Toxicants in Terrestrial Ecosystems
A Guide for the Analytical and Environmental Chemist

T.R. Crompton

Toxicants in Terrestrial Ecosystems

A Guide for the Analytical and Environmental Chemist

With 18 Figures and 45 Tables

 Springer

T. R. Crompton
Hill Cottage (Bwthyn Yr Allt)
Anglesey, Gwynedd
United Kingdom

Library of Congress Control Number: 2006924585

ISBN-10 3-540-33694-X Springer Berlin Heidelberg New York
ISBN-13 978-3-540-33694-5 Springer Berlin Heidelberg New York
DOI 10.1007/b95908

Springer is a part of Springer Science+Business Media
springer.com
© Springer-Verlag Berlin Heidelberg 2006
Printed in Germany

Cover design: *design & production* GmbH, Heidelberg
Typesetting and production: LE-TEX Jelonek, Schmidt & Vöckler GbR, Leipzig, Germany
Printed on acid-free paper   2/3141/YL - 5 4 3 2 1 0

Preface for Toxicants in Terrestrial Ecosystems

This book discusses the methods currently available in the world literature up to 2005 for the determination of organic, organometallic and metallic impurities in soil and plant materials, vegetables and fruit. Radioactive substances and anions are also discussed.

In the case of soils, the presence of deliberately added or adventitious toxicants can cause contamination of the tissues of crops grown on the land or animals feeding on the land and, consequently, can cause adverse toxic effects on man, animals, birds and insects. Drainage of theses substances from the soil can also pollute adjacent streams, rivers and eventually the oceans. Some of the organic substances included in this category are pesticides, herbicides, growth regulators, and organic fertilisers.

Individual chapters deal with the determination of metals, non-metals, organic compounds and organometallic compounds in soil and in plants that grow in soil. A separate chapter deals with sampling procedures. A relationship between toxicant levels in soil and plants that grow in that soil has been established and is the subject of the concluding chapter.

Examining for toxicants combines all the exciting features of analytical chemistry. First, the analysis must be successful and in many cases, must be completed quickly. Often the nature of the substances to be analysed is unknown, might occur at exceedingly low concentrations and might, indeed, be a complex mixture. To be successful in such an area requires analytical skills of a high order and the availability of sophisticated instrumentation.

The work has been written with the interests of the following groups of people in mind: agricultural chemists, agriculturists concerned with the ways in which inorganic and organic chemicals are used in crops and soil treatment that permeate through the ecosystem, biologists and scientists involved in plant life, and also people in the medical profession, such as toxicologists, public health workers and public analysts. Other groups or workers to whom the work will be of interest include environmentalists and not least members of the public who are concerned with the protection of our environment.

Finally it is hoped that the work will act as a spur to students of all subjects mentioned above and assist them in the challenge that awaits them in ensuring that the pollution of the environment is controlled so as to ensure we have a worthwhile environment to protect.

Anglesey, United Kingdom, July 2006 *T. R. Crompton*

Contents

1 Sampling for the Analysis of Soil and Plant Sample

1.1
Soil

1.1.1
Soil Sampling

An HMSO publication covers the subject of soil sampling methods very extensively, including a detailed discussion on matters such as regular sampling methods, random sampling methods, grab sampling, systematic square sampling of fields, alternative random field sampling methods, auger sampling, depth profiling, and sampling by pedogenetic horizons. As this detailed information is readily available elsewhere it will not be discussed further here [1].

The whole field of the sampling of soils has been reviewed by Stoeppler [2], Lijtha et al. [3] and Fortunati et al. [4]. Fortunati et al. discuss the strategies of soil sampling. What is good practice in soil sampling has been discussed by Epps [5], including the effect of sampling variation on test results and the need for standardisation of soil sampling methods.

For individual investigations of contaminated soil, sampling errors are much greater than analytical errors, so the theory and practice of sampling contaminated soils needs to be developed further and much remains to be done in this field [6].

Van Der Veen and Alink [7] have reviewed methods for evaluating the performance of sampling, sample preparation and subsampling. Several new methods and apparatus for sampling solid matrices have been described recently [5–12] and, in particular, a new sampling method has been developed that is especially adapted to the specific conditions of sampling contaminated bulk soil masses [8].

Eccles and Redford [9] have investigated the use of dynamic (window) sampling in site investigations of soil.

Lancaster and Keller-McNulty [10] showed that reduced costs and improved statistical performance can be achieved by applying composite sampling methods to soil.

Various new devices that have been used in soil sampling include a gravity-driven, hydraulically sampled multi-piston corer for fine-grained soils [11] and time-series trap that can collect 21 samples of soil at programmed intervals [12].

Dong et al. [13] evaluated sampling and analytical errors for the determination of manganese in soils, and no doubt the conclusions reached in this work could be applied to other elements.

On the organic chemicals side, Thiboutot et al. [14] devised protocols for a sampling campaign for sites contaminated by explosives.

Brown and Reinsch [15] have discussed the collection and preparation of soil samples for the US Federal Soil Survey Laboratory Programme.

Burton [16] discussed factors affecting the realism of the collected samples of soil.

Various aspects of the sampling and analysis of soils for total petroleum hydrocarbons and benzene, toluene, ethyl benzene and xylenes have been discussed [202–205].

1.1.2
Soil Preparation for the Analysis and Determination of Metals

It is very important that the subsample analysed represents the original sample, otherwise the analytical result will be of little value. Each sample must be treated according to the analysis required. A very good guide covering the initial preparation of samples which is applicable to most samples has been published [1]!

The former UK Ministry of Agriculture and Food has also published [17] recommended soil preparation techniques for the determination of a wide range of metals and for the preparation of plant samples for analysis by dry combustion and the determination of ash.

Contamination problems can arise during the preparation and analysis of soils. Sources of trace elements can be atmospheric dust contamination after the initial sample was taken, laboratory equipment, adventitious contaminants such as cosmetics, and reagents used during the analysis. The analyst should take suitable precautions to reduce these to a minimum. Some of these problems have been reviewed by Mitchell [18].

Special precautions are necessary during the initial preparation of soil for certain analyses, such as for boron, mercury and selenium, and details of these precautions are given in any well-written published analytical method. Any analytical method for such elements that does not include such information is not worthy of further consideration.

For volatile and labile determinands, particularly in the field of the analysis of nonvolatile organic or organometallic substances, special attention should be given to methods of drying or reducing the sample. Drying must not be used in the case of soils in which it is necessary to determine volatile organic compounds.

Samples can be dried and moisture content determined by special methods [19] and soil samples can be homogenised with a blender or a similar device.

For many analyses, soil is brought to the air-dried condition. This term refers to soil conditioned to ambient temperature and humidity, although in the case of the determination of organic or inorganic nonvolatile-containing samples, artificial heating at a temperature not exceeding 30 °C may be used in the drying process.

The length of time required to dry a sample to produce a friable material for subsequent sieving will depend on the nature and type of the soil.

In order to sieve samples that are to have their inorganic constituents determined, the soil is ground to pass through a nylon sieve meeting the requirements of BS 410i77. This avoids sample contamination associated with the use of metallic sieves. When sieving samples prior to the determination of organic constituents, a metal sieve may be used provided it does not react with the determinand of interest.

Methods of sample drying, sieving and sample volume reduction as well as long piles and quartering, sifting, core quartering, rotating pie wedge sampling, particle size reduction, sample storage, sample blending and blending homogeneity determination have been discussed in detail elsewhere [1] and will not be discussed further here.

Recent work on soil sample preparation is reviewed below. Rubio and Une [20] have discussed the risks of soil sample contamination using inappropriate materials, containers and tools as well as possible analyte loss during sample loading.

Houba et al. [21] studied the influence of grinding procedures and found that for some soils the availability of some analytes was significantly influenced by the degree of grinding.

1.1.3
Extraction of Inorganic Substances from Soil

Reynolds [25] has reviewed digestion procedures for the analysis of metal-contaminated soils.

Extraction of Metals

Microwave extraction methods are now being developed [22–25]. Krishna-murti et al. [22] found that the microwave extraction of cadmium in a soil reference material gave results comparable to those obtained by conventional soil extraction methods. In another study, Kingston and Walter [23] compared microwave *versus* conventional dissolution of soils. About 90% of the lead and cadmium were extracted from soils and dusts by a microwave digestion procedure [24].

An extraction procedure based on ethylene diamine tetraacetic acid has been evaluated for the extraction of metals from soils [26]. In a collaborative study between 54 different laboratories, all of the laboratories produced some extreme outlying results, but most results were in good agreement once the outliers had been removed.

An ultrasonic bath extraction procedure gave acceptable accuracy and precision in the determination of metals [27].

In an interlaboratory study involving 160 accredited hazardous materials laboratories reported by Kimbrough and Wakakuwa [28], each laboratory performed a mineral acid digestion on five soils spiked with arsenic, cadmium, molybdenum, selenium and thallium. Analysis of extracts was carried out by atomic emission spectrometry, inductively-coupled plasma mass spectrometry, flame atomic absorption spectrometry and hydride generation atomic absorption spectrometry.

At most concentrations, inductively coupled plasma mass spectrometry exhibited higher precision and accuracy than the other techniques, but also the highest rates of false positives and negative results.

Much work has been reported on the evaluation of sequential extraction procedures. The three-stage sequential extraction procedure for speciation of heavy metals proposed by the Commission of the European Communities Bureau of References (BCR) was found to be acceptable and reproducible with some modifications [29]. In another study, when applied to real soils and sediments, this (unmodified) BCR method was queried [30]. Lopez-Sanchez et al. [31] found that significant results can be obtained when different sequential extraction procedures are used.

Shan and Chen [32] reported that various proportions of metals released from exchangeable, carbonate-bound iron, manganese oxide-bound and organic-bound fractions were readsorbed onto the other solid geochemical phases during sequential extractions.

Some work on sediments is reported here in the belief that it may also be useful in the analysis of soil samples. Thus Asikainen and Nikolaides [33] have carried out a sequential extraction study of chromium from contaminated aquifer sediments and found that 65% of the chromium was extractable. Of this amount 25% was exchangeable, 11% was bound to organic matter and 30% was bound to iron and manganese oxide surfaces. Thomas et al. [34] also investigated the use of BCR sequential extraction procedures for river sediments, and found the method to work well. Real et al. [35] improved sequential extraction by optimising microwave heating.

Martens and Suarez [37] employed sequential extraction and hydride generation atomic absorption spectrometry to analyse soil for arsenic and selenium and achieved excellent precision.

Ren and Salin [36] showed that direct analysis of solid samples is possible, by using furnace vaporisation with Freon modification and inductively coupled plasma mass spectrometry. The relative standard deviations obtained for several metals in marine reference sediments varied from 3 to 15%.

A number of studies on sequential extractions of trace metals have been reported [42–49]. Metal distributions were significantly different among three compared sequential extraction procedures [43]. Silty soils may have a relatively high heavy metal retention capacity due to the presence of carbonate, and this retention capacity can be comparable in magnitude to that of certain clayey soils [47]. Soil amended with sewage sludge exhibited a different distribution of metals in the soil [48]. Bodog extended the sequential extraction procedure of Ure, and observed good agreement with a total acid extraction procedure [49].

The effects of soil sample preparation procedures for the determination of chromium in soils have been reported [38]. The optimum conditions included the use of an homogeneous sample with a mass of less than 4 kg, a grain diameter of less than 0.25 mm, digestion with a solution of nitric acid plus perchloric acid (3:2) and hydrochloric acid after dry ashing, with the addition of 1% lanthanum or 1% ammonium chloride to eliminate interferences.

Mierzwa and Dobrowolski [39] determined selenium using combined slurry sampling, microwave-assisted extraction and hydride atomic absorption spectrometry. Lopez-Garcia et al. [40] also used slurry sampling in the determination of arsenic and antimony in soil.

Direct solvent extraction of soil by an organic solvent containing an organic complexing has been used. Thus Reddy and Reddy [41] showed that extraction of soil with a chloroform solution of xanthate completely extracted cadmium. Compared to untreated soil or sediment, none of the three drying methods studied – freeze drying, air drying and oven drying – completely preserved the distribution of selected metals in the various geochemical fractions [50].

Reviews have been conducted on the problems associated with techniques and strategies of soil sampling [4] and on the collection and preparation of soil samples for the Federal Soil Survey Laboratory Program [51]. Factors affecting the realism of the collected sample were discussed by Burton [16]. Various sampling schemes for soils have been described [52–56]. Different sampling designs are needed, depending upon whether the soil contamination is expected to be "spread" over the whole area or to exist in localised "hot spots" [52]. A decision support system for the sampling of aquatic sediments in lakes was described by Wehrens and was applied to a real environmental problem [53]. Lame showed that the *fundamental* sampling error for soils only affects the analytical variance when sample sizes are less than 10 g [54]. For larger samples, the variance is determined by the *segregation* error. A sampling board method of estimating the *segregation* error has been described. Skalski showed that a two-way compositing strategy could be used to attribute detected contamination in composited samples directly to constituent samples without further analyses [55].

Evaluations of various soil and sediment samplers have been reported [56, 57]. The sediment shovel proved highly practical, but was limited because small particles tend to be lost when the shovel is lifted [56]. A cryogenic sediment sampler was less convenient to use, but allowed the collection of almost undisturbed samples. Houba described a different device for the automatic subsampling of soil, sediment and plant material for proficiency testing [57]. In another study, Thoms showed that freeze-sampling collects representative sediment samples, whereas grab-sampling introduces a bias into the textural composition of the 120 mesh fraction, due to washout and elutriation of the finer fractions [58].

1.1.4
Extraction of Organic Substances from Soils

Analysis of organic pollutants in environmental soil samples is an important task with respect to the protection of the environment.

Conventionally, organic contaminants in solid samples are examined by Soxhlet extraction, followed by separation and identification. Several methods have been proposed to reduce the use of organic solvents and to increase the speed of analysis, such as supercritical fluid extraction [59, 60], accelerated solvent extraction [61], subcritical fluid solvent extraction [62, 63] and headspace solid-phase microextraction [64–66]. Separation and phase identification methods such as gas chromatography–mass spectrometry are typically used to examine the extracts. Attenuated total reflectance–infra-red spectroscopy [67] provides a direct method for detecting organic species in samples of varying physical composition and is very suitable for handling aqueous solutions because the evanescent wave penetrates into the adjoining medium for a short distance. Examples of these techniques are reviewed below.

Conventional Solvent Extraction

Miellet [80] and Lopez-Avila et al. [81] have reviewed the applications of Soxhlet extraction to the determination of pesticides in soil. This technique has been applied extensively to the extraction of polycyclic aromatic hydrocarbons, volatile organic compounds, pesticides, herbicides and polychlorodibenzo-*p*-dioxins in soils. Details of the extraction procedures and the analytical finish employed are reviewed in Table 1.1.

Accelerated Solvent Extraction

This relatively new technique has been proposed as an alternative to the Soxhlet procedure [92–94]. In this technique the soil sample is packed into an extraction cartridge and the analytes are extracted from the matrix with conventional low

Table 1.1. Conventional solvent extraction procedures for organic compounds in soils (from author's own files)

Compound–pesticides and herbicides	Extractant	Clean-up	Analytical finish	Reference
Organochlorine insecticides	Polar solvents	Florisil	–	[68]
Alachlor	Methanol enrichment on C_{18} cartridge		–	[69,70]
Bromoxynil, ioxynil residues	Solvent partitioning	Solvent partitioning	Gas chromatography with ion-trap detector	[71]
Triazine type herbicides	Methanol enrichment on C_{18} cartridge	–	Gradient C_{18} high-performance liquid chromatography	[72]
Imazapyr	Methanol	None	High-performance liquid chromatography, detection at 250 nm	[73]
Terbuthylazine and degradation products	Hot acetone, then cation exchange solid-phase cartridge	–	High-performance liquid chromatograph with photodiode array detection	[74,75]
Norflurazon herbicide	Methanol	Filtration	C_{18} high-performance liquid chromatography	[76]
Isoproturon, dichloroprop-P. bifenox	Solvent	Size exclusion chromatography	High-performance liquid chromatography	[77]
Fenopropimorph and metabolite fenopropimorphic acid fungicide	Acetone – water, partitioning with methylene chloride	Gel permeation chromatography	Methylation – gas liquid chromatography with NP detection and GC-MS	[78]
Fenoxaprop, Fenoxapropethyl	Solvent	Florisil or alumina	High-performance liquid chromatography	[79]
Polycyclic aromatic hydrocarbons Polycyclic aromatic hydrocarbons	Soxhlet extraction and saponification	Silica gel	1 ppm detected in soil	[82]

Table 1.1. (continued)

Compound–pesticides and herbicides	Extractant	Clean-up	Analytical finish	Reference
Polycyclic aromatic hydrocarbons	BF_3 in methanol	–	–	[83]
Polycyclic aromatic hydrocarbons	Miniaturised liquid–liquid extractor (100 μl solvents)	–	–	[84]
Polycyclic aromatic hydrocarbons	Automated Soxhlet extraction with ethyl acetate	–	–	[85]
Polycyclic aromatic hydrocarbons	Organic solvent, methanolic hydrolysis	–	–	[86]
Volatile organic compounds	Hot (70 °C) methanol	–	–	[87]
Volatile organic compounds	Comparison of solvent extraction, headspace analysis and vapour partitioning, methanol extraction	–	–	[88]
Volatile organic compounds	Comparison of methanol extraction and purge and trap method	–	Hot methanol extraction is the most efficient	[89]
Polychlorodibenzo p-dioxins and benzofurans	Comparison of Soxhlet extraction using (a) toluene and (b) methylene dichloride–acetone	–	–	[90, 103]
Medium polar and polar analytes	Hot phosphate buffered water extractant	–	–	[91]
Aromatic hydrocarbons	Solvent extraction with methanolic hydrolysis	–	–	[86]
Chlorophenols	Soxhlet extraction	–	–	[99, 100]

boiling point solvents or solvent mixtures at elevated temperatures of up to 200 °C and pressures of up to 20 MPa [93–95] to maintain the solvent in a liquid state.

Two comparative studies have shown that accelerated solvent extracted quantities of pesticides from soils equal to or larger than those found by other extraction techniques [120, 121]. However, only 36 to 72% of phenoxyacetic acid herbicides were recovered by this technique from clay, loam and sand [43]. A further limitation of the accelerated solvent extraction technique, which is shared by several of the other newer extraction techniques reviewed here, is that selective extraction of organics based on their polarities is difficult. For example, in the case of the extraction of soil with a high organic content (9.6%) at 100 °C with methanol or acetone as such or acidified with phosphoric acid with each of these extractants, large amounts of wax-like substances – presumably cellulose, lignin and waxes from plant cells – were coextracted with the herbicides considered. The presence of these high molecular weight compounds in soil extracts caused interference in the final analytical finish employed to determine the herbicides, and can only be avoided in some, not all, cases by tedious and time-consuming clean-up procedures. To a lesser extent these species are also present in soils with a lower organic content. Subcritical water extraction overcomes this difficulty and will be discussed further below [123].

The range of materials for which the technique is proposed includes semivolatile compounds, including polycyclic aromatic hydrocarbons, organochlorine pesticides, organophosphorus pesticides, chlorinated herbicides and polychlorinated biphenyls [92].

Saim et al. [96] investigated the interdependence of selected operating parameters on the recovery of 16 polycyclic aromatic hydrocarbons from nine highly contaminated soils, including a range of pressures from 1000–2400 psi, operating temperatures from 40–200 °C, and extraction times from 2 to 16 minutes.

At the 95% confidence interval, no significance in terms of the three operating parameters was found when considering the total polycyclic aromatic hydrocarbon recovery. However, recoveries of some individual polycyclic aromatic hydrocarbons were found to be dependent on operating variables. In particular, low operating temperatures of 40 °C were very significant for naphthalene, chrysene and benzo(b)fluoranthene.

Wennrich et al. [97] have described a method for the determination of nine chlorophenols in soil using accelerated solvent extraction with water as the solvent combined with solid-phase microextraction and gas chromatography – mass spectrometry. An extraction temperature of 125 °C and ten-minute extractions were optimal.

Hofler et al. [98] also studied the application of accelerated solvent extraction with an organic solvent, followed by clean-up and preconcentration procedures.

Hubert et al. [101] state that accelerated solvent extraction compared to alternatives such as Soxhlet extraction, steam distillation, microwave extraction, ultrasonic extraction and, in some cases, supercritical fluid extraction is an exceptionally effective extraction technique. Hubert et al. [101] studied the effect of operating variables such as choice of solvent and temperature on the solvent extraction of a range of accelerated persistent organic pollutants in soil, including chlorobenzenes, HCH isomers, DDX, polychlorobiphenyl cogeners and polycyclic aromatic hydrocarbons. Temperatures of between 20 and 180 °C were studied. The optimum extraction conditions use two extraction steps at 80 and 140 °C with static cycles (extraction time 35 minutes) using toluene as a solvent and at a pressure of 15 MPa.

Pyle and Marcus [102] achieved low ppb detection limits for the determination of organochlorine insecticides in soil using accelerated solvent extraction followed by gas chromatography ion trap tandem mass spectrometry. Richter et al. [103] showed that accelerated solvent extraction gave essentially equivalent recoveries of chlorinated dibenzo-p-dioxins and dibenzofurans from soil compounds to Soxhlet extraction, but in less time and using much less solvent.

Pressurised Liquid Extraction

Pressurised hot water extraction has been used to isolate polycyclic aromatic hydrocarbons from soil [104, 105]. Ramos et al. [106] reported an rapid (ten minutes) miniaturised pressurised liquid extraction method using only 100 μl solvent for extracting polycyclic aromatic hydrocarbons from soil.

Microwave-Assisted Extraction

Lopez-Avila et al. [107] showed that microwave-assisted extraction of pesticides and polycyclic aromatic hydrocarbons from soil is a viable alternative to Soxhlet extraction and needs a smaller sample volume and extraction time [108, 109]. These techniques have also been compared in the case of chlorophenols. Lopez-Avila et al. compared microwave-assisted extraction with electron capture gas chromatography to ELISA for the determination of polychlorinated biphenyls in soils. Both techniques are applicable to field screening and monitoring applications. Microwave-assisted extraction [111, 112] and solid-phase microextraction [113] have been applied to the extraction of pesticides from soil. It was observed by these and other workers [114] that the selectivity of microwave-assisted extraction is highly dependent on the soil composition.

Microwave-assisted extraction [115] has been compared with ultrasonic extraction [116] in the context of soil extraction. Microwave-assisted extraction [117, 195–198] and supercritical fluid extraction coupled with on-line infrared spectroscopy detection [118, 119] have been compared as methods for the extraction of hydrocarbons from soil.

Subcritical Water Extraction

This technique, as discussed above under "Accelerated Solvent Extraction", has the outstanding advantage that extraction with water as opposed to organic solvents does not cause contamination of the extract with potentially interfering organic components such as cellulose, lignin and waxes originating in plant cells or interference due to contamination by the solvent or impurities therein.

Crescenzi et al. [122] evaluated the feasibility of selectively extracting phenoxyacetic acid herbicides with subcritical hot water and collecting the analytes on a Carbograph-4 solid-phase extraction cartridge set on-line with the extraction cell. Final analysis was by liquid chromatography–mass spectrometry with an electrospray ion source. With few exceptions, recoveries were in the range 81 to 93% (with the exception of 24 DB and MCPB, which gave 63%) recovery and detection limits of between 1.7 and 10 ng/g. Other applications of subcritical water extraction are reviewed in Table 1.2.

Table 1.2. Applications of subcritical water extraction to the determination of organic compounds in soil (from author's own files)

Determinand	Subcritical water extractant	Sorbent trap	Analytical finish	Reference
Mixtures of herbicides	Water	Miscellaneous traps		[123]
Terbuthylazine and metabolites	Phosphate buffered water	Graphitised carbon block cartridge		[123]
Herbicides and breakdown products	Water	Miscellaneous traps		[124]
Polycyclic aromatic hydrocarbons	Static subcritical water	Styrene-divinyl-benzene discs		[124]
Polycyclic aromatic hydrocarbons	Water	Solid phase	High-performance liquid chromatography, post column fluorimetric detection	[125]
Polychlorobiphenyls	Water at 250–300 °C and 50 atmospheres pressure	–		[126]
Polychlorobiphenyls	Water	Solid-phase microextraction		[63]

Solid-Phase Microextraction

This technique seems to have been introduced in late 1998, and consists of extracting organic contaminants from the soil with a solvent, generally subcritical water, and then passing the extract through a small disc of solid sorbent. The solid sorbents discussed to date include graphitised carbon black [123], styrene-divinyl benzene [124], Carbograph-4 [122] and polyisobutylene [193].

An example of the application of subcritical water extraction–solid-phase microextraction is that of Crescenzi et al. [122] (see above).

Water extraction is also occasionally combined with solid-phase microextraction. Thus Wennrich et al. [97] determined chlorophenols in soil by using accelerated water extraction to remove the chlorophenols from the soil followed by adsorption onto a solid sorbent for ten minutes at 125 °C. Low ppb detection limits were thus achieved.

Other applications of subcritical water extraction–solid-phase microextraction are the determination of terbuthylazine and its metabolites [123], polycyclic aromatic hydrocarbons [124, 125] and polychlorobiphenyls [63]. Yang and Her [193] collected 1-chloronaphthylene, nitrobenzene and 2-chlorotoluene in soil on a hydrophobic polyisobutylene disc prior to analysis by attenuated total reflectance Fourier transform infrared spectroscopy.

Supercritical Fluid Extraction

This is an attractive technique for recovering organic compounds from soils. Carbon dioxide is currently the fluid of choice, due to its low toxicity and environmental acceptability. The physicochemical properties of supercritical fluids, including low viscosity, variable solvent strength and high diffusivity, contribute to faster extractions compared to conventional extraction techniques such as Soxhlet extraction or sonication. Supercritical fluid extraction methods have been successfully developed for nonpolar compounds that exhibit high solubilities in carbon dioxide, such as polycyclic aromatic compounds [127–133], polychlorobiphenyls [134–136], chlorodioxins [137–142], amines [143], pyridine [144], triaryl and trialkyl phosphates [145], hydrocarbons [146], volatile organic compounds [147–150], phenols [151], organic acids [152], ketones [152], enteroviruses [153], organochlorine pesticides [154] and miscellaneous herbicides and pesticides [155–165]. With methanol as a modifier, supercritical carbon dioxide becomes more amenable to the extraction of moderately polar pesticides including triazines [160, 166], organophosphorus insecticides [167, 168], sulfonyl ureas [161–163], organochlorine insecticides [167, 170], flumetron [171], and other herbicides [172]. Further details of the methods are given in Table 1.3. The work of Field et al. [165] is quoted as an example of the application of supercritical carbon dioxide and subcritical (hot) water extraction of the widely used pre-emergent herbicide dacthal and its mono and diacid metabolites in soil. These compounds were sequentially extracted from soils by first performing a supercritical hot water extraction for 15 minutes at 150 °C and 400 bar to recover dacthal, followed by a subcritical hot water extraction to recover the metabolites, which were then trapped in situ on a strong anion exchange disc placed over the exit frit of the extraction cell. Dacthal was combined with the metabolites by placing the disc into a gas chromatograph autosampler vial containing the supercritical fluid extract. The metabolites are then simultaneously eluted from the disc and derivatised to their ethyl esters by reaction with ethyl iodine at 100 °C.

Meyer et al. [173] showed that supercritical fluid extraction results can give recoveries comparable to Soxhlet extraction methods, even for soils with high carbon contents. McNally et al. [174] have studied factors affecting the supercritical fluid extraction of soils. It was shown that soil type affects the recovery of moderately polar analytes. In general the organic carbon content of the soil governs analytical recovery.

Online coupling of supercritical fluid extraction and high-performance liquid chromatography considerably decreases sample preparation time and analysis time [175]. Dunkers [128] showed that by using dilute dichloromethane as a static modifier, 20–30 minute supercritical fluid extractions gave results comparable to those obtained by conventional four-hour sampling methods in soil extractions.

Fahing et al. [176] studied the effect of the addition of modifiers such as methanol and water on the SCFE of organic solutes from soils and clays. Hawthorne et al. [177] compared the application of sub- and supercritical water in the extraction of organics from soil, and found that both were effective extractants.

Headspace Analysis

This technique is, of course, only applicable to organic compounds in soil that are sufficiently volatile at room temperature or slightly above that they exist in the headspace above the samples. For such samples, the technique is elegant in that it is solventless, i.e., there is no solvent interference, is amenable to automation, and can be directly coupled to a gas chromatograph and/or alternate techniques such as mass spectrometry to ensure equivocal identification of the organics.

Headspace analysis is the method of choice for determining volatile organic compounds in soil [178–183]. A limitation of this method is the incomplete desorption of the contaminants in soil–water mixtures, but this problem can be overcome through the addition of methanol to the sample [181]. Good recoveries of volatile organic compounds in soils were obtained via thermal vaporisation of the sample followed by Tenax GC trapping and gas chromatography–mass spectrometry.

Stuart et al. [184] studied the analysis of volatile organic compounds in soil using an automated static headspace method. Recoveries increased in the or-

Table 1.3. Determination of organic compounds from soil by supercritical fluid extraction (from author's own files)

Determinand	Extractant	Conditions/comments	Analytical finish	Reference
Polycyclic aromatic hydrocarbons	Dichoromethane modified CO_2	30 minute extraction	–	[128]
Polycyclic aromatic hydrocarbons	CO_2	–	High-performance liquid chromatography	[129]
Polycyclic aromatic hydrocarbons	Water–methanol dichoromethane	Comparison with CO_2	–	[130]
Polycyclic aromatic hydrocarbons	CO_2–dichoromethane	Comparison with CO_2 and CO_2–methanol. CO_2 recommended	–	[131]
Polycyclic aromatic hydrocarbons	CO_2	Use of liquid–solid traps compared to analyte trapping	–	[133]
Polycyclic aromatic hydrocarbons	Pretreatment of sample with (a) 15% water, (b) 5% (ethylenedinitrilo) tetraacetic acid tetrasodium salt or (c) 50% methanol then extraction with CO_2	60–98% recovery using Na_4 EDTA–CO_2. Only 7–63% recovery with CO_2 alone		
Diesel hydrocarbons	CO_2	–	Gas chromatography	[146, 147, 195–198]
Petroleum hydrocarbons	CO_2	Comparison of sorbent trapping with solvent trapping	Gas chromatography	[149]
Petroleum hydrocarbons	Argon	500 atm at 150 °C	Infrared spectrometry	[148]
Volatile organic compounds	CO_2	–	–	[150]
Phenols, cresols	CO_2	–	–	[151]
Organic acids and ketones	CO_2	–	–	[152]

Table 1.3. (continued)

Determinand	Extractant	Conditions/comments	Analytical finish	Reference
Polychlorobiphenyls	Comparison of CO_2, $CHClF_2$, N_2O	$CHClF_2$ gave the best recovery of PAH	FT offline IR spectroscopy	[135]
Polychlorobiphenyls	CO_2	–		[136]
2,3,7,8-Tetrachloro-p-benzo dioxin	CO_2	–		[138–141]
Trialkyl and triaryl phosphates	CO_2	Microwave extraction	–	[145]
Aromatic amines	CO_2	–	–	[143]
Pyridine	CO_2	–	–	[144]
Enteroviruses	CO_2	–	–	[153]
Insecticides and herbicides				
Dacthal herbicide and acid metabolites	Supercritical CO_2 at 150 °C and 400 bar, then supercritical water at 50 °C and 200 bar	Metabolites derivatised to ethyl esters	–	[165]
Diuron, Linuron	CO_2	–	–	[155]
Chlordane	CO_2	Compared to Soxhlet and accelerated solvent extraction	–	[156]
Chlorpyrifos metabolite	Polychlorobiphenyls	Comparison with subcritical water extraction	–	[157]
Organochlorine insecticides	CO_2	–	Comparison with Soxhlet extraction	[170]
Imazaquin	CO_2	–	–	[158]

Table 1.3. (continued)

Determinand	Extractant	Conditions/comments	Analytical finish	Reference
Atrazine	CO_2	–	High-performance liquid chromatography–mass spectrometry	[159]
Flumetron	CO_2	–	–	[171]
Atrazine, cyanazine, desethylatrazine	CO_2	–	–	[160]
Metochlor	CO_2–methanol SF extractions			
Sulfonylurea herbicides and metabolites	CO_2	–	Superfluid chromatography	[161]
Sulfonylurea herbicides and metabolites	CO_2–methanol	Using C_{18} solid phase extraction disk	High-performance liquid chromatography with UV detection. Dimethyl sulfonyl ureas determined by GC with detection by electron capture or N-P or MS.	[162, 163]
Triazine herbicides	CO_2–methanol	–	–	[166]
Organochlorine, organophosphorus insecticides	CO_2–3% methanol	Comparison with classical sonication and Soxhlet extratction	–	[165, 167]
Flumetron, fenpropiomorph, perimicarb, parathion, ethyl triallate, fenvalerate	CO_2–5% methanol	Extraction at 60 °C and 3.8×10^7 Pa	–	[172]

der: water, pure sand, sandy soil, clay and top soil. A full evaporation technique that uses little or no aqueous phase and higher equilibration temperature gave the most reproducible analyte recoveries. Hewitt [185] compared three vapour partitioning methods, three solvent extraction methods and headspace analysis for the preparation of soil samples for the determination of volatile organic compounds.

Samples Used for the Determination of Volatile Organic Compounds in Soils

Methanol extraction was the most efficient method of recovering volatile organic compound spikes from soils, but results depended on the organic carbon content of the soil.

Various other workers have reported on the determination of volatile organic compounds in soils [186, 187] and landfill soils [188]. Soil fumigants such as methyl bromide have also been determined by this technique [189]. Trifluoroacetic acid is a breakdown product of hydrofluorocarbons and hydrochlorofluorocarbon refrigerant products in the atmosphere and, as such, due to the known toxicity of trifluoroacetic acid, it is important to be able to determine it in the atmosphere, water and in soil from an environmental point of view [190]. In this method the trifluoroacetic acid is extracted from the soil sample by sulfuric acid and methanol, which is then followed by the derivatisation of it to the methyl ester. The highly volatile methyl ester is then analysed with a recovery of 87% using headspace gas chromatography. Levels of trifluoroacetic acid in soil down to 0.2 ng/g can be determined by the procedure.

Purge and Trap Analyses

Kester [191] has reviewed the application of this technique to the determination of a wide range of organic compounds in soil, including ketones, aldehydes, aromatic hydrocarbons, halogenated aliphatic compounds, alcohols and vinyl acetate.

Basically, in these methods the volatiles released by heating the sample are collected on a Tenax GC and subsequently desorbed from the Tenax and determined by gas chromatography–mass spectrometry.

Roche and Miller [192] have shown that ultrasonic extraction gives more accurate results when compared with a heated nitrogen purge in the determination of volatile organic compounds in soils.

Yang and Her [193] have described a rapid method for the determination of down to 200 ppt of semi-volatile compounds such as 1-chloronaphthalene, nitrobenzene and 2-chlorotoluene in soils by coupling solid-phase microextraction with attenuated total reflectance Fourier transform infrared spectroscopy.

Pervaporation

Papaefstathion and Luque de Castro [194] used pervaporation as an alternative to headspace analysis for the determination of down to 1 ng/g of volatile organic compounds in soils.

1.2
Plants and Crops

1.2.1
Plant and Crop Sampling

Many of the comments made in Sect. 1.1 regarding sampling of soil samples apply equally well to the sampling of plant and crop materials. Some analysis of plant materials must be carried out on the fresh plant material. However, most analysis for nonvolatile organic constituents is carried out after the fresh material has been subsampled, dried and ground. At all stages of the preparation, suitable precautions need to be taken to avoid metallic or organic contamination of the soil. Where trace element analysis is required, samples must be taken with great care to avoid soil contamination of the plant sample. Actively growing, fresh material free from dust or surface contamination should be sampled. Water washing of plant material to remove surface contamination should be viewed with caution and performed only after thorough checks.

Various authorities have published guidelines on the washing, subsampling [17, 199], drying [17], grinding, storage and prior to storage.

Table 1.4. Preparation of plant extracts by acid digestion for determination of metals (from author's own files)

Type of sample	Determinand	Dry ash acid digestion reagent	Analytical finish	Reference
Plant	Cu	HNO_3–$HClO_4$	AAS	[201]
Plant	Se	HNO_3–$HClO_4$	AAS	[202]
Plant	Cu, Mn, Zn	$KHSO_4$–HNO_3	AAS	[203]
Plant	Co, Mo	$KHSO_4$–HNO_3	Spectrophotometry	[204]
Plant	As, Al, Fe, Zn, Cr, Cu	HNO_3–H_2SO_4	AAS	[205]
Plant	Mo	HNO_3–H_2O_2 HNO_3	GF AAS spectrophotometry	[206, 208]
Plants	Fe	HNO_3–$HClO_4$	AAS	[207]
Plants	Al	HNO_3–HCl–H_2O 25 + 25 + 50 μl	Spectrophotometry	[209]

Table 1.5. Preparation of plant extracts by solvent extraction for the determination of organic compounds (from author's own files)

Type of sample	Determinand	Solvent extraction	Analytical finish	Reference
Grass	m-s methyl (phenylthio) carbamate	$CHCl_3$ or CH_2Cl_2	–	[210]
Crop	Hexamethylene pyrophosphate	$CHCl_3$	GLC	[211]
Carrots	Linuron, trifluralin	C_6H_6–$C_2H_5OC_2H_5$	GLC–EC	[212]
Fruit, vegetables	Synthetic pyrethroids	C_6H_6–CH_3COCH_3	Chromatography	[213]
Crops	Phenyl urea herbicides	MeOH	–	[214]
Fruit, vegetables	Natural pyrethrins	CH_3COCH_3 acetonitrile	GLC–EC	[215]
Fruit, vegetables	Pyrethroids	Acetone–PET ether	GLC–EC	[216]
Vegetables	Pyrethroids, organophosphorus insecticides	ETOH, partition with with toluene	GLC	[217]
Plant tissues	Dalapon, 2,2-dichloro-propionic acid	$C_2H_5OOC_2H_5$	GLC	[218]
Vegetables	Nitrophen	Acetonitrile	Spectrophotometry	[219]
Plants	Triazidepyrimidine growth retarder	CH_2CH_2	GLC	[220]
Fruit, vegetables	Organophosphorus insecticides	H_2O–ethyl acetate on-line microextraction	–	[221]
Fruit, vegetables	Carbamate insecticides and metabolites	Ethyl acetate	GLC–EC	[222]
Vegetables, crops	Buprotezin	Ethyl acetate	GLC–MS	[223]
Plants	Pesticides	Soxhlet extraction	–	[80]
Plants	S-methyl methionine	0.1 N HCl in 70% C_2H_5OH	Automated amino analyser	[224]
Plants	Alkyl mercury	Miscellaneous	–	[225]

1.2.2
Preparation of Plant Extracts

Metals

Generally speaking, digestion with mineral acids is used to prepare extracts for subsequent analysis for metals. Some typical examples are reviewed in Table 1.4. It is seen that digestion of the plant sample with nitric acid–perchloric acid is still used extensively.

Preparation of an ultrasonic slurry of the sample is occasionally used, as for example in the determination of cobalt, nickel and copper [200], selenium [39] and arsenic and antimony [40]. Extraction of leaves with a chloroform solution of xanthate completely extracted cadmium [41, 103]. X-ray fluorescence spectroscopy is a nondestructive method of analysing plant materials if they can be converted into a suitable form for presentation to the instrument.

Organic Compounds

Solvent extraction of the plant material is still by far the most popular method for extracting organic compounds or organometallic compounds from plant materials, crops and fruit and vegetables prior to final analysis, see Table 1.5.

Applications of supercritical fluid extraction and headspace analysis are, however, now creeping in. Thus supercritical fluid extraction with carbon dioxide–methanol has been used to extract 2,4-chlorophenol from crops [231], sulfonylurea herbicides from plants [161], and organophosphorus pesticides from fruit and vegetables [226].

Headspace analysis has been employed in the extraction of dithiocarbamate insecticide in vegetables [227]. Other techniques occasionally used are vacuum distillation followed by gas chromatography–mass spectrometry in the determination of volatile organic compounds in leaves, steam distillation in the determination of organochlorine insecticides in fruit and vegetables [229], and water distillation followed by high-performance liquid chromatography in the determination of 2-aminobutane in potatoes [102, 230].

References

1. Standing Committee of Analysts (DoE)(1986) *The Sampling and Initial Preparation of Sewage and Waterworks Sludge, Soils, Sediments, Plant Materials and Contaminated Wildlife Prior to Analysis*, Second Edition, HMSO, London, UK.
2. Stoeppler M (1997) Sampling: An Introduction. In: Stoeppler M (ed) *Sampling and Sample Preparation*, Springer, Berlin, p. 1–16.
3. Lijtha K, Jarrell WM, Johnson DW, Sollins P (1999) *Long Term Ecol Res Network Ser* 2:166.
4. Fortunati M (1994) *Fresen J Anal Chem* **348**:86.

5. Epps RJ, Leonard RE, Robinson SJ (1997) *Proc Conf Geoenviron Eng: Contam Ground*, Telford, London, p. 17.
6. Neesse T, Breiter R (1996) *Aufbereit Tech* **37**:565.
7. Van Der Veen AMH, Alink A (1998) *Accredit Qual Assur* **3**:20.
8. Neesse T, Dueck J, Breiter R (1997) *Aufbereit Tech* **38**:653.
9. Eccles CS, Redford RP (1997) In: Yong RN, Thomas HR (eds) *Proc Conf Geoenviron Eng: Contam Ground*, Telford, London, p. 11–16.
10. Lancaster V, Keller-McNulty S (1998) *Environ Test Anal* **7**:15.
11. Jahnke RA, Knight LH (1997) *Deep-Sea Res Pt 1* **44**:713.
12. Kremling K, Lentz U, Zeitzschel B, Schulz-Bull DE, Duinker JC (1996) *Rev Sci Instrum* **67**:4360.
13. Dong D, Zhu X, Yang B, Liu M (1998) *Microchem J* **59**:356.
14. Thiboutot S, Ampleman G, Dube P, Hawari J, Jenkins TF, Walsh ME (1998) *29th Int Annu Conf ICT (Energetic Materials)*, 30 June–3 July 1998, Karlsruhe, Germany, p. 127.1–127.14.
15. Brown LE, Reinsch TG (1993) *ASTM STP* 1162:32–40.
16. Burton GA (1992) In: Burton GA (ed) *Sediment Toxicity Assessment*, Lewis, Boca Raton, FL, USA, p. 37–66.
17. Ministry of Agriculture, Fisheries and Food (1979) *The Analysis of Agricultural Materials*, Technical Bulletin RB427, HMSO, London, UK
18. Mitchell RL (1960) *Sci Food Agric* **10**:553.
19. Standing Committee of Analysts (DoE) (1984) The Conditionability, Filterability, Selectability and Solids Contents of Sludges. In: *Methods for the Examination of Waters and Associated Materials*, HMSO, London, UK.
20. Rubio R, Ure A (1993) *Int J Environ Anal Chem* **51**:205.
21. Houba VLG, Chardon WJ, Roelse K (1993) *Commun Soil Sci Plant Anal* **24**:1591.
22. Krishnamurti GSR Huang PM, Van Rees KCJ, Kozak LM, Rostad HPW (1994) *Commun Soil Sci Plant Anal* **25**:615.
23. Kingston HM, Walter PJ (1992) *Spectroscopy* **7**:20.
24. Feng Y, Barrett RS (1994) *Sci Total Environ* **143**:157.
25. Reynolds CM (1992) In: Iskander IK, Selim HM (eds) *Engineering Aspects of Metal-Waste Management*, Lewis, Boca Raton, FL, USA, p. 49.
26. Crosland AR, McGrath SP, Lane PW (1993) *Int J Environ Anal Chem* **51**:153.
27. Sanchez J, Garcia R, Millan E (1994) *Analusis* **22**:222.
28. Kimborough DE, Wakakuwa J (1994) *Analyst* **119**:383.
29. Davidson CM, Thomas RP, McVey SE, Perala R, Littlejohn D, Ure AM (1994) *Anal Chim Acta* **291**:277.
30. Whalley C, Grant A (1994) *Anal Chim Acta* **291**:287.
31. Lopez-Sanchez JF, Rubio R, Rauret G (1993) *Int J Environ Anal Chem* **51**:113.
32. Shan X, Chen B (1993) *Anal Chem* **65**:802.
33. Asikainen JM, Nikolaidis NP (1994) *Ground Water Monit Rem* (1994) **14**:185.
34. Thomas RP, Ure AM, Davidson CM, Littlejohn D, Ruret G, Rubio R, Lopez-Sanchez JF (1994) *Anal Chim Acta* **286**:423.
35. Real C, Barreiro R, Carballeira A (1994) *Sci Total Environ* **152**:135.
36. Ren JM, Salin ED (1994) *Spectrochim Acta Part B* **49B**:567.
37. Martens DA, Suarez DL (1997) *Environ Sci Tech* **31**:133.
38. Kalembkie-Wicz J, Filar L (1997) *Chem Anal (Warsaw)* **42**:87.
39. Mierzwa J, Dabrowolski R (1998) *Spectrochim Acta Part B* **53**:117.
40. Lopez-Garcia L, Sanchez-Merloz M, Hernandez-Cordoba M (1997) *Spectrochim Acta Part B* **52**:437.

41. Reddy PRK, Reddy SJ (1997) *Fresen Environ Bull* 6:661.
42. Marin B, Valladon M, Polve M, Monaco A (1997) *Anal Chim Acta* 342:91.
43. Usero J, Gamero M, Morillo J, Gracia I (1998) *Environ Int* 24:487.
44. Vanni A, Gennaro MC, Cignetti A, Petronio BM, Petruzelli G, Liberatori A (1997) *J Environ Sci Heal A* A32:1467.
45. Mave MR, Wragg J (1997) *Analyst* 122:1211.
46. Ma Y, Uren NC (1998) *Geoderma* 84:157.
47. Cabral AR, Lefebvre G (1998) *Water Air Soil Poll* 102:329.
48. Nyamangara J (1998) *J Agric Ecosyst Environ* 69:135.
49. Bodog I, Polyak K, Hlavay J (1997) *Int J Environ Anal Chem* 66:79.
50. Bordas F, Bourg ACM (1998) *Water Air Soil Poll* 103:137.
51. Brown LE, Reinsch TG (1993) *ASTM STP* 1162:32.
52. McBratney AB, Laslett GM (1993) *Soil Environ* 1:435.
53. Wehrens R, van Hoof P, Buydens L, Kateman G, Vossen M, Mulder WH, Bakker T (1992) *Anal Chim Acta* 271:11.
54. Lame FPJ, Defize PR (1993) *Environ Sci Tech* 27:2035.
55. Skalski JR, Word JQ (1994) *Environ Toxicol Chem* 13:15.
56. Ristenpart E, Gitzel R, Uhl M (1992) *Water Sci Tech* 25:63.
57. Houba VJG (1993) *Fresen J Anal Chem* 345:156.
58. Thoms MC (1994) *J Geochem Explor* 51:131.
59. Bowadt S, Hawthorne SB (1995) *J Chromatogr A* 703:549.
60. Deuster R, Lubahn N, Friedrich C, Kleiböhmer W (1997) *J Chromatogr A* 785:227.
61. Richter BE, Jones BA, Ezzell JL, Poter NL, Avdalovic N, Phol C (1996) *Anal Chem* 68:1033.
62. Hawthorne SB, Yang Y, Miller DJ (1994) *Anal Chem* 66:2912.
63. Hawthorne SB, Grabanski CB, Hageman KJ, Miller DJ (1998) *J Chromatogr A* 814:151.
64. Zhang Z, Pawliszyn J (1993) *Anal Chem* 65:1843.
65. Zhang Z, Pawliszyn J (1995) *Anal Chem* 67:34.
66. Santos FJ, Sarrión MN, Galceran MT (1997) *J Chromatogr A* 771:181.
67. Harrick NJ (1967) *Internal Reflection Spectroscopy*, Wiley, New York, USA.
68. Kimbraugh DE, Chin R, Wakahuwa J (1994) *Analyst* 119:1283.
69. Schwab AP, Splichal P, Sonon LS (1993) *ASTM STP* 1162:86.
70. Churnside AEM, Ritter WF (1993) *ASTM STP* 1162:92.
71. Sanchez-Brunete C, Garcia-Valcarcel AI, Tadeo JL (1994) *Chromatographia* 38:624.
72. Qiao X, During R, Humnel HE (1991) *Meded Fac Landbouwwet Rijksuniv Gent* 56:949.
73. Liu W, Pusino A, Gena C (1992) *Sci Total Environ* 39:123.
74. Schwes R, Maide FX, Fishbeck G, Lepschy von Gleisenthall J, Suess A (1993) *J Chromatogr* 641:89.
75. Schwes R, Wuset S, Lepschy Lepschy von Gleisenthall J, Maide FX, Suess A, Hock B, Fishbeck G (1994) *Anal Lett* 27:487.
76. Willianm WT, Mueller TC (1994) *J AOAC Int* 77:752.
77. Licycois E, Dehon Y, De Brabant B, Perry P, Porteteller D, Copin A (1992) *Sci Total Environ* 17:123.
78. Dieckmann H, Stockmaier M, Frenzig R, Bahandir M (1993) *Fresen J Anal Chem* 345:784.
79. Celi L, Negre M, Gennari M (1993) *Pest Sci* 38:43.
80. Miellet A, Faisil A (1986) *Expert Chim Toxicol* 72:245.
81. Lopez-Avila V, Young R, beckert WF (1993) *J AOAC Int* 76:864.
82. Coidina G, Vaqucro MT, Comellas I, Broto-Duig F (1994) *J Chromatogr* 673:21.
83. Lutermann C, Dott W, Hollender J (1998) *J Chromatogr A* 811:151.

84. Ramos L, Vreuls JJ, Brinkmann UAT (2000) *J Chromatogr A* **89**:275.
85. Oliver HS, Szolar OHJ, Rost H, Braun R, Loiber AP (2002) *Anal Chem* **74**:2379.
86. Eschenback A, Kaestner M, Bierl R, Schaefer G, Mahro B (1994) *Chemosphere* **28**:683.
87. Ball WP, Xia G, Durfee DP, Wilson RD, Brown MJ, Mackay DM (1997) *Ground Water Monit Rev* **17**:104.
88. Hewitt AD (1998) *Environ Sci Technol* **32**:143.
89. Askari MDE, Maskarinec MP, Smith SM, Bean DM, Trevis CC (1996) *Anal Chem* **68**:3431.
90. Kjeler LO, Kulp SE, Jonsson B, Rappe C (1993) *Toxicol Environ Chem* **39**:1.
91. Crecenzi C, Di Corcia A, Nazzari M, Samperi R (2000) *Anal Chem* **72**:3050.
92. US EPA (1995) *Test Methods for Evaluating Solid Waste, Method 3545*, US EPA SW-846, 3rd Edition, Update 111, USGPO, Washington, DC.
93. Ezzell JL, Ritter BE, Felix WD, Black SR, Meikle JE (1995) *LC-GC* **13**:390.
94. Richter BE, Jones BA, Ezzell JL, Porter NL, Audalovic N, Pohl C (1996) *Anal Chem* **68**:1033.
95. Dean JR (1996) *Anal Comm* **33**:191.
96. Saim N, Dean JD, Abdulla MP, Zakaria Z (1998) *Anal Chem* **70**:420.
97. Wennrich L, Popp P, Modee M (2000) *Anal Chem* **72**:546.
98. Hofler F, Ezzell J, Richter B (1995) *Labor Praxis* **4**:58.
99. Tavendale MH, Wilkins AL, Langdon AG, Mackie KL, Stuthrigde TR, Mc Farlane PN (1995) *Environ Sci Technol* **29**:1407.
100. Paasivirta J, Hakala H, Knutinene J, Otollinen T, Sarkkla J, Welling L, Paukku R, Lammi R (1990) *Chemosphere* **21**:1355.
101. Hubert A, Wenzel KD, Manz M, Weissflog L, Engewald W, Schüürmann G (2000) *Anal Chem* **72**:1294.
102. Pyle S, Marckus AB (1997) *J Mass Spec* **32**:897.
103. Richter BE, Ezzell JL, Knowles DE, Hoefler F, Mattulat AKR, Schentwinkel M, Waddall DS, Gobran T, Knurona U (1997) *Chemosphere* **34**:975.
104. Lundstedt S, Von Bavel B, Hagland D, Tysklind M, Oberg L (2000) *J Chromatogr A* **883**:151.
105. Morales-Munoz S, Luque-Garcia JL, De Castro MDL (2002) *Anal Chem* **74**:4213.
106. Ramos L, Vreuls JJ, Brinkman UAT (2000) *J Chromatogr A* **891**:275.
107. Lopez-Avila V, Young R, Beckert WF (1994) *Anal Chem* **66**:1097.
108. Alonso MC, Puig D, Silgoner I, Grasserbauer M, Barceló D (1998) *J Chromatogr A* **823**:231.
109. Danis TG, Albanis TA (1996) *Toxicol Environ Chem* **53**:9.
110. Lopez-Avila V, Benedicto J, Cheron C, Young R, Beckert VF (1995) *Environ Sci Technol* **29**:2709.
111. Silgoner L, Kriska R, Lombas E, Gans O, Rosenberg E, Grasserbauer M (1998) *Fresen J Anal Chem* **362**:120.
112. Stout SJ, Babbit BW, Dacunha AR, Safarpaur MM (1998) *J AOAC Int* **81**:1054.
113. Bac ML, Pantani F, Barbieri K, Buarini D, Grittine O (1996) *J Environ Anal Chem* **64**:23.
114. Molins C, Hogendorf EA, Heusinokveld HAG, Van Zoner P, Bauman RA (1997) *J Environ Anal Chem* **68**:155.
115. Depeyron S, Duderwel PM, Conturier D (1997) *Analysis* **25**:286.
116. Sun F, Littlejohn D, Daved Gibson M (1998) *Anal Chim Acta* **364**:1.
117. Pastor A, Vazquez E, Ciscar R, de la Guardia M (1997) *Anal Chim Acta* **344**:241.
118. Current RW, Tilotta DC (1997) *J Chromatogr A* **785**:269.
119. Hawari J, Halasz A, Beiruty A, Sas I, Tra HV (1997) *J Environ Anal Chem* **66**:299.
120. Frost SP, Dean JR, Evans KP, Harridine K, Cary C, Comber MHI (1997) *Analyst* **122**:895.

121. Conte E, Milani R, Morali G, Abballe F (1997) *J Chromatogr* 765:121.
122. Crescenzi C, D'Ascenzo G, DiCorcia A, Nazzari H, Marchese S, Samperi R (1999) *Anal Chem* 71:2157.
123. Di Corcia A, Caracciolo AB, Crescenzi C, Guiliano G, Murtas S, Samperi R (1999) *Environ Sci Technol* 33:3271.
124. Hawthorne SB, Trembley S, Moniot CL, Grabanski CR, Miller DT (2000) *J Chromatogr A* 886:237.
125. Fernandez-Perez V, Luque de Castro MD (2000) *J Chromatogr A* 902:357.
126. Yang Y, Bowadt S, Hawthorne SB, Miller DJ (1995) *Anal Chem* 67:4571.
127. Hawthorne SB, Miller DJ (1994) *Anal Chem* 66:4005.
128. Dunkers J, Groenenboom M, Scholtes LHA, Van der Heiden C (1993) *J Chromatogr* 641:357.
129. Reinde S, Hoefler F (1994) *Anal Chem* 66:1808.
130. Lee H-B, Peart TE, Hong-You RL, Gere RD (1993) *J Chromatogr* 653:83.
131. Burford MD, Hawthorne SB, Miller DJ (1993) *Anal Chem* 65:1497.
132. Meyer A, Kleiböhmer W (1993) *J Chromatogr* 657:327.
133. Guo F, Li QX, Alcantara-Licudine JP (1999) *Anal Chem* 71:1309.
134. Alexandrou N, Pawliszyn J (1989) *Anal Chem* 61:2770.
135. Hawthorn Sb, Langenfeld JJ, Miller JJ, Burford MD (1992) *Anal Chem* 64:1614.
136. Fuoco R, Griffiths PR (1992) *Ann Chim (Rome)* 82:235.
137. Onuska FI, Terry KA (1991) *J High Res Chromatogr* 14:830.
138. Onuska FI, Terry KA (1989) *J High Res Chromatogr* 12:357.
139. Wright Bw, Wright CW, Fruchter JS (1989) *Energy Fuels* 3:474.
140. Onuska FI, Terry KA, (1989) *J High Res Chromatogr* 12:527.
141. Hawthrone SB, Miller KA, Lagenfeld JJ (1990) *J Chromatogr Sci* 28:2.
142. Lohleit M, Beechmann K (1990) *J Chromatogr* 505:227.
143. Oostdyk TG, Grob RL, Snyder JL , McNally ME (1993) *Anal Chem* 65:596.
144. Peters RJB, Von Renesse M, Duivenbode JAD (1994) *Fresen J Anal Chem* 348:249.
145. Geus H, Zegers BN, Lingeman H, Brinkman UAT (1994) *Int J Environ Anal Chem* 56:119.
146. Burford MD, Hawthorne SB, Miller JD (1996) *Am Environ Lab* 8:1.
147. Emery AP, Chesler SN, MacCrehan WA (1992) *J Chromatogr* 606:221.
148. Liang S, Tilotta DC (1998) *Anal Chem* 70:616.
149. Yang Y, Hawthorne Sb, Miller DJ (1995) *J Chromatogr A* 699:265.
150. Janku J, Kubes V, Machokova Z, Kuras M (1994) *Anal Chem Bull* 3:345.
151. Futter JC, Wall P (1993) *J Planar Chromatogr* 6:372.
152. Langbehn A, Steinhart H (1994) *J High Res Chromatogr* 17:293.
153. Staub TM, Pepper IL, Abbazzadegon M, berba C (1994) *Appl Environ Microbiol* 60:1014.
154. Wong JM, Li Q, Hammock BD, Seiber JN (1991) *J Agric Food Chem* 39:1802.
155. McNally MEP, Wheeler JR (1988) *J Chromatogr* 447:53.
156. Brumley WC, Latorre E, Kelliner V, Marcus A, Knowles DE (1998) *J Liq Chromatogr R T* 21:1199.
157. Jiminez-Carmona, MM, Mancins JJ, Montoya A, Luque de Castro MD (1997) *J Chromatogr* 785:329.
158. Reddy KN, Locke MA (1994) *Weed Sci* 42:249.
159. Papillod S, Hoerdi W, Chiron S, Barcelo D (1996) *Environ Sci Tech* 30:1822.
160. Steinheimer TR, Pfeiffer RL, Scoggin KD (1994) *Anal Chem* 66:645.
161. McNally MEP, Wheeler JR (1988) *J Chromatogr* 435:63.
162. Klaffenbach P, Holland PS (1993) *J Agric Food Chem* 41:396.
163. Klaffenbach P, Holland PJ (1993) *Biol Mass Spec* 22:565.

164. Lopez-Avila V, Dodhiwela NS, Beckert WF (91993) *J Agric Food Chem* **41**:2038.
165. Field JA, Monohan K, Reed R (1998) *Anal Chem* **70**:1956.
166. Pappilond S, Haerdi W (1994) *Chromatographia* **38**:514.
167. Snyder JL, Grob RL, McNally Me, Oosterdyk TS (1993) *J Chromatogr Sci* **31**:183.
168. Snyder JL, Grob RL, McNally ME, Oostdyk TS (1992) *Anal Chem* **64**:1940.
169. McNally ME, Wheeler JR (1988) *J Chromatogr* **435**:53.
170. Schanz M, Bowadt S, Benner BA, Wize SA, Hawthorne SB (1998) *J Chromatogr* **816**:213.
171. Locke MA (1993) *J Agric Food Chem* **41**:1081.
172. Koinecke A. Krenzig R, Bahider M (1997) *J Chromatogr* **786**:155.
173. Meyer A, Kleiboehwer W, Cammann K (1993) *J High Res Chromatogr* **16**:491.
174. McNally MEP, Derdorff CM, Fahing TM, Hoffman D (1993) *Natl Meet Am Soc Div Environ Chem* **33**:327.
175. Mangin C, Dubroca J, Barriuso M (1996) *J Chromatogr A* **19**:700.
176. Fahing TM, Paulatic ME, Johnson DM, McNally MEP (1993) *Anal Chem* **65**:1462.
177. Hawthorne SB, Yang Y, Miller DJ (1994) *Anal Chem* **66**:2912.
178. Hewitt AD, Miyares DH, Leggett DC, Jenkins TF (1992) *Environ Sci Tech* **26**:1932.
179. Milana MR. Maggio A, Denaro M, Feliciani R, Gramiccioni L (1991) *J Chromatogr* **552**:205.
180. Maggio A, Milana MR, Denaro M, Feliciani R, Gramiccioni L (1991) *J High Res Chromatogr* **14**:618.
181. Paviostathis SG, Mathaven GN (1992) *Environ Technol* **13**:23.
182. Bianchi AP, Varney MS, Phillips J (1991) *J Chromatogr* **542**:413.
183. Yokouchi Y, Sano M (1991) *J Chromatogr* **555**:297.
184. Stuart JD, Miller ME, Williams-Burnett ML (1997) *J Soil Contam* **6**:439.
185. Hewitt AD, (1998) *Environ Sci Tech* **32**:143.
186. Roc UD, Lacy MJ, Stuart JD, Robbins GA (1989) *Anal Chem* **61**:2584.
187. Kawata K, Tanabe A, Saito S,Sakai M, Yasuhara A (1997) *Bull Environ Contam Toxicol* **58**:893.
188. James KJ, Stack MA (1996) *J High Res Chromatogr* **19**:515.
189. Gan J, Papiernik S, Yates SR (1998) *J Agric Food Chem* **46**:986.
190. Cahill TM, Benesch JA, Gustin MA, Zimmerman EJ, Seiber JN (1999) *Anal Chem* **71**:4465.
191. Kester PE (1975) *The Analysis of Volatile Organic Compounds in Soils by Purge and Trap Gas Chromatography*, Tekman Co., Cincinnati, OH, USA.
192. Roche AC, Miller CT (1993) *Fresen J Anal Chem* **732**:19.
193. Yang J, Her JW (1999) *Anal Chem* **71**:4690.
194. Papaefstathion I, Luque de Castro MD (1997) *J Chromatogr A* **779**:352.
195. Jones S (1993) In: Calabrese EJ, Kostecki PT (eds) *Hydrocarbon Contaminated Soils and Groundwater*, CRC, Boca Raton, FL, USA, **3**:111.
196. Ilias AM, Jaeger C (1993) In: Calabrese EJ, Kostecki PT (eds) *Hydrocarbon Contaminated Soils and Groundwater*, CRC, Boca Raton, FL, USA, **3**:147.
197. Karp KE (1993) *Ground Water Monit Rem* **13**:101.
198. Picer M, Hccenski V (1994) *Water Res* **28**:619.
199. Houba VJG (1993) *Fresen J Anal Chem* **345**:156.
200. Takuwa DT. Sawula G, Wibetoe G, Lund W (1997) *J Anal Atomic Spec* (1997) **12**:849.
201. Simmons WJ (1978) *Anal Chem* **50**:870.
202. Clinton DE (1977) *Analyst* **102**:187.
203. Heans DL (1981) *Analyst* **106**:182.
204. Heans DL (1981) *Analyst* **106**:172.
205. Arafat NM, Glooschenko WA (1981) *Analyst* **106**:1174

206. Hoenig M, Elsen YV, Couter RV (1988) *Anal Chem* **58**:777.
207. Mortatti FJ, Krug FS, Pessendra LCR, Zagatto EAG, Jorgensen SS (1982) *Analyst* **107**:659.
208. Bradfield EG, Stickland JF (1975) *Analyst* **100**:1.
209. Jayman TCZ, Sivasubramanian S, Wijedasa MA (1975) *Analyst* **100**:716.
210. Westlake WE, Monick J, Gunther F (1972) *Bull Environ Contam Toxicol* **8**:109.
211. Crossley JJ (1970) *J AOAC Int* **95**:935.
212. DiAmato A, Semerara I, Biechi C (1993) *J AOAC Int* **76**:657.
213. Baker PG, Bottomley P (1982) *Analyst* **107**:206.
214. Lagana A, Marino A, Fago G, Pardo-Martinez B (1994) *Analusis* **22**:63.
215. Nakamura Y, Tonogai Y, Tsumuar Y, Ito Y (1993) *J AOAC Int* **768**:1348.
216. Fang GF, Fan CL, Chao YZ, Zhao TS (1994) *J AOAC Int* **77**:738.
217. Wan HB, Wong MK, Lim PY, Mok CY (1994) *J Chromatogr* **662**:147.
218. Getzendaner ME (1969) *J AOAC Int* **52**:824.
219. Kvalväg J (1974) *Analyst* **99**:666.
220. Read AN (1988) *J Chromatogr* **438**:383.
221. Steinwand H (1992) *Fresen Z Anal Chem* **343**:887.
222. Branckhoff S, Thier HPZ (1987) *Lebensm Unters Forschung* **184**:91.
223. Vaverde-Garcia S, Fernandez Alba AR, Herrara JC, Radon E (1994) *J AOAC Int* **77**:1041.
224. Tovatcheva EG (1979) *Analyst* **104**:79.
225. Liang L, Horvat M, Cernichia Z, Gerlain B, Bolagh S (1996) *Talanta* **43**:1883.
226. Wuchner K, Ghijsen RT, Brinkman UAT, Grob R, Mathieu J (1993) *Analyst* **118**:11.
227. Society of Analytical Chemistry Committee (1981) *Analyst* **106**:782.
228. Hiatt MH (1998) *Anal Chem* **70**:851.
229. Parrends M, Larson BJ (1993) *Trace Microprobe Technology* **11**:133.
230. Scudamore KA (1980) *Analyst* **105**:1171.
231. Thompson CA, Chesney DJ (1992) *Anal Chem* **64**:848.

2 Determination of Metals in Soils

The presence of deliberately added or adventitious metallic compounds in soils can cause contamination of the tissues of crops grown on the land or animals feeding on the land, and can consequently cause adverse toxic effects on man, animals, birds and insects. Drainage of these substances from the soil can also cause pollution of adjacent streams, rivers and eventually the oceans. Some of the metal-containing substances included in this category are fertilisers, crop sprays, sheep dips, etc. A major source of metal contamination of soil arises from the addition of sewage sludge to land, especially if the sewage originates in a sewage treatment plant handling industrial effluent.

Most of the elements in the periodic table have been found in soils, some naturally occurring and others deliberately added.

Analyses used for these elements are discussed next, in alphabetical order.

2.1
Actinides

See under "Multi-Cation Analysis" (Sect. 2.55).

2.2
Aluminium

An early spectrophotometric method [1] for aluminium in soil involves the use of a Technicon sample changer, proportioning pump and automatic colorimeter. The method is based on the measurement of the rate of colour development in the reaction between aluminium and xylenol orange in ethanolic media. The calibration graph is rectilinear up to 2.7 mg/l aluminium and the coefficient of variation is 4.5%.

Flow injection analysis has been used to determine aluminium in soil. Reis et al. [2] studied the spectrophotometric determination of aluminium in soil using merging zones and sequential addition of pulsed reagents.

Tecator [3] has described a flow injection method for the determination of 0.5–100 mg/l aluminium in 0.1 M potassium chloride extracts of soils in which the acidified soil extract is injected into a carrier stream which has the same composition as the sample matrix, (i.e., 0.1 M KC1) and merged with a masking solution for iron (hydroxylamine and 1,10-phenanthroline monohydrate or o-phenanthroline hydrochloride) and subsequently with the colour reagent for aluminium (pyrocatechol violet and aqueous hexamethylene tetramine buffer). The coloured complex formed between aluminium and pyrocatechol violet is measured at 585 nm. Repeatability is 1% RSD.

In addition to the above method, based on the use of pyrocatechol violet, Tecator also describes a flow injection analysis for determining 0.5–0.5 mg/l aluminium in soil extracts based on the measurement of the chromazurol–aluminium complex at 570 nm [4, 5].

Ross et al. [6] analysed samples of soil leachates from laboratory columns and of soil pore water from field porous cup lysimeters for aluminium by atomic absorption spectrometry under two sets of instrumental conditions. Method 1 employed uncoated graphite tubes and wall atomisation; method 2 employed a graphite furnace with a pyrolytically coated platform and tubes. Aluminium standards were prepared and calibration curves used for the colorimetric quantification of aluminium. Method 1 gave results which compared favourably with method 2 in terms of both sensitivity and interference reduction for samples containing 1–15 uM aluminium.

The determination of aluminium is also discussed under "Multi-Cation Analysis of Soils" in Sects. 2.55 (inductively coupled plasma atomic emission spectrometry) and 2.55 (emission spectrometry).

Mitrovic et al. [7] and Kozuh et al. [8] have carried out aluminium speciation studies on soil extracts. Various workers [9–11] have discussed the determination of aluminium in soils. Using isotachoelectrophoresis, Schmidt and coworkers [12] were able to differentiate aluminium(III) and aluminium species in soil leachates.

2.3
Ammonium

Keay and Menage [13] have described an automated method for the determination of ammonium and nitrate in 2 M potassium chloride extracts of soil. In this method, a sample of soil (2 g) is shaken for one hour with 2 M potassium chloride (20 ml) and the filtrate is diluted, in the AutoAnalyser, with a 0.25% suspension of magnesium oxide; the ammonia evolved is absorbed in 0.1 M hydrochloric acid and determined spectrophotometrically at 625 nm by the indophenol method. The sum of ammonium plus nitrate is determined similarly, but with addition of 4.5% titanous chloride solution before distillation; this reduces nitrate but not nitrite.

Waughman [14] has described a microdiffusion method for the determination of ammonium and nitrate in soils. Nitrate in the sample solution is reduced to ammonia by titanous sulfate and the ammonia is then released from the solution and diffused and absorbed onto a nylon square impregnated with dilute sulfuric acid. The nylon is then put into a solution which colours quantitatively when ammonia is present, and a spectrophotometer is used to measure the colour.

Adler et al. [15] describe a method for determining low levels of ammonium ions in solution, in which the ammonium ion is oxidised with sodium hypobromite in alkaline medium and the evolved nitrogen is passed into an argon plasma:

$$2NH_3 + 3NaBr = 3NaBr + 3H_2O + N_2$$

The nitrogen–hydrogen emission intensity produced in the plasma at 336 nm is monitored. A practical detection limit of 0.1 ug nitrogen per ml was obtained for 5 ml aqueous sample solutions. The method has been applied to determine the exchangeable ammonium contents of soil samples.

The instrumental system employed utilised a 2 kW crystal-controlled radiofrequency generator operating at 27 MHz (International Plasma Corporation, model 120-27, Hayward, CA, USA) and a 1 m plane grating scanning monochromator (Monospek 1000, Rank Hilger Ltd., Margate, UK). A demountable plasma torch with tangential argon inlets and sample introduction from a central injector tube was used. The outer quartz tubing was extended to a height of 40 mm above the work coil to prevent entrainment of atmospheric nitrogen into the discharge.

Tecator Ltd. [16] have described a flow injection analysis method for the determination of 0.2 – 1.4 mg/l (as NH_3N) of ammonia nitrogen in soil samples extractable by 2 M potassium chloride. The soil suspension in 2 M potassium chloride is centrifuged and filtered and introduced into the flow injection system for the analysis of ammonia (and nitrate) one parameter at a time. Ammonia is determined by the gas diffusion principle, in which a PTFE membrane is mounted in the gas diffusion cell.

HMSO (UK) [17] have published a method for the determination of ammonia, nitrate and nitrite in potassium chloride extracts of soil extracts. An aliquot of the extract is made alkaline and the ammonia released, originating from ammonium ions, is determined either with an ammonia-selective probe or, after removal by distillation, by titration (Crompton TR, private communication).

2.4
Antimony

Chikhalikar et al. [18, 19] have discussed the speciation of antimony in soil extracts and soils. Asami et al. [20] have reviewed methods for the determination of antimony in soils.

Various other techniques that have applied to the determination of antimony in multi-cation analysis include atomic absorption spectrometry (Sect. 2.55), inductively coupled plasma atomic emission spectrometry (Sect. 2.55), neutron activation analysis (Sect. 2.55) and photon activation analysis (Sect. 2.55).

2.5
Arsenic

Arsenic occurs naturally in the Earth's crust, but a considerable amount of arsenic is added to the environment through its use in wood preservatives, sheep dips, fly paper, arsenical soaps, rat poison, glass additives, dye pigment for calico prints, wallpaper, lead shot and pesticides. During 1971, the estimated production of organoarsenical herbicides such as monosodium methanearsenate, disodium methanearsenate and hydroxydimethylarsine oxide (cacodylic acid) in the USA was 10.7×10^8 kg [13]. Generally, soils contain about 5 ppm of arsenic, but soils with a known history of arsenic application average about 165 ppm [21]. In some places such as Buns, Switzerland and Wiatapu Valley, New Zealand, the arsenic level in the soil may reach 10^4 ppm [22]; a substantial portion of the arsenic in soil and soil-like material (sediment, clay, sand, etc.), is expected to be found in a soluble form and can probably can be easily dislodged by the action of water moving through the soil. Soluble forms of arsenic are relatively more mobile in the environment and pose a greater potential for contaminating both ground water and surface water. Soluble forms of arsenic from soil and soil-like material are likely to enter a bioconversion chain through their initial uptake by vegetation.

An early method for the determination of arsenic in soils is that of Forehand et al. [23]. This method is based on the selective extraction of arsenic(III) by benzene and analysis of the extract by atomic absorption spectrometry. Firstly the soil is allowed to stand with 9.9 M hydrochloric acid for 12 hours, and then the arsenic is reduced from arsenic(V) to arsenic(III) with stannous chloride and potassium iodide. Following adjustment to pH 9 with hydrochloric acid, the aqueous phase is extracted with benzene. The benzene extract is then treated with water and the water extract analysed by atomic absorption spectrometry at 193.7 nm. An average recovery of 88% of the arsenic present in sandy soils was achieved by this procedure.

More recently Lopez-Garcia et al. [24] determined arsenic and antimony in soil by slurry sampling and graphite furnace atomic absorption spectrometry.

To avoid problems previously encountered with flame atomic absorption spectrometry of arsenic, and also with flameless methods such as that in which the element is converted to arsine, Ohta and Suzuki [25] proposed an alternative method based on electrothermal ionisation with a metal microtube atomiser. Effective atomisation can be achieved by the addition of thiourea to the arsenic solution or by preliminary extraction of the arsenic–thionalide complex. The second method is recommended for soil samples so as to avoid interference due to the presence of trace elements.

A UK standard method also discusses the determination of arsenic in soil by atomic absorption spectrometry [26].

Licht and Skogerboe [27] determined arsenic down to 5 mg/l in plants and soils by atomic emission spectroscopy.

Extractable arsenic in soil has been determined by slurry sampling on-line, microwave extraction and hydride generation [28].

The determination of arsenic by atomic absorption spectrometry with thermal atomisation and with hydride generation using sodium borohydride has been described by Thompson and Thomerson [29], and it was evident that this method could be modified for the analysis of soil. Thompson and Thoresby [30] have described a method for the determination of arsenic in soil by hydride generation and atomic absorption spectrophotometry using electrothermal atomisation. Soils are decomposed by leaching with a mixture of nitric and sulfuric acids or fusion with pyrosulfate. The resultant acidic sample solution is made to react with sodium borohydride, and the liberated arsenic hydride is swept into an electrically heated tube mounted on the optical axis of a simple, laboratory-constructed absorption apparatus.

The advantages of high sensitivity, rapid analysis and simplicity of equipment are discussed, and the results for both types of sample material are compared with values obtained through use of the molybdenum blue method.

Haring et al. [31] determined arsenic and antimony by a combination of hydride generation and atomic absorption spectrometry. These workers found that, compared to the spectrophotometric technique, the atomic absorption spectrophotometric technique with a heated quartz cell suffered from interferences by other hydride-forming elements.

The recommended procedure for the determination of arsenic and antimony involves the addition of 1 g of potassium iodide and 1 g of ascorbic acid to a sample of 20 ml of concentrated hydrochloric acid. This solution should be kept at room temperature for at least five hours before initiation of the programmed MH 5-1 hydride generation system, i.e., before addition of ice-cold 10% sodium borohydride and 5% sodium hydroxide. In the hydride generation technique the evolved metal hydrides are decomposed in a heated quartz cell prior to determination by atomic absorption spectrometry. The hydride method offers improved sensitivity and lower detection limits compared to graphite furnace atomic absorption spectrometry. However, the most important advantage of hydride-generating techniques is the prevention of matrix interference, which is usually very important in the 200 nm area.

Jiminez de Blas et al. [32] have reported a method for the determination of total arsenic in soils based on hydride generation atomic absorption spectrometry and flow injection analysis. The method gave good recoveries and had a detection limit below 1 µg/l for an injection volume of 160 µl.

Merry and Zarcinas [33] have described a silver diethyldithiocarbamate method for the determination of arsenic and antimony in soil. The method involves the addition of sodium tetrahydroborate to an acid-digested sample which has been treated with hydroxylammonium chloride to prevent the formation of insoluble antimony compounds. The generated arsine and stibine react with a solution of silver diethyldithiocarbamate in pyridine in a gas washtube. Absorbance is measured twice at wavelengths of 600 and 504 nm.

The concentration of arsenic can be determined at 600 nm because the Sb–Ag DDTC complex does not absorb light of this wavelength. The molar absorptivity of the antimony complex with Ag DDTC reaches its maximum value at 504 nm, but there is also appreciable light absorbance from the As–Ag DDTC complex at this wavelength. The antimony concentration can be calculated from the total extinction value measured at 504 nm by subtracting the extinction value (at 504 nm) that corresponds to the previously determined arsenic concentration. It is clear that calibration curves of arsenic at 504 and 600 nm and of antimony at 504 nm are needed to perform the calculation.

The limitations of the Gutzeit method for determining arsenic are well-known. The spectrophotometric molybdenum blue or silver diethyldithiocarbamate procedures tend to suffer from poor precision. Sandhu [34] has described a spectrophotometric method for the direct determination of hydrochloric acid-releasable inorganic arsenic in soils and sediments. The method provides reliable data on the quantitative recovery of 2.0 μg of arsenic(V) added to 5.0 g (0.4 mg/kg) of soil, clay, sand and sediment samples. The method is simple, reliable and relatively rapid; 24 samples can be analysed in about an hour. It does not require elaborate equipment and can be routinely used for the quantitative determination of arsenic in soil and soil-like material. The detection limit has been established as 0.5 μg of arsenic. The extent of ionic interference when this method is used for arsenic determination in soil was also quantitatively evaluated.

Thomas et al. [35] used coupled high-performance liquid chromatography with inductively coupled plasma mass spectrometry to determine various forms of arsenic in soil.

Gas chromatography and high-performance liquid chromatography have both been combined with the introduction of hydride generation into inductively coupled plasma mass spectrometry for the speciation determination of arsenic in soils [36].

Van Laecke et al. [37] determined arsenic in solid plant material by electro-vaporisation–inductively coupled plasma mass spectrometry. Use of an internal standard (antimony) is important when obtaining accurate results.

Hydride generation inductively coupled plasma atomic emission spectrometry has been used to determine arsenic in soils. This technique was found to greatly reduce sample preparation time [38].

Lasztity et al. [39] have reported on inductively coupled plasma mass spectrometric methods for the determination of total arsenic in soils.

Barra et al. [40] have described a microwave-assisted procedure based on atomic fluorescence for the quantitative determination of down to 0.006 μg/g of inorganic arsenic in soils.

Demesmay and Olle [41] showed that the use of microwave extraction procedures work well for mineral arsenic but may affect the relative amounts of various arsenic species.

Agemian and Bedak [42] have described a semi-automated method for the determination of total arsenic in soils. Chappell et al. [43] have described an inexpensive but effective method for the quantitative determination of arsenic species in contaminated soils. Chappell found that the extraction efficiency varied with the ratio of soil to acid and with the concentration of the acid. Rurikova and Beno [346] accomplished speciation of arsenic(III) and arsenic(V) in soils by cathodic stripping voltammetry. Wenclawiak and Krah [347] used reactive supercritical fluid extraction in speciation studies of inorganic and organic arsenic in soils. In this method, derivatisation with thioglycollic acid methyl ester was performed in supercritical carbon dioxide. Various other workers have discussed the determination of arsenic in soils [44–46].

Naidu et al. [9] showed that the separation of arsenic species from soil solutions could be performed in less than five minutes by using capillary electrophoresis. Levels of arsenic down to 0.1–0.5 ng/l can be detected.

The determination of arsenic is discussed under "Multi-Metal Analysis of Soils" in Sect. 2.55.

2.6
Barium

The determination of barium is discussed under "Multi-Metal Analysis of Soils" in Sect. 2.55.

2.7
Beryllium

The determination of beryllium is discussed under "Multi-Metal Analysis of Soils" in Sect. 2.55.

2.8
Bismuth

The determination of bismuth is discussed under "Multi-Metal Analysis of Soils" in Sect. 2.55.

Asami et al. [20] have reviewed methods for the determination of bismuth in soils.

2.9
Cadmium

Cadmium is readily taken up by most plants. The occurrence of cadmium in motor oils, car tyres, phosphorus fertilisers and zinc compounds explains its accumulation in soils. The cadmium contents of soils in unpolluted areas are below 1 ppm, but values as high as 50 ppm can be found [50].

Nitric perchloric acid soluble cadmium has been determined in soils by an official method, which involves examination of the acid digest at 228.8 nm by atomic absorption spectrometry [48].

Extractable cadmium in soil has been determined by extraction with 0.5 M acetic acid followed by extraction with a chloroform solution of pyrrolidine dithiocarbamate, then decomposition of the cadmium complex with hydrochloric acid and determination of cadmium by atomic absorption spectrometry at 228.8 nm [49].

The determination of cadmium by graphite furnace atomic absorption spectrometry is especially difficult because cadmium is a volatile element, and matrix constituents cannot be removed by charring without a loss of cadmium. The use of selective volatilisation often makes it possible to obtain a cadmium peak before the background has risen to such a high value that it interferes with the cadmium measurement. Another unrecognised source of interference is char loss resulting from the salt matrix. Although uncoated graphite tubes can be used for the determination of cadmium because of its volatility, some workers have found that pyrolytically coated tubes give better results when cadmium is determined in the presence of high contents of alkali and alkaline-earth elements [51]. Many studies of the determination of cadmium in soil extracts have been reported, but a chelation–extraction step has always been used prior to determination by graphite furnace atomic absorption spectrometry in order to reduce matrix interferences and to improve detection limits.

Atomic absorption spectrometry with or without preliminary solvent extraction of metal has been applied extensively to the determination of cadmium in soils.

Berrow and Stein [52] have described a procedure based on digestion with aqua regia followed by atomic absorption spectrometry for the determination of cadmium (also iron and zinc) in soils. The soil sample was air dried at a maximum temperature of 30 °C and sieved through a 2 mm sieve. The sieved soil was mixed coned and quartered and about 30 g ground in an agate planetary mill for 30 minutes to < 150 μ size. Three grams of soil were weighed into a 100 ml flask and 2–3 ml water added, and then 7.5 ml concentrated hydrochloric acid and 2.5 ml concentrated nitric acid were added per gram of dry sample. The flask is covered and left to digest at 20 °C for 16 hours. A 30 cm reflux condenser is attached to the top of the flask which is then boiled gently for two hours on a temperature-controlled electrothermal extraction apparatus.

After cooling, the condenser is rinsed with 30 ml water and the solution filtered into a 100 ml calibrated flask. The filter paper is rinsed five times with a few millilitres of warm (250 °C) 2 M nitric acid. After cooling, the flask contents are made up to 100 ml with 2 M nitric acid. The solution was then analysed for cadmium by atomic absorption spectrometry equipped with a single-slot burner and an air–acetylene flame using a 228.8 nm hollow cathode lamp. A relative standard deviation for cadmium of 3.4–5.2% was obtained.

Baucells [53] applied graphite furnace atomic absorption spectrometry to the determination of cadmium in soils with a precision of 0.4% at the 69 μg/g cadmium level. The loss of cadmium during the charring cycle was high, preventing the use of any char in the atomisation process in order to remove the organic matrix or minimise interference effects.

The application of inductively coupled plasma atomic emission spectrometry and graphite furnace atomic absorption spectrometry to the determination of cadmium (and molybdenum) in soils has been discussed by Baucells et al. [53]. Baucells et al. chose the 228.802 nm cadmium line because it is well resolved from the 228.763 nm iron line with the spectrometer used in this work. Background measurements could only be carried out at +0.05 nm. These workers obtained good agreement between cadmium values obtained by direct graphite furnace atomic absorption spectrometry and inductively coupled plasma atomic emission spectrometry. Chelation extraction procedures that require extensive sample handling are avoided.

Problems in the direct determination of cadmium in soil extracts by graphite furnace atomic absorption spectrometry are overcome by the use of a low atomisation temperature of 1200 °C (mini-furnace or high heating rate of > 2000 °C/s), the addition of molybdenum, hydrogen peroxide and nitric acid as a matrix modifier, and accurate optimisation of the instrumental parameters.

The addition of ammonium dihydrogen phosphate and ammonium sulfate, normally used as matrix modifiers in the determination of cadmium, is not recommended with this type of sample because of the appearance of multiple peaks.

Inductively coupled plasma atomic emission spectrometry has proved to be an excellent technique for the direct analysis of soil extracts because it is precise, accurate and not time-consuming, the level of matrix interference being very low. Of course, the graphite furnace technique yields better detection limits than the inductively coupled plasma procedure.

Reddy and Reddy [54] used differential pulse anodic stripping voltammetry on chloroform extracts of the xanthate to determine cadmium in soil.

Microwave extraction of cadmium in soil gave results comparable to those obtained by conventional extraction procedures [55].

Lewin and Beckett [56] have shown that cadmium added to soils treated with sewage will quickly divide between a number of different forms of combination from some of which it can become available to plants during crop growth. These workers investigated reagents which can extract cadmium and make it available for analysis. Acidified fluoride and EDTA were effective extractants.

Christensen and Lun [57] developed a speciation procedure using a cation-exchange resin (Chelex 100) in a sequential batch/column/batch system for determining free divalent cadmium and cadmium complexes of various stabilities at the cadmium concentrations typically found in landfill leachates (less than 100 pg/l). Results obtained on standardised solutions containing cadmium and on two actual leachates are included. The leachates had only a small percentage of free divalent cadmium and a large percentage of labile complexes.

Turner et al. [58] discussed the limitations in research on adsorption of trace metals on soils owing to inadequate control of composition and pH of the equilibrium solution. The use of chelating resins is suggested to establish and maintain constant pH and metal activity in a solution of constant ionic strength and composition. Details are given of the preparation of suitable resins and the experimental procedure used to investigate the adsorption of cadmium on iron gel and on organic matter over a range of cadmium:calcium ratios similar to those found under normal soil solution concentrations. The suggested method was more difficult and more time-consuming than conventional equilibration methods. In some cases, however, its use may make determination of an entire adsorption isotherm unnecessary, since adsorption may be determined in response to one or more predetermined metal activities. It could also be used to evaluate possible mechanisms of metal adsorption.

Roberts et al. [59] have discussed the simultaneous extraction and concentration of cadmium and zinc from soil extracts. Extractions were conducted with calcium chloride adjusted to various pH values between 3 and 11. The simultaneous recovery of cadmium and zinc was essentially quantitative over the pH range 4–7, with values ranging from 92 to 102%. An extraction at pH 4.5 was adopted. Adequate recoveries were obtained when the procedure was applied to spiked soils.

Carlosena et al. [60] and Hirsch and Banin [61] have conducted studies on the speciation of cadmium in soil. Feng and Barrett [62] showed that microwave dissolution of soil and dust samples with nitric–hydrofluoric acid gave recoveries of cadmium (and lead) of over 90% for a 30-minute digestion. Various other workers [65–68] have reviewed methods for the determination of cadmium in soils.

The determination of cadmium is also discussed under "Multi-Metal Analysis of Soils" in Sect. 2.55.

2.10
Caesium

Caesium contamination of soil is a system for which spectroscopic imaging investigations of contaminants would be of high interest [67, 68]. Caesium isotopes comprise one of the lasting health problems of the Chernobyl accident [69,70]. The 134 and 137 isotopes decay by γ-emission and are formed in high fission yields. The 137 isotope has a moderately long half-life. Caesium can be highly mobile in some environments, and geochemically it has many of the same characteristics as potassium because of its similar ionic radius [68].

Hence, there is a motivation to understand the interaction of caesium with naturally occurring mineral surfaces at the molecular level.

Caesium sorption has been investigated extensively, primarily by using sequential extractions together with γ-spectroscopy (for radioisotopes) or atomic absorption (for detection) [72, 73]. This approach has been applied to the study of caesium contamination of soils [73]. Caesium was shown to prefer the mineral soil horizons in high organic soils [74]. From these studies, it has been possible to infer mechanistic details: caesium will tenaciously adhere to adsorption sites and can be supplanted only by K^+ and NH_4^+. It appears to prefer surface "defects", which have been termed frayed edges, and wedge sites [75–77]. However, understanding of caesium soil systems would benefit from direct spectroscopic information.

To study caesium speciation on soil particles, Groenewold et al. [78] used imaging time-of-flight secondary ion mass spectrometry (ToFSIMS) and also scanning electron microscopy/energy-dispersive X-ray spectroscopy (SEM/EDS). The results showed that Cs^+ could be readily detected and imaged on the surface of the soil particles at concentrations down to 160 ppm, which corresponds to a 0.04 monolayer. Imaging revealed that most of the soil surface consisted of aluminosilicate material. However, some of the surface was more quartzic in composition, primarily silica with little aluminium. It was observed that adsorbed Cs^+ was associated with the presence of aluminium on the surface of the soil particles. In contrast, in high-silica areas of the soil particle where little aluminium was observed, little adsorbed Cs^+ was observed on the surface of the soil particle. Using EDS, Cs^+ was observed only in the most concentrated Cs^+-soil system, and Cs^+ was clearly correlated with the presence of aluminium and iodine. These results are interpreted in terms of multiple layers of caesium iodine forming over areas of the soil surface that contain substantial aluminium. These observations are consistent with the hypothesis that the insertion of aluminium into the silica lattice results in the formation of anionic sites, which are then capable of binding cations.

The determination of caesium in soil by multi-element analysis is discussed in Sect. 2.55 (isotope dilution analysis).

2.11
Calcium

Xing-Chu and Yu Sheng [79] have described a spectrophotometric method for the determination of exchangeable calcium in soil. In this method, a portion of an aqueous extract of the soil is treated with ammoniacal ammonium acetate and an aqueous solution of chlorophosphonazo–mA. The solution is evaluated spectrophotometrically at 630 nm. Recoveries of calcium are 99% and relative standard deviations of between 0.9% at the 11 mequiv/100 g of soil level and 3.1% at the 2 mequiv/100 g soil level are obtained.

Soil cation exchangeable capacity is an index used both to evaluate the nutrient and water retention ability of the soil and as an important basis for the amelioration of soil and to apply, rationally, fertiliser. Exchangeable cations absorbed by soil colloid include K^+, Na^+, Ca^{2+}, Mg^{2+}, Al^{3+} and H^+. K^+, Na^+, Ca^{2+} and Mg^{2+} are exchangeable bases. Al^{3+} and H^+ are exchangeable acids and the sum of these ions is known as the cation exchangeable capacity. Exchangeable Cu^{2+}, Zn^{2+} and Mn^{2+} are present at negligible concentrations.

Among the numerous methods used to determine the total amount of exchangeable metal cations in acidic soils, the 1 M ammonium acetate leaching method and 0.1 M hydrochloric acid extraction–titrimetric methods are the best known and most applied. The former involves complete evaporation of the solution obtained after eluting the soil, and ignition of the residue to change all of the exchangeable metal salts into their carbonate form. The carbonates are dissolved in standard hydrochloric acid and the excess of acid back-titrated with standard sodium hydroxide solution. The drawback of this method is that the Al^{3+} and Fe^{3+} ions precipitate as hydroxides during the titration and absorb the indicator. The endpoint of the titration is indistinct and no exact result can be obtained. The same problem also occurs in the 0.1 M hydrochloric acid extraction–titrimetric method. Moreover, some nonexchangeable metal also dissolves when soil is extracted with hydrochloric acid, so the result includes exchangeable and some nonexchangeable metal ions in the soil.

According to the pH balance method recommended by Jackson [80], acetic acid can be used to extract the total amount of exchangeable metal cations from an acidic soil, and the pH change is measured carefully. This is then compared with the pH calibration graph of acetic acid so as to obtain the decrease in H^+ in solution. However, this method requires a very precise acidity measurement (generally an accuracy of ± 0.01 pH units when the pH value is within 2.3–2.8). To overcome this difficulty, Xing-Chu and Ying-Quan [79] have developed the bromophenol blue spectrophotometric method for the determination of exchangeable calcium in soils.

The determination of chromium is also discussed under "Multi-Metal Analysis of Soils" in Sect. 2.55 (atomic absorption spectrometry), Sect. 2.55 (inductively coupled plasma atomic emission spectrometry), Sect. 2.55 (emission spectrometry), Sect. 2.55 (photon activation analysis), Sect. 2.55 (neutron activation analysis), and Sect. 2.55 (differential pulse anodic stripping voltammetry).

2.12
Cerium

The determination of cerium in multi-element analysis is discussed in Sect. 2.55 (neutron activation analysis).

2.13
Chromium

Qi and Zhu [81] investigated a highly sensitive method for the determination of chromium in soils. In this method, chromium(VI) is reacted with o-nitrophenyl-fluorone in the presence of cetyltrimethyl ammonium bromide to form a purplish-red complex at pH 4.7 to 6.6 by heating at 50 °C for ten minutes. The composition of the complex was determined as 1:2:2-chromium(VI):NPF:CTAB. The wavelength of maximal absorbance was 582 nm and the molar absorptivity was 111 000 litres per mole/cm. Beer's law was obeyed up to 0.2 ug/l chromium(VI). Interference due to copper(II), iron(III) and aluminium(III) was eliminated by the addition of a masking reagent containing potassium fluoride, trans-1,2-diaminocyclohexanetetraacetic acid and potassium sodium tartrate. This method was more sensitive than the diphenylcarbazone method.

Fodor and Fischer [84] have investigated problems of chromium speciation in soils. When employing spectrophotometric detection, only a method based on the diphenylcarbazide reaction was found suitable for chromium speciation analysis.

Smith and Lloyd [82] determined chromium(VI) in soil by a method based on complexation with sodium diethyldithiocarbamate in pH 4 buffered medium followed by extraction of the complex with methylisobutylketone and analysis of the extract by atomic absorption spectrometry (Evans R, City Analyst, Dundee, UK, private communication) [86]. Using this method, levels of chromium(V) of between 90 and 176 mg/l were found in pastureland on which numerous cattle fatalities had occurred.

Chakraborty et al. [87] determined chromium in soils by microwave-assisted sample digestion followed by atomic absorption spectrometry without the use of a chemical modifier.

The sequential extraction of chromium from soils has been studied [89]. A three-step sequential extraction scheme has been proposed using acetic acid, hydroxylamine hydrochloride and ammonium acetate as extracting agents. Steps 1 and 2 were measured by electrothermal atomic absorption spectrometry (ETAAS). Step 3 was measured by flame atomic absorption spectrometry. Interfering effects when measuring chromium in soils were circumvented through the use of a 1% δ-hydroxyquinoline suppressor agent.

Prokisch et al. [85] described a simple method for determining chromium speciation in soils. Separation of different chromium species was accomplished by the use of acidic activated aluminium oxide. Polarographic methods have been applied in speciation studies on chromium(VI) in soil extracts [86]. Milacic et al. [88] have reviewed methods for the determination of chromium(VI) in soils.

Mierzwa et al. [90] determined chromium in soil by modifier-free slurry sampling with an overall analytical repeatability of better than 20%.

X-ray fine-structure spectroscopy has been used to determine the Cr(VI):Cr(III) ratio in soils [91]. The Cr(VI):Cr(III) ratio in extracts of soils has been determined using X-ray absorption near-edge structure spectroscopy [92].

Kalembkiewicz and Filar [93] have reported on the effect of the soil sample preparation procedure on the determination of chromium in soils.

Marques [94] has reviewed literature on chromium speciation in soils. The determination of chromium is also discussed under "Multi-Metal Analysis of Soils" in Sect. 2.55.

2.14
Cobalt

A method based on measurement of the ammonium pyrollidine dithiocarbamate complex at 240.7 nm has been described for the determination of 0.5 m acetic acid [96] and nitric–perchloric acid-soluble cobalt in soils [95].

The determination of cobalt by atomic absorption spectrometry is discussed under "Multi-Metal Analysis of Soils" in Sect. 2.55.

The determination of cobalt is also discussed under "Multi-Metal Analysis of Soils" in Sect. 2.55.

2.15
Copper

Official methods have been published for the determination of nitric–perchloric acid-soluble copper in soil [97] and ethylenediaminetetraacetic acid-soluble copper in soil [98]. The former method involves atomic absorption spectrometric evaluation of the acid digest and the second method involves extraction of the soil with an aqueous solution of ammonium EDTA and atomic absorption spectrometric evaluation of the extract.

Mesuere et al. [99] and Gerringa et al. [100] have reviewed methods for the determination of copper in soils. Residual copper(II) complexes have been determined in soil by electron spin resonance spectroscopy. Fast neutron activation analysis has been studied [101] as a screening technique for copper and (zinc) in waste soils. Experiments were conducted in a sealed tube neutron generator and a germanium γ-ray detector.

The determination of copper is also discussed under "Multi-Metal Analysis of Soils" in Sect. 2.55 (atomic absorption spectrometry), Sect. 2.55 (emission spectrometry), Sect. 2.55 (inductively coupled plasma atomic emission spectrometry), Sect. 2.55 (photon activation analysis), Sect. 2.55 (neutron activation analysis), Sect. 2.55 (electron probe microanalysis) and Sect. 2.55 (differential pulse anodic stripping voltammetry).

2.16
Curium

The determination of curium in soil by α-spectrometry is discussed under "Multi-Metal Analysis of Soils" in Sect. 2.55.

2.17
Europium

The determination of europium in soil by neutron activation analysis is discussed under "Multi-Metal Analysis of Soils" in Sect. 2.55.

2.18
Hafnium

The determination of hafnium in soil by neutron activation analysis is discussed under "Multi-Metal Analysis of Soils" in Sect. 2.55.

2.19
Indium

The determination of indium in soil by atomic absorption spectrometry is discussed under "Multi-Metal Analysis of Soils" in Sect. 2.55.

2.20
Iridium

Stefanov and Daieva [102] determined micro amounts of iridium in soil by neutron activation analysis. The soil is activated for 30 hours in a neutron flux of $\approx 5 \times 10^{12}$ neutrons/cm/s, then left for 2–3 days to decay, and then iridium was determined from the 317 keV peak of ^{123}iridium. Down to 30 ng of iridium in soil could be determined.

2.21
Iron

The determination of total iron in soils has been discussed by Jayman et al. [103]. This method is based on the formation of the 1,10-phenanthroline complex of iron. Unfortunately, aluminium also forms a similar complex which exhibits identical absorption characteristics. However, iron can be determined without interference following the removal of aluminium and phosphates. In this method, finely ground soil is ignited overnight at 450 °C and the residue dissolved in nitric acid–hydrochloric acid–water (25 + 25 + 50v/v). Following the separation of iron from aluminium and phosphate and the formation of the 1,10-phenanthroline complex, iron is determined spectrophometrically at 490 nm in the aluminium and phosphate-free extract.

The determination of iron is also discussed under "Multi-Metal Analysis of Soils" in Sects. 2.55 (atomic absorption spectrometry), 2.55 (inductively coupled plasma atomic emission spectrometry), 2.55 (neutron activation analysis), 2.55 (photon activation analysis), 2.55 (differential pulse anodic stripping voltammetry) and 2.55 (emission spectrometry).

2.22
Lanthanum

The determination of lanthanum in soil by neutron activation analysis is discussed under "Multi-Metal Analysis of Soils" in Sects. 2.55 (neutron activation analysis) and 2.55 (column chromatography).

2.23
Lead

Most of the lead in soil exists in sparingly soluble forms. When 2784 ppm of lead nitrate were added to soil, it was found that after three days the soluble lead content was only 17 ppm [104]. It is to be expected that all ions will accumulate in nature as their less soluble compounds, such as oxides, carbonates, silicates and sulfates, the relative proportions of each depending on the nature of the soil and on solubility.

Several acids and acid mixtures have been used for the digestion of soil samples prior to the analysis of lead, including nitric acid–perchloric acid (1 + 1) [105], hydrochloric acid [106], perchloric acid [107], nitric acid–hydrofluoric acid (1 + 1) [108, 109] and aqua regia [52].

Savvin et al. [110] have discussed a spectrophotometric method for the determination of lead in soils.

Official methods have been published for the determination of nitric–perchloric acid-soluble lead [111] and ammonium pyrrolidine dithiocarbamate-extractable lead [112] in soil. Atomic absorption spectrometric evaluations of the digest or extract is conducted at the 217 nm emission line from a lead hollow cathode lamp. Rigin and Rigina [122] determined lead in soil by flameless atomic fluorescence using electrolytic preconcentration. The limit of detection is 15 pg lead and the standard deviation is not greater than 0.04.

Tills and Alloway [113] investigated the speciation of lead in soil solution using a fractionation scheme, ion exchange chromatography and graphite furnace atomic absorption spectrophotometry. Soils from four sites were selected (Snertingdal in Norway, Pen Craig-ddu in Dyfed, Wales, Velvet Bottom in Somerset, England and Beaumont Leys Sewage Farm, Leicester, England). The sources of contamination for each soil and its chemical properties are described. The percentages of lead in cationic, anionic, neutral and less polar organic complexes were determined and discussed with respect to organic matter content and pH. No direct relationship was established between total lead content of the soil and total lead content in soil solution.

Using palladium–magnesium nitrate mixtures as chemical modifiers, Hinds and Jackson [114] effectively delayed the atomisation of lead until atomic absorption spectrometer furnace conditions were nearly isothermal. This technique was used to determine lead in soil slurries. Zhang et al. [115] investigated the application of low-pressure electrothermal atomic absorption spectrometry to the determination of lead in soils.

Hinds et al. [116] investigated the application of slurry electrothermal atomic absorption spectrometry to the determination of lead in soils. Hinds and Jackson [117] also investigated the application of vortex mixing slurry graphite furnace atomic absorption spectrometry to the determination of lead in soils.

Somer and Aydin [118] determined the lead content of soil adjacent to roads in Turkey using anodic stripping voltammetry. These workers found that aqua regia was the most suitable acid for extracting lead from roadside soil. The lead salt that may be trapped in the silicate crystal lattice of the soil was brought into solution by keeping the soil in acid overnight. To avoid the possibility of the presence of undissolved lead salts even after digestion, EDTA was added to the digested sample in order to ensure quantitative dissolution of lead. The lead content of this solution was determined by anodic stripping voltammetry.

Differential pulse anodic scanning voltammetry has been applied to the determination of lead in soils [119]. Sakharov [120] determined lead in soil polarographically by digesting the sample with sodium carbonate, followed by dissolution in hydrochloric acid. He found that when hydrochloric, sulfuric or nitric acids were used as digestion media instead of sodium carbonate, no lead could be detected in the resulting solution. Lead was determined in the digest by anodic scanning voltammetry [121].

Wegrzynek and Holynska [127] have developed a method for the determination of lead in arsenic-containing soils by energy-dispersive X-ray fluorescence spectroscopy. The correction for arsenic interference is based on the use of an arsenic-free reference sample.

Bedrosian et al. [123] have described a spectrographic method for the determination of down to 1 mg/kg of lead in 50 mg samples of soil.

Xan et al. [124] described a highly selective method for the determination of traces of lead in soils and sediments. It is based on the preconcentration of lead on a microcolumn packed with a macrocycle immobilised on silica gel.

Several investigations have reviewed the determination of lead in soils [125–129]. Lead has been determined in soil using a slurry sampling technique with lead nitrate and magnesium nitrate as a chemical modifier [130]. Results were in good agreement with known concentrations of a standard reference material. Feng and Barrett [62] showed that wave dissolution of soil and dust samples with nitric–hydrofluoric acid gave recoveries of lead (and cadmium) of over 90% for 30 minutes of digestion.

Chen and Hong [126] found that 5-carboxy methyl-L-cysteine was especially effective for the chelating extraction of lead from contaminated soils. The chelator could be recovered and reused over consecutive runs with no loss in performance.

The determination of lead in soil is also discussed under "Multi-Cation Analysis" in Sects. 2.55 (inductively coupled plasma atomic emission spectrometry), 2.55 (atomic absorption spectrometry), 2.55 (photon activation analysis), 2.55 (emission spectrometry), 2.55 (anodic stripping voltammetry) and 2.55 (neutron activation analysis).

2.24
Magnesium

Official methods have been published for the determination of exchangeable and extractable magnesium in soils [131]. Magnesium is extracted from the soil with 1 M ammonium acetate and determined by atomic absorption spectrometry. The determination of magnesium in soils is also discussed under "Multi-Metal Analysis of Soils" in Sects. 2.55 (atomic absorption spectrometry), 2.55 (inductively coupled plasma atomic emission spectrometry), 2.55 (photon activation analysis) and 2.55 (ion chromatography).

2.25
Manganese

In an official method [132], exchangeable and easily reducible manganese is determined in 1 M ammonium acetate for exchangeable manganese and 1 M ammonium acetate containing 0.2% quinol for exchangeable plus easily reducible manganese. Manganese in the extracts is determined by atomic absorption spectrometry at the 403 nm emission from a hollow cathode lamp.

Alekseeva and Davydova [133] determined micro amounts of manganese(II) in sulfuric acid–hydrofluoric acid digests of clays by a kinetic method involving the oxidation of o-dianiside by potassium periodate. The extinction of the solution is measured at 460 nm over a period of 15 minutes. The manganese content is found from the rectilinear portion of a plot of extinction *versus* time with the aid of a calibration graph which is rectilinear in the range from 0.0245 to 25 µg of manganese.

The determination of manganese in soils is also discussed under "Multi-Metal Analysis of Soils" in Sects. 2.55 (inductively coupled plasma atomic

emission spectrometry), 2.55 (differential pulse anodic stripping voltammetry), 2.55 (X-ray fluorescence spectroscopy), 2.55 (emission spectrometry),
2.55 (neutron activation analysis), 2.55 (photon activation analysis) and 2.55
(ion chromatography).

Kamburova [134] has reported a spectrophotometric method based on
the formation of the mercury–triphenyltetrazolium chloride complex for the
determination of mercury in soils.

2.26
Mercury

Kimura and Miller [135] determined mercury in amounts down to 0.1 μg
mercury per sample charge. In this method, a concentrating aeration procedure
at 20 °C is used following digestion of the soil with sulfuric acid, hydrogen
peroxide and potassium permanganate. Mercury is swept from the acid digest
with air for 30 minutes into an absorbing solution consisting of potassium
permanganate and sulfuric acid. The mercury content of the absorbing solution
is determined by a dithizone procedure with spectrophotometric evaluation at
605 nm. Both inorganic and organic forms of mercury in sandy loams or turf
samples are determined in this procedure.

Various atomic absorption spectrophotometric procedures have been described for the determination of mercury in soils. Methods based on attacking
the mercury in soil samples with mineral acids and permanganate have been
shown to give low mercury recoveries. In recent years methods based on decomposition of the sample by heating have gained favour in that they obviate
any tendency to produce low results.

Methods based on acid digestions of the soil with 7 M nitric acid [136] or
sulfuric acid–nitric acid [137] have been described. Released mercury is absorbed in stannous chloride–sulfuric acid–hydroxylamine [136] or potassium
permanganate–potassium persulfate–hydroxylamine–sodium chloride [137]
prior to cold vapour atomic absorption spectrometry.

Kuwae et al. [138] have described a rapid determination of mercury in
soils by high-frequency induction heating (rf) followed by cold vapour atomic
absorption spectrometry. The mercury released from the sample is absorbed
in stannous chloride–hydroxylamine prior to atomic absorption spectrometry.
Recovery of 99.4 to 99.8% mercury was obtained by this method from portions
of sample containing between 0.025–0.15 μg of mercury.

Nicolson [139] has described a rapid thermal decomposition technique for
the atomic absorption determination of mercury in soils. In this method, air
is used to sweep mercury vapour from the heated (650–750 °C) sample onto
gold foil. In the second stage, heating of the gold foil releases mercury vapour
into a cold vapour atomic absorption spectrometer.

Cold vapour (or flameless) atomic absorption spectrometry is *the method* of
choice for the determination of mercury in soils [136–147]. Ure and Shand [141]

investigated various procedures for the digestion of soil samples prior to analysis by cold vapour atomic absorption spectrometry. They found good agreement between two digestion methods involving digestion of the soil with a mixture of nitric and sulfuric acids and potassium permanganate, and oxygen flask combustion over acid potassium permanganate solution.

Floyd and Sommers [142] evaluated a simple one-step digestion procedure for extracting total mercury from soils. The sample was digested with concentrated nitric acid and 4 N potassium dichromate for four hours at 55 °C and the mercury in the extract determined by flameless atomic absorption spectrometry. The method can be applied to soils containing up to 20% organic matter.

Cold vapour atomic absorption spectrometry and atomic fluorescence spectrometry (253 nm emission) have been applied to the determination of down to 0.01 mg/kg of mercury in soils and sediments [144].

Sakamoto et al. [148] have shown that the differential determinations of different forms of mercury in soil can be accomplished by successive extraction and cold vapour atomic absorption spectrometry.

Azzaria and Aftabi [149] showed that stepwise (as compared to continuous) heating of soil samples before determination of mercury by atomic absorption spectrometry gives increased resolution of the different phases of mercury. A gold-coated graphite furnace atomic absorption spectrometer has been used to determine mercury in soils [150].

Bandyopadhyay and Das [151] extracted mercury from soils with the liquid anion exchanger Aliquat-336 prior to determination by cold vapour atomic absorption spectrometry.

A study by Rasemann et al. demonstrated to what extent mercury concentrations depend on the method of handling soil samples between sampling and chemical analysis for samples from a nonuniformly contaminated site [152]. Sample pretreatment contributed substantially to the variance in results and was of the same order as the contribution from sample inhomogeneity. Welz et al. [153] and Baxter [154] have conducted speciation studies on mercury in soils. Lexa and Stulik [155] employed a gold film electrode modified by a film of tri-*n*-octylphosphine oxide in a PVC matrix to determine mercury in soils. Concentrations of mercury as low as 0.02 ppm were determined.

Cherian and Gupta [156] have described a simple field test for the determination of mercury in soil. Saouter et al. [157] showed that the use of hydrogen peroxide as an oxidising agent for organics in soils can result in the loss of mercury. This is because hydrogen peroxide can act as a reducing agent for mercury compounds.

Voltammetric methods have been used to determine mercury in soil composts. The amount of mercury leaching from composts was very low [158]. Neutron activation analysis has been used to determine mercury in soil [159].

Carpi and Lindberg [160] have developed a Teflon dynamic flux chamber for measuring mercury soil emissions.

Easterling et al. [161] has reported a rapid field screening method for the determination of elemental mercury in soil. This method involves thermal desorption of the mercury onto gold, followed by thermal desorption from the gold film mercury analyser.

Atomic absorption spectrometry has also been used to determine mercury in multi-metal mixtures (see Sect. 2.55).

2.27
Molybdenum

Subclinical effects are often observed when molybdenum levels in soil exceed $3\,\mu g/g$; an excess of molybdenum in forage is toxic to livestock. Deficiency diseases have been observed in livestock when soil molybdenum levels are below $0.5\,\mu g/g$.

Official methods have been published for the determination of extractable [162] and total [163] molybdenum in soil.

In the method for extractable molybdenum [162], the molybdenum is extracted with ammonium oxalate–oxalic acid solution. The oxalate ion irreversibly exchanges with the molybdate ion, which then forms a stable complex with excess oxalate. Organic matter (including oxalates) in the extract is destroyed by dry combustion, and the soluble mineral constituents in the ash are dissolved in hydrochloric acid. The concentration of molybdenum in a diisopropyl ether extract of this solution is determined spectrophotometrically as the orange complex formed when molybdenum reacts with iron and thiocyanate in the presence of a reducing agent (stannous chloride). Sodium fluoride is added to prevent interference by titanium. To determine total molybdenum [163], organic matter in the soil is first destroyed by dry combustion at $500\,^\circ$C. The residual mineral matter is digested with hydrochloric acid and hydrofluoric acids and the residue dissolved in hydrochloric acid. The concentration of molybdenum in this solution is determined spectrophotometrically as the orange complex formed when molybdenum reacts with iron and thiocyanate in the presence of the stannous chloride reducing agent. The coloured complex is extracted into diisopropyl ether for spectrophotometric evaluation.

Molybdenum(VI) has also been determined spectrophotometrically in soil extracts using toluene-3,4-dithiol chromogenic reagent. However, copper(II) interferes in this procedure. Milham et al. [169] discussed an isoamylacetate extraction procedure for overcoming such interference.

Earlier atomic absorption methods [164–167] from the determination of molybdenum in soils employed a preliminary solvent extraction step to improve sensitivity in view of the low concentrations of molybdenum occurring in most soils. Baucells et al. [5] developed a graphite furnace atomic absorption procedure which was capable of determining down to 8.4 pg of molybdenum in a soil matrix solution with a precision of 4% for $100\,\mu g/l$ molybdenum. These workers showed that a char temperature of $1500\,^\circ$C and an atomisation temperature of 2400 °C are optimum for molybdenum. Under these conditions, the background absorbance is 0.015. However, the use of a char temperature of 700 °C in part prevents attack of the graphite, gives a better precision, and the background absorbance in the atomisation step is only 0.030.

During the extraction method described by Baucells et al. [53] for the determination of molybdenum, the dry residue was solubilised with nitric acid. To observe the influence of nitric acid concentration on the absorbance signal of molybdenum, different acid concentrations were used. There was a decrease of 22.86% in the peak height when 10% nitric acid was present compared with no concentrated nitric acid.

The interference study was carried out with 200 µg/l of molybdenum at various interferent concentrations (10, 100, 1000 and 4000 µg/l). The most important interferences were given by aluminium, iron and magnesium.

Five replicate determinations of molybdenum in a siliceous soil sample obtained by this proposed method gave a precision of 17.1% with a mean concentration of 35 µg/l. The main problem concerning the determination of molybdenum is the corrosion of the pyrolytic layer in the graphite tubes by the acid and the atomisation temperature used. After 30 firings (and the corresponding cleanings), the sensitivity decreased by 50%, so the minimum acceptable atomisation temperature and time must be used. Recalibration must be carried out frequently, and a computer program was developed in order to correct for the possible variations in the readings for standards and samples during the analysis.

Baucells et al. [53] applied ICPAES to the determination of very low levels of molybdenum (and cadmium) in soils. Among the most sensitive molybdenum lines, 202.030 nm proved to be an excellent analytical line; although there were two iron lines close-by, at 201.99 and 202.074 nm, it was still found to be free from interference. The other molybdenum-sensitive lines were subject to interference from iron, chromium and vanadium.

Calcium and magnesium are the most serious interferents when they are present at high levels (more than 1000 µg/cm^3). A depressant effect is observed in both instances.

The effect of the calcium/molybdenum combination was studied with the molybdenum line; their joint effect was virtually identical to the sum of their separate effects (95% at the 1000 µg/ml level and 99% at the 4000 µg/ml level). These results agree with those of Maessen et al. [167]. A background correction is necessary in the determination of molybdenum when aluminium is present in the samples because of the enhancement of the background.

Atomic emission spectrometry is not sufficiently sensitive to determining molybdenum at the levels at which it occurs in soils. Due to its greater intrinsic sensitivity, inductively coupled plasma atomic emission spectrometry is capable of achieving the required sensitivity. Manzoori [168] has utilised inductively coupled plasma optical emission spectrometry to determine down to 0.01 mg/l molybdenum in 1 M ammonium acetate extracts of soils.

Thompson and Zao [170] have described a solvent extraction–inductively coupled plasma atomic emission spectrometric method for the determination of down to 0.02 – 0.03 μg/g of molybdenum in soils. The soil sample is pressure-leached with 6 M hydrochloric acid and at 120 °C for 15 minutes. The digest is then extracted with heptan-2-one to separate molybdenum from potentially interfering elements such as iron, aluminium, calcium and magnesium. This organic extract is then directly sprayed into an inductively coupled plasma atomic emission spectrometer operated at 1.65 to 1.7 kW power.

The determination of molybdenum in soil is of interest because molybdenum is necessary for normal crop growth, but an excess in forage has a toxic effect on ruminants. The absorption of molybdenum by plants is influenced by other soil components, especially extractable iron, pH and organic matter. The average abundance of molybdenum in soils is about 2 ppm, but deficient soils can have much less than 1 ppm [171]. Jiao et al. [172] and Rowbottom [173] have reviewed methods for the determination of molybdenum in soils.

Other techniques applied to the determination of molybdenum include neutron activation analysis (Sect. 2.55), emission spectrometry (Sect. 2.55), inductively coupled plasma mass spectrometry (Sect. 2.55), and atomic absorption spectrometry (Sect. 2.55).

2.28
Nickel

Standard official methods have been described for the determination of nitric–perchloric acid-soluble nickel [174] and acetic acid-extractable nickel [175] in soil. To determine nitric acid–perchloric acid-soluble nickel [174], the acid digest is dissolved in hydrochloric acid and the nickel is determined by atomic absorption spectrometry. To determine extractable nickel, the nickel is first extracted from the soil with 0.5 M acetic acid and the nickel is then converted to the ammonium pyrrolidine dithiocarbamate complex. Extraction of the complex with chloroform provides an extract for the determination of nickel by atomic absorption spectrometry.

The determination of nickel is also discussed under "Multi-Metal Analysis of Soils" in Sects. 2.55 (atomic absorption spectrometry), and 2.55 (inductively coupled plasma mass spectrometry), 2.55 (differential pulse anodic stripping voltammetry), 2.55 (photon activation analysis), 2.55 (emission spectrometry) and 2.55 (neutron activation analysis).

2.29
Palladium

Manceau et al. [129] studied the application of X-ray absorption fine-structure analysis (EXAES) to the speciation and quantification of the forms of trace metals in solid materials. Palladium was studied in particular.

2.30
Platinum

Inductively coupled plasma atomic absorption spectrometry (Sect. 2.55) and neutron activation analysis (Sect. 2.55) have both been applied to the determination of platinum in multi-metal mixtures.

2.31
Plutonium and Americium

Sekine et al. [176] studied the liquid–liquid extraction separation and sequential determination of plutonium and americium in soils by alpha-spectrometry. The chemical recovery of plutonium from standard soil samples was 51–99% (average 81%) of the analytical level and for americium 60–70% of the analytical level.

Microwave digestion and anion exchange chromatography have also been used to determine plutonium in soil [177].

2.32
Potassium

An official method has been published for the determination of nickel in 1 M ammonium nitrate extracts of potassium from soil [178]. The level of potassium in the extract is determined by flame photometry. Inductively coupled plasma atomic emission spectrometry (Sect. 2.55) and stable isotope dilution (Sect. 2.55) have been applied to the determination of potassium in multi-metal analyses.

2.33
Rubidium

Techniques that have been applied to the determination of rubidium in multi-cation analyses include stable isotope dilution (Sect. 2.55).

2.34
Scandium

Neutron activation analysis has been used to determine scandium, as discussed under "Multi-Metal Analysis of Soils" in Sect. 2.55. Scandium has also been determined by atomic absorption spectrometry (Sect. 2.55).

2.35
Selenium

The fate of selenium in natural environments such as soils and sediments is affected by a variety of physical, chemical and biological factors which are associated with changes in its oxidation state. Selenium can exist in four different oxidation states (–II, 0, IV and VI), and as a variety of organic compounds. The different chemical forms of selenium can control selenium solubility and availability to organisms. Selenate (Se(VI)) is the most oxidised form of selenium; it is highly soluble in water and generally considered to be the most toxic form. Selenite (Se(IV)) occurs in oxic to suboxic environments and is less available to organisms because of its affinity to sorption sites of sediment and soil constituents. Under anoxic conditions, elemental selenium and selenide(–II) are the thermodynamically stable forms. Elemental selenium is relatively insoluble, and selenide(–II) precipitates as metal selenides(–II) of very low solubility. Organic selenium(–II) compounds such as selenomethionine and selenocystine can accumulate in soil and sediments or mineralise to inorganic selenium. Therefore, Se(VI), Se(IV) and organic selenium(–II) are the most important soluble forms of selenium in natural environments.

A widely used method for the routine determination of selenium in soils is based on the reaction between selenium and 2,3-diaminonaphthalene to form a fluorescent piazoselenol product [179–184]. While methods based on this principle give satisfactory results, they require very careful technique with strict attention to detail, especially in the sample dissolution stage. Complete destruction of organic matter is necessary in order to avoid the possibility of fluorescent interference from these amounts of residual material, and is achieved by treatment of the sample with hot acidic oxidising mixtures. However, excessive temperature or prolonged heating bring about the loss of selenium by volatilisation and considerable ingenuity is required to devise methods that will satisfy these conflicting requirements. These considerations and problems associated with the purity of the 2,3-diaminonaphthalene reagent are now tending to preclude recommendation of fluorometric methods.

An atomic fluorescence spectrometric determination of selenium was first reported by Dagnall et al. [185] using a dispersive spectrometer equipped with an air–propane flame, giving a detection limit of 0.25 µg/ml of selenium on aspiration of aqueous solutions using a pneumatic nebuliser. Fluorescence from the 204 nm selenium resonance line was observed when the flame was irradiated by radiation from a selenium electrodeless discharge lamp, the optical axis of which was aligned at 90 °C to the optical axis of the monochromator.

Azad et al. [186] used a similar technique for the determination of selenium in soil extracts using a nondispersive spectrometer, with which it was possible to observe fluorescence from the 196.1, 214.3 and 204.0 lines simultaneously, thus enabling a detection limit of 10 ng/ml to be observed using discrete sample introduction via the hydride generation technique. In this method, soil samples were digested using a mixture of nitric and perchloric acids at 200 °C, taking care to avoid selenium volatilisation. Potassium bromide is added to the digest to convert selenium to the selenium(VI) state, which is necessary in order to apply the hydride generation technique. Although the hydride generation technique is normally subject to interference from copper, this effect can be eliminated by employing chemical pretreatment of the samples, using lanthanum hydroxide as a coprecipitant on the addition of tellurium(IV) to form stable copper telluride during reduction. Azad et al. [186] applied both methods successfully to the determination of selenium in soil digests.

Hydride generation methods are finding increasing favour for the determination of selenium in soils and sediments. This method consists of measuring the atomic absorption of selenium hydride formed as a result of the reduction of selenium and its compounds with different reducing mixtures such as sodium borohydride or, occasionally, zinc–stannous chloride–potassium iodide. Hydride generation techniques are about three orders of magnitude more sensitive for determining selenium than are classical flame ionisation techniques; a detection limit of 0.2 ng/g is achievable. They have an additional advantage of separating selenium from the matrix before atomisation, thus avoiding interferences inherent to the conventional atomic absorption technique. Practical working ranges for selenium are 3 – 250 µg/ml, 0.03 – 0.3 µg/ml and up to 0.12 µg/ml for flame atomic absorption, atomic absorption and vapour generation methods, respectively [187–194].

The most intense resonance line of selenium (196.03 nm) corresponds to a range near to the vacuum ultraviolet. Moreover, the most frequently applied air–acetylene flame absorbs about 55% of the radiation intensity of the light source. When using electrodeless discharge lamps and an air–acetylene flame, appreciably lower detection limits can be achieved by application of a deuterium lamp for a background correction. The argon–hydrogen flame is often used to augment the sensitivity, but it increases interference too. Extraction has also been attempted [195] as a means of improving sensitivity, but in selenium determinations a re-extraction to a water solution is necessary.

Flameless atomic absorption spectrometric techniques offer high sensitivity (5×10^{11} g Se) but are not simple nor free from interference, due to the high volatility of selenium. This technique is particularly suitable for the direct analysis of samples, and an additional advantage lies in the possibility of "chemically treating" samples in the graphite cell in order to diminish chemical interference.

The addition of nickel significantly enhances the sensitivity to selenium by about 30% and allows higher ashing temperatures (1000 °C) without losses [196–198]. Other elements capable of forming selenides (i.e., barium, copper, iron, magnesium and zinc) did not interfere and arsenic interference was minimised. A detection limit of 10 – 12 µg/kg selenium has been achieved using a graphite electrothermal furnace and background correction with a deuterium lamp [208].

A method has been reported [200] for determining total arsenic (and selenium) in soils based on atomic absorption spectrometry and flow injection analysis. The method exhibits good recoveries and detection limits below 1 μg/l for an injection volume of 160 μl.

Hydride generation atomic absorption spectrometry is widely used to determine the speciation of selenium in natural water and soil–sediment extracts because of its low detection limits. The speciation of selenium is determined by subdividing sample solutions for selective treatments. Selenite is determined by directly analysing aliquots of samples without any treatment or by analysing samples acidified to pH 2 with concentrated hydrochloric acid or samples in 4–7 M hydrochloric acid solutions. Selenate plus Se(IV) are determined after reduction of Se(VI) to Se(IV) in 4–7 M hydrochloric acid at high temperatures (80–100 °C) and analysis for selenium to obtain Se(VI+IV) concentrations. Selenate is determined by the difference between a determination of Se(VI+IV) and a determination of Se(IV) in another subsample. Total selenium is determined by oxidising all selenium species (organic Se(–II) and Se(IV)) to Se(VI) with hydrogen peroxide or persulfate ($K_2S_2O_8$ or $(NH_4)_2S_2O_8$) then reducing Se(VI) to Se(IV) with 4–7 M hydrochloric acid at a high temperature (80–100 °C) and analysing the samples for total selenium. The determination of organic Se(–II) is obtained as the difference between the Se(VI+IV) and the total selenium analyses. To separate organic Se(–II) from inorganic selenium, a technique was developed by passing an acidified sample (pH 1.6–2.2) through an XAD-8 resin column to remove hydrophobic and neutral organic Se(–II) compounds before selenium species analysis. These methods have provided valuable information about selenium speciation in natural water and soil–sediment extracts. However, the following comments should be noted.

Some drawbacks for the speciation of selenium using hydride generation atomic absorption spectrometry have been found by some researchers. Thus Se(VI) is recovered poorly from many samples after a reduction with 6 M hydrochloric acid at 100 °C. The addition of ammonium persulfate increased the recovery of Se(VI). However, part of the organic Se(–II) was included in the value reported for Se(VI) due to the oxidation of organic Se(–II) by persulfate. The reduction of Se(VI) to Se(IV) in soil extracts with 6 M hydrochloric acid oxidised organic Se(–II) present in the sample, resulting in an overestimation of the Se(VI) concentration. XAD-resin has been used to separate hydrophobic and neutral organic Se(–II) compounds. However, hydrophilic organic Se(–II) compounds in solution, such as selenomethionine which are found in soil extracts, will be detected as part of Se(VI). Also, a considerable fraction of Se(IV) is removed due to a complexion of Se(IV) with humic substances when an acidified sample is passed through an XAD-8 column, thus resulting in an overestimation of the organic Se(–II) or Se(VI) concentration in the samples. The net consensus view is that many of the published methods of determining selenium speciation using hydride generation atomic absorption spectrometry may be possible only in solutions with little or no organic Se(-II), but this situation is rarely found in natural environments.

To overcome these drawbacks, Zhang et al. [201] developed a new method of determining organic selenium(-II) in soils and sediments. In this method, persulfate is used to oxidise organic selenium(-II), and manganese oxide is used as an indicator of oxidation completion. This method was used to determine selenium speciation in soil–sediments and agricultural drainage water samples collected from the western United States. Results showed that organic selenium can be quantitatively oxidised to selenite without changing the selenate concentration in the soil–sediment extract and agricultural drainage water, and then quantified by hydride generation atomic absorption spectrometry. Recoveries of spiked organic selenium and selenite were 96–105% in the soil–sediment extracts and 96–103% in the agricultural drainage water. Concentrations of soluble selenium in the soil-sediment extracts were 0.05–2.45 µg/g of which organic selenium accounted for 4.5–59.1%. Selenate is the dominant form of selenium in agricultural drainage water, accounting for about 90% of the total selenium. In contrast, organic selenium(-II) was an important form of selenium in the wetlands. These results showed that wetland sediments are more active in reducing selenite compared to evaporation pond sediments.

Martens et al. [202] and McCurdy et al. [203] have employed hydride generation atomic absorption spectrometry and inductively coupled plasma mass spectrometry, respectively, to determine selenium in soils.

Pahlavanpour et al. [204] determined trace concentrations of selenium in soils by conversion to hydrogen selenide with sodium tetrahydroborate and the introduction of hydrogen selenide into an inductively coupled plasma source for emission spectrometry. These workers found that nitric acid–perchloric acid was the most suitable for the digestion of soil samples prior to selenium determination. This is because selenium needs strongly oxidising conditions during the acid attack, not only to destroy organic matter, but also to prevent its reduction to metallic selenium by organic matter or by iron(II) in the sample. A maximum temperature of 180 °C is essential during digestion to avoid loss of selenium by volatilisation.

Again, as in the method described by Azad et al. [186], lanthanum hydroxide coprecipitation is employed to remove interfering elements such as copper. Selenium in the soil extracts is converted to hydrogen selenide by reduction with sodium tetrahydroborate(III) and the hydrogen selenide introduced into the argon–hydrogen air-entrained flame.

With few exceptions, the results given by fluorometry, chromatography and neutron activation analysis compare well with those obtained by inductively coupled plasma atomic emission spectrometry. The precisions obtained for the various samples were very good for between- and within-batch samples.

Square-wave cathode stripping voltammetry [212] and PIXIE [213] and neutron activation analysis have been used to determine selenium in soil [205–212].

Agemian and Bedek [209] have described a semi-automated method for the determination of total selenium in soils.

Dong et al. [211] used a mixture of phosphoric acid, nitric acid and hydrogen peroxide in the digestion of soils prior to the determination of selenium.

Bem [210] has reviewed methods developed up to 1981 for the determination of selenium in soil. These methods include neutron activation analysis, atomic absorption spectrometry, gas chromatography and spectrophotometric methods. Square-wave cathodic stripping voltammetry has been used to determine selenium in soils [212].

Further work on the determination of selenium in soil is reported under "Multi-Cation Analysis Methods" including atomic absorption spectrometry (Sect. 2.55), inductively coupled plasma atomic emission spectrometry (Sect. 2.55), and neutron activation analysis (Sect. 2.55).

2.36
Silver

The determination of silver by atomic absorption spectrometry is discussed in Sect. 2.55.

2.37
Sodium

An official standard method has been published for the determination of 1 M ammonium nitrate-extractable sodium in soils. The sodium content of the extract was determined by atomic absorption spectrometry [214].

Sodium has been determined by inductively coupled plasma atomic emission spectrometry, as discussed in Sect. 2.55.

2.38
Strontium

Akcay et al. [215] have shown that extraction of total strontium using an ultrasonic extraction procedure was not as good as was achieved using conventional extraction methods. Further work on the determination of strontium in soil is reported under "Multi-Cation Analysis Methods" including inductively coupled plasma atomic emission spectrometry (Sect. 2.55), emission spectrometry (Sect. 2.55), stable isotope dilution (2.55), and photon activation analysis (2.55).

2.39
Thullium

The determination of this element in soil is described under "Multi-Cation Analysis" in Sect. 2.55 (emission spectrometry).

2.40
Tantalum

The determination of tantalum by neutron activation analysis is discussed under "Multi-Metal Analysis of Soils" in Sect. 2.55.

2.41
Technetium

Tagami and Uchida [216] have described a combustion method for determining ^{99}technetium in soil. Tagami and Uchida [217] have also reported a method for the determination of ^{99}technetium in soil using inductively coupled plasma mass spectrometry. Morita et al. [218] and Harvey et al. [219] have reviewed methods for the determination of technetium in soils.

2.42
Tellurium

The determination of tellurium by emission spectrometry is discussed under "Multi-Metal Analysis of Soils" in Sect. 2.55.

2.43
Terbium

The determination of terbium is discussed under "Multi-Metal Analysis of Soils" in Sects. 2.55 (emission spectrometry) and 2.55 (neutron activation analysis).

2.44
Thallium

Atomic absorption spectrometry has been used to determine thallium in soil [220]. This element has also been determined in multi-metal mixtures by emission spectrometry (Sect. 2.55).

Chikhalikar et al. [18] have studied the speciation of thallium (and antimony) in soil. Lukaszewski and Zembrzuski [221] and Sagar [222] have discussed the determination of thallium in soils.

Opydo [224] used anodic stripping voltammetry to determine thallium in soil extracts in the presence of a large excess of lead.

Van Laar et al. [223] have reviewed methods for the determination of thallium in soils. The determination of thallium is also discussed under "Multi-Metal Analysis of Soils" in Sect. 2.55 (emission spectrometry).

2.45
Thorium

Toole et al. [225], Shaw and Francois [226] and Zbiral et al. [227] determined thorium (and uranium) in soils by inductively coupled plasma mass spectrometry.

Parsa [228] has described a sequential radiochemical method for the determination of thorium (and uranium) in soils. Mukhtar et al. [229] have described a laser fluorometric method for the determination of thorium (and uranium) in soils. Steam digestion has been employed in the preparation of soil samples for the determination of thorium (and uranium) [230]. Thorium (and uranium) were determined by X-ray fluorescence using a germanium planar detector and by chemometric techniques. No sample preparation was required in this method [231].

Various other workers have been reported for the determination of thorium (and uranium) in soils [232, 233].

The determination of thorium is also discussed under "Multi-Metal Analysis of Soils" in Sects. 2.55 (neutron activation analysis), and 2.55 (emission spectrometry).

2.46
Tin

Li et al. [235] have reviewed methods for the determination of tin in soils.

2.47
Titanium

Abbasi [234] has described a spectrophotometric method employing $N-p$-methoxyphenyl-2,-furohydroxic acid for the determination of titanium in soils. In this method, the soil sample was subjected to alkali fusion. The ash was treated with nitric acid to adjust it to pH 2.0 and filtered prior to adjustment to 10 pM with respect to hydrochloric acid. Stannous chloride (5 M) was added to the filtrate, and the chromogenic reagent was dissolved in chloroform. The chloroform extract was evaluated spectrophotometrically at 385 nm against the reagent solution as blank. Approximately 61 ppm of titanium was found in a soil sample by this method.

The determination of titanium is also discussed under "Multi-Metal Analysis of Soils" in Sects. 2.55 (inductively coupled plasma atomic emission spectrometry), 2.55 (emission spectrometry), and 2.55 (photon activation analysis).

2.48
Tungsten

Quinn and Brooks [236] have described a rapid method for the determination of down to 0.01 ppm of tungsten in soils. The sample is fused with potassium hydrogen sulfate and the melt leached with 10 M hydrochloric acid, then heated with stannous chloride. This solution is heated with a solution of dithiol in isoamyl acetate, and then dissolved in petroleum ether prior to spectrophotometric evaluation at 630 nm.

Tungsten has been determined by emission spectrometry, as discussed under "Multi-Metal Analysis of Soils" in Sect. 2.55.

2.49
Uranium

Earlier methods for the determination of uranium in soils employed spectrophotometry of the chromophore produced with arsenic(III) at 655 nm [237] and neutron activation analysis [238]. More recently, laser fluorescence [239] and in situ laser ablation–inductively coupled plasma atomic emission spectrometry [240] have been employed to determine uranium in soil. D'Silva et al. [241] compared the use of hydrogen chloride gas for the remote dissolution of uranium in soil with microwave digestion.

Inductively coupled plasma mass spectrometry has been used for the analysis of uranium. However, the technique suffers from spectral interferences and it has relatively poor detection limits.

Inductively coupled plasma mass spectrometry is a relatively new technique for elemental analysis and has superior limits of detection over optical methods. Also, this technique has an order of magnitude better detection limit than that obtained by the conventional fluorometric method. Uranium has many stable and unstable isotopes but ^{238}U has the largest percentage abundance (99.274%).

Boomer and Powell [242] have developed an analytical technique using inductively coupled plasma mass spectrometry to estimate the concentration of uranium in a variety of environmental samples including soil. The lower limit for quantitation is 0.1 ng/ml. Calibration is linear from the low limit to 100 ng/ml. Precision, accuracy and a quality control protocol were established. Results are compared with those obtained by the conventional fluorometric method.

In this method the soil sample is dried overnight at 85 °C and ground into an homogeneous mixture. A 1 g soil sample is placed into a beaker and 10 ml of concentrated nitric acid added. The solution is heated to dryness and 5 ml of concentrated nitric acid is added. The uranium is redissolved in 5 ml of 8 N nitric acid and diluted to 25 ml with distilled water. The inductively coupled plasma mass spectrometry system used was an ELAN Model 250. The ion source consists of a modified plasma Thermal Model 2500 control box. The forward power was set at 1200 W with the plasma flow, auxiliary flow and nebuliser pressure set at 13 l/min, 1.0 l/min and 0.27 MPa, respectively. The focusing lenses B, El, P and S2 are set at +5.3 V, –12.5 V, –18.0 V and –7.6 V, respectively. The m/z238 ion was monitored for two seconds with five replicates of this measurement carried out for each determination.

Toole et al. [225] and Shaw and Francois [226] determined uranium (and thorium) in soils by inductively coupled plasma mass spectrometry.

Nass et al. [238] used a delayed neutron counting technique to determine down to 50 ng of ^{235}uranium in soils. Steam digestion has been employed in the preparation of soil samples for the determination of uranium (and thorium).

Mukhtar et al. [229] have described a laser fluorimetric method for the determination of uranium (and thorium) in soils. To determine uranium (and thorium) in soils, fluorescent X-rays were measured by the use of a germanium planar detector and chromometric techniques [227]. No sample preparation was required in this method.

The determination of uranium is also discussed under "Multi-Metal Analysis of Soils" in Sects. 2.55 (emission spectrometry), 2.55 (photon activation analysis), and 2.55 (neutron activation analysis).

2.50
Vanadium

Vanadium leaches soil from a large number of diverse sources, including waste effluents from the iron and steel industries and chemical industries. Phosphate industries are also a major source of vanadium pollution because vanadium becomes soluble along with phosphoric acids when rock phosphates are leached with sulfuric acid. Vanadium is present in all subsequent phosphoric acid preparations, including ammonium phosphate fertilisers, and is released into the environment along with them. Other sources of vanadium pollution are fossil fuels, such as crude petroleum, coal and lignite. Burning these fuels releases vanadium into the air, which then settles in the soils.

Abbasi [243] described a spectrophotometric method based on N(-pN, N-dimethyl anilo-3-methoxy-2-naphtho)hydroxaminic acid for the determination of vanadium in soil. The soil sample was digested with 8 M hydrochloric acid and potassium permanganate (to oxidise vanadium to the pentavalent state). This extract was retrieved with a chloroform solution of the chromogen and the violet colour was evaluated spectrophotometrically.

Vanadium has been determined in soil by inductively coupled plasma atomic emission spectrometry (Sect. 2.55), and by emission spectrometry (Sect. 2.55).

2.51
Yttrium

Yttrium has been determined in soils by emission spectrometry, as discussed under "Multi-Metal Analysis of Soils" in Sect. 2.55.

2.52
Zinc

A standard official method has been published for the spectrometric determination of nitric–perchloric acid-soluble zinc [244] and 0.5 M acetic acid-extractable zinc [245] in soils.

External beam photon-induced X-ray emission spectrometry has been used to determine total zinc in soils [246].

Roberts et al. [59] have discussed the simultaneous extraction and concentration of zinc and cadmium from calcium chloride soil extracts, as discussed in Sect. 2.9. Adequate recoveries of zinc were obtained when the pH of the extractant was adjusted to the range 4–7.

Fast neutron activation analysis has been studied as a screening technique for zinc (and copper) in waste soils [247]. Experiments were conducted in a sealed tube neutron generator and a germanium X-ray detector.

Inductively coupled plasma atomic emission spectrometry and inductively coupled plasma mass spectrometry have been applied to the determination of zinc, as discussed under "Multi-Metal Analysis of Soils" in Sects. 2.55 (inductively coupled plasma atomic emission spectrometry) and 2.55 (inductively coupled plasma mass spectrometry). Other techniques include atomic absorption spectrometry (Sect. 2.55), X-ray fluorescence spectroscopy (Sect. 2.55), electron probe microanalysis (Sect. 2.55), photon activation analysis (Sect. 2.55), emission spectrometry (Sect. 2.55), neutron activation analysis (Sect. 2.55), spectrophotometry (Sect. 2.55) and ion chromatography (Sect. 2.55).

2.53
Zirconium

Emission spectrometry has been applied to the determination of zirconium in soils, as discussed under "Multi-Metal Analysis of Soils" (Sect. 2.55), and photon activation analysis has also been applied (Sect. 2.55).

2.54
Selective Extraction of Metal Ions Associated with Humic Acid from Soil

2.54.1
Extraction with Aqueous Reagents

Only a fraction of the total metal content of soils and sediments tends to be available for uptake by plants or biota. This fraction is generally associated with the colloidal material (i.e., clay minerals, hydrous oxides and organic matter), but views differ on the relative effects of the individual components.

Opinions also vary with respect to the most suitable procedure(s) for evaluating "available" levels. A widely adopted approach is extraction with a chemical solution, and a wide variety of active constituents have been proposed, ranging from acids (strong or weak) to complexing agents or salt solutions of different types. Further, the correlations obtained between extraction values and plant uptakes tend to be sensitive to the species of plant grown and type of soil used [248]. There is also uncertainty about whether different reagents release metal ions from the same or from different types of binding sites.

By making arbitrary assumptions about the behaviour of metal ions associated with different components of a sample when exposed to solutions of varying reactivity, it is possible to propose sequential procedures which theoretically fractionate the total metal content into subcategories. Some investigators have been satisfied with a simple division (e.g., between detrital and nondetrital) while others [249–253] have been more ambitious and have sought to identify several fractions, loosely classified as ion-exchangeable, weakly adsorbed, associated with organic matter, precipitated, retained by hydrous oxides, etc. The order of attack adopted varies between authors, owing in part to a limited understanding of some or all of the competing equilibria involved.

The relative efficiency of various reagents in retrieving Cu, Pb, Cd or Zn ions presorbed onto clay suspension has been investigated [254]. It was found that a few extractants (e.g., EDTA, oxalic acid) recovered all of the adsorbed metal ion. However, in most of the systems examined the extraction yield varied with the type of clay, the metal ion and the pH and concentration of the extractant, as well as with the pH during the initial sorption stage.

Recent studies have shown that the adsorption capacity of a common organic component (humic acid) can exceed that of clay minerals. A change in pH can cause marked changes in the uptake of metal ions by such humic acids [255] or humic acid–clay mixtures [256]. In this connection, Slavek et al. [257] examined the effect of various electrolytes on the organic acid–metal ion equilibria, with a view to clarifying the situation.

Extraction of Zinc and Cadmium

It can be observed from Fig. 2.1 that most of the extractants tested displaced more than 80% of the presorbed zinc or cadmium ions.

Humic acid I (symbol ▨) retained the sorbed material to a slightly greater extent in acid media, but released marginally greater amounts to the complexing agents. The alkaline complexing solutions (i.e., citrate, pyrophosphate, EDTA, DTPA) appeared to dissolve or colloidally disperse the humic acids, but there was a residual > 0.45 μm fraction which retained a measurable amount of the sorbed ion (this observation is consistent with the adsorption/pH trends observed in earlier studies).

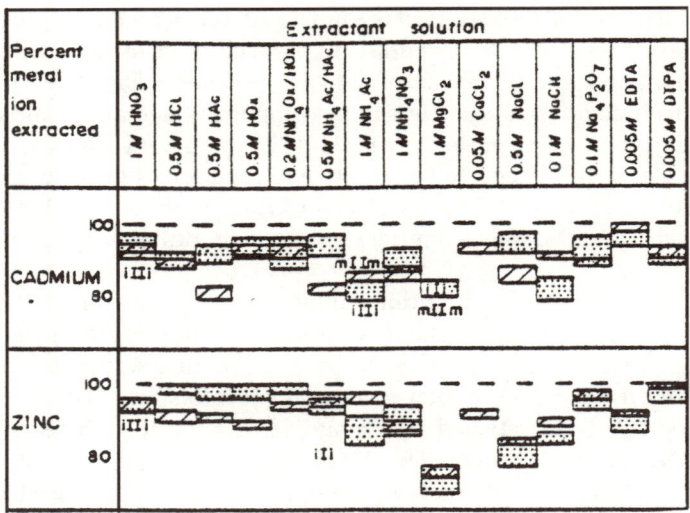

Figure 2.1. Percentage of adsorbed Cd or Zn ions extracted from humic acid suspensions by electrolyte solutions. ▨ Humic acid I; ▦ Humic acid II. Symbols indicate mean±SD. The lettered entries indicate systems in which the addition of clay particles (i = illite, m = montmorillonite) altered the recovery level. From [257]

No attempt was made to reduce the ash content of the humic acid samples, since drastic purification methods can cause abnormal changes in humate characteristics, and it was believed that most organic acids in their natural environment would be in salt form and associated with other colloidal matter. Though competition from displaced cations may have contributed to the smaller uptake by humic acid (HA II) in the adsorption stage (as shown in Table 2.1), any residual counterions should have had little effect on the metal-ion extraction step.

The two humic acids released different proportions of the adsorbed metal ions into most of the extractants studied (see Figs. 2.1 and 2.2), but these variations have been attributed to differences in chemical structure; that is, the type and spatial location of the functional groups attached to the organic moeity. The chemical reactivity of humic acid samples is usually attributed to constituent carboxylic acid groupings, in particular these located *ortho* or *meta* to phenolic groups or a second carboxylic acid group, with contributions from other functional groups such as $-NH_2$ or $-SH$. Interactions between metal ions and humic acids have been shown to yield both 1:1 species and 1:2 complexes [258–260]. In the 1:1 forms, the metal ion is either retained through a salt-type linkage (i.e., $RCOO^-M^{2+} \dots$) or coordinated to appropriately located pairs of functional groups.

In the 1:2 complexes, the metal ion becomes fully coordinated to functional groups:

Table 2.1. Effect of pH and metal salt concentration on the amount of metal ion sorbed

Soil suspension		Initial	Zinc		Cadmium	
Humic acid	Clay types	$[M^{2+}]$, mM	pH	% sorbed	pH	% sorbed
HA I	–	0.167	4.7	35	4.0	30
	–	0.167	5.2	40	4.3	35
	–	0.167	–	–	4.9	45
	+ Illite	0.167	6.5	50	5.9	60
	+ Montmorillonite	0.167	6.8	80	6.8	90
HA II	–	0.050	5.2	50	5.2	65
	–	0.100	5.1	35	5.1	45
	–	0.167	5.5	25	5.4	30
	–	0.167	6.8	40	6.7	50
	+ Illite	0.167	6.2	40	6.7	75
	+ Montmorillonite	0.167	6.6	65	6.5	75

with $[M^{2+}]$ initialy 1.67×10^{-6} M, 100% absorption with HA alone corresponds to CA, 1.85 moles M^{2+} per kg of HA. Each 10% increase in sorption due to presence of clay (similar PH) corresponds to ca. 45 mmole M^{2+} sorbed per kgd clay

Extraction of Copper and Lead

The greater affinity of humic acids for copper or lead ions is clearly reflected in the extraction values reported in Fig. 2.1. In particular, far less of the adsorbed material was displaced by the salt solutions. The amount of lead retrieved by these extractants, and also with the mineral acids and buffer solutions, exceeded the copper recoveries, so copper appears to be more firmly bound than lead on the humic acid samples studied.

As with cadmium and zinc, the amount released by chelating agents such as DTPA and EDTA was > 90%, but not total, which again indicates strong bonding of some of the metal ion to residual solids. Increasing the chelate concentration tenfold (i.e., to 0.05 M EDTA) had little measurable effect. Behaviour in the presence of acids was somewhat variable, with lead recovery in the presence of oxalic acid being quite low. This has been attributed to conditions favouring the formation of sparingly soluble lead oxalate. The addition of ammonium oxalate, however, favoured the formation of soluble oxalato-complexes and the lead extraction values then exceeded 90%.

The electrolyte extractants (NH_4NO_3, NaCl, $MgCl_2$, $CaCl_2$) displaced only about half of the sorbed ions (with less from HA I than HA II), which indicates that a higher proportion of Cu and Pb ions become coordinated to the organic solid.

Hydrochloric acid (0.1 M), 1 M ammonium chloride or acetate, and 0.05 M calcium or 1 M magnesium chloride [252, 261–263] have all been proposed

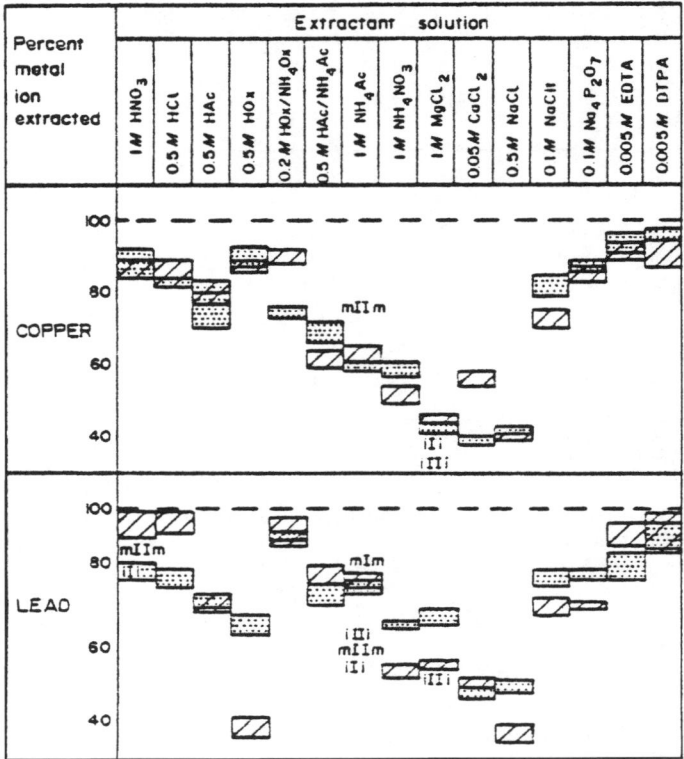

Figure 2.2. Percentage of adsorbed Cu or Pb ions extracted from humic acid suspensions by electrolyte solutions. Symbols as for Fig. 2.1. From [257]

as suitable extractants for the evaluation of the "available" or "exchangeable" fractions of the total metal content of soils.

It can be seen from Figs. 2.1 and 2.2, however, that reagents such as these extract differing amounts of metal ion from humic acids, a result similar to that noted earlier for clay suspensions [254]. The preferred reagents, magnesium and calcium chlorides, generally displace less than the other proposed reactants, but the amount released by each system tends to be quite variable, depending inter alia on the type of humic acid, surface loading, equilibrium pH, and type of clay present (if any).

Variations between reagent responses may be of minimal concern if the analytical data are intended solely to indicate major differences in soil behaviour (i.e., for comparative studies) and no special significance needs to be placed on the assignment terminology used, although the influence of other variables on extraction values probably contributes to the correlation problems associated with plant pot-trials.

However, where sequential extraction steps are to be adopted, with the aim of fractionating the total content into specific categories (e.g., exchangeable, weakly adsorbed, associated with organic matter), an improved understanding of the chemical equilibria associated with the extractant–solid interactions becomes highly desirable.

A summary of the sequential approach proposed by four different groups [249, 252–254] is provided in Table 2.2, and this can be used to illustrate the complexity of the problem.

In summary, examination of metal ion recoveries from humic acid samples by different chemical extractants has confirmed that in any extraction procedure very careful consideration must be given to the associated chemical equilibria and the impact of competing reactions.

Table 2.2. Extraction sequences for the subdivision of the total metal content of soils

Cu [261]	Zn, Mn, Cu [265]	Cd, Co, Cu, Ni, Pb, Zn, Fe, Mn [264]	Cu, Zn, Fe, Mn, Mo [266]
'Available' or *'Exchangeable'* ($CaCl_2$) ($MgCl_2$) or (NaAc)			Sulfides and bound to *organic matter (NaOCl)*
Weakly bound, specific sites (HAc)		Bound to carbonates (HAc/NaAc)	Adsorbed ions, soluble carbonates (HCl)
Bound to *organic matter* ($Na_4P_2O_7$) (H_2O_2)			
	Bound to *hydrous oxides of iron and manganese* (HO_x/NH_4O_x)		($NH_2OH.HCl$) ($Na_2S_2O_4$/citrate/HCO_3^-)
		Bound to *organic matter (H_2O_2)*	
Residual (lattice) components (HF)	Sand (totals) Silt (totals) Clay (totals) Soil (totals) (HCl/HNO_3/HF)	Residual solids (HCl/HF) (HNO_3/$HClO_4$)	Silicate minerals

2.55
Multi-Metal Analysis of Soils

Frequently methods for the determination of metals in soil deal not with the determination of a single element as discussed in Sects. 2.1 and 2.53, but instead deal with more than one element. This work is reviewed in Table 2.3.

Crosland et al. [264] reported on the application of the ethylenediaminetetraacetic acid extraction procedure to soil analysis. A round robin involving

Table 2.3. Multi-metal analysis of soil (from author's own files)

Technique	Element determined	Limit of detection	Reference
Spectrophotometry	Zn, Cd		[296]
AAS	Heavy metals	0.07 – 2.5 µg/g	[56, 301, 302, 306, 339–344]
AAS	As, Cd, Tl, Cd, Mo		[303] [304]
AAS	Ag, Bi, In, Mo, Se, heavy metals		[305]
AAS	Heavy metals, Mo, Hg, As, Se		[307, 308]
AAS	Heavy metals, Mg		[309]
Inductively coupled plasma atomic emission spectrometry	As, Sb, Bi, Se		[310–312]
Inductively coupled plasma atomic emission spectrometry	Miscellaneous		[313]
Inductively coupled plasma atomic emission spectrometry	Heavy metals		[314–316]
Inductively coupled plasma atomic emission spectrometry	Heavy metals, Al, Ba, Ca, K, Mg, Na, Si, Se, Ti, V		[317]
Inductively coupled plasma atomic emission spectrometry	Mo, Co, B	0.01 ppm Mo 0.05 ppm Co, B	[312]
Inductively coupled plasma atomic emission spectrometry	17 elements		[318, 319]
Spark source or glow mass spectrometry	Miscellaneous		[320]
Spark source or glow mass spectrometry	Miscellaneous	10 µg/g	[316]
Spark source or glow mass spectrometry	50 elements		[321]
Emission spectrometry	Heavy metals, Al, Ba, Be, Mo, Sr, V, Tb, Te, TH, Ti, Tl, Tm, U, W, Y, Zr	10 µg/g	[314–316]
Stable isotope dilution mass spectrometry	K, Rb, Cs, Ca, Sr, Ba	µg/kg	[131]
Photon activation analysis	Heavy metals, Al, Mg, Sr, Ca, Ti, As, Zr, Sb, U		[322]
Neutron activation analysis	Heavy metals, Sc, As, Se, Mo, Sb, La, Ce, Eu, Hf, Ta, Th, U	< 0.002 ppm	[322]
Neutron activation analysis	Cu, Zn		[106]

Table 2.3. (continued)

Technique	Element determined	Limit of detection	Reference
Neutron activation analysis	As, Se		[216]
Neutron activation analysis	Lanthanides		[323]
X-ray fluorescence spectrometry	Miscellaneous		[324]
X-ray fluorescence spectrometry	Miscellaneous	10 – 50 ppm	[325]
Differential pulse anodic stripping voltammetry	Heavy metals, As		[326–329]
Electron probe microanalysis	Zn, Cu		[330]
Column chromatography	Actinides, lanthanides		[331–334]
Gas chromatography–combustion isotope dilution ratio mass spectrometry	Miscellaneous		[335]
Ion chromatography	Mg, Ca, Mn, Zn	20 μg/l – Ca 10 μg/l – Mg	[336]
Laser-induced breakdown spectroscopy	Miscellaneous	1 – 300 ppm	[337]
α-spectrometry	Curium		[338]

six laboratories showed that all laboratories produced some outliers but most results were in good agreement once the outliers were removed.

Kingston and Walter [270] compared microwave extraction with conventional dissolution procedures for soils.

Reynolds [271] reviewed microwave digestion procedures for the analysis of metal-contaminated soils, and this technique has been found to give better than 80% accuracy for a wide range of metals [268].

La Guardia and Garrigues [273] and Hall [274] have reviewed methods for the determination of metals in soil. Waymaugh [275] has reviewed the monitoring of historical changes in soil and atmospheric trace metal levels using dendrochemical analysis.

Kimbrough and Wakakuwa [276,330] reported on an interlaboratory comparison study involving 160 accredited hazardous materials laboratories. Each laboratory performed a mineral acid digestion on five soils spiked with arsenic, cadmium, molybdenum, selenium and thallium. The instrumental detection methods used were inductively coupled plasma atomic emission spectrometry, inductively coupled plasma mass spectrometry, flame atomic absorption spectrometry, electrothermal atomic absorption spectrometry and hydride generation atomic absorption spectrometry. At most concentrations, the results obtained with inductively coupled plasma atomic emission spectrometry gave higher precision and accuracy than results obtained by other techniques, but also gave the highest rate of false positives and negatives.

A new type of pedologically based soil sampling technique (i.e., by soil horizon rather than incrementally in depth) has been suggested by Meriweather et al. [277]. He gives an example where classical sampling approaches would lead to erroneous conclusions about anthropogenic contamination. Considerable simplification of sample preparation for trace elements in sediments and soils has been reported by using ultrasonic slurry sampling and graphite furnace atomic absorption spectrometry. Before analysis, samples were ground to a grain size of < 63 µm.

Lopez Garcia et al., [24] suspended soil samples in water containing 5% (v/v) concentrated hydrofluoric acid before injection into an electrothermal atomic absorption spectrophotometric system. No modifier other than the hydrofluoric acid was required for the determination of lead, cadmium and thallium.

Jerrow et al. [278] have reviewed methods for the determination of alkaline earths in soils.

Ure [279] and Spevachova and Kucera [280] studied the determination and speciation of trace metals in soil. Sagar [281] has reviewed the chemical speciation and environmental mobility of metals in soils. Stachel et al. [282] performed comparative studies using different digestion procedures in the determination of heavy metals and arsenic in the fine grain particle size fraction of suspended particulate matter. The highest metal concentrations were found for microwave heating in a closed system using an acid mixture of nitric acid–hydrofluoric acid. In a separate study, only digestion procedures using nitric acid and hydrofluoric acid with subsequent evaporation to dryness and dissolution in hydrochloric acid gave appropriate results for a wide range of elements.

Del Castilho and Rix [265] reviewed the suitability of the ammonium acetate extraction method to predict heavy metal availability in soils.

For the sample digestion of sediments, Scancar et al. found that significantly lower results were obtained for chromium, nickel and cadmium when aqua regia digestion was employed, compared to total acid dissolution and acetic acid extraction methods [266].

Another study comparing various digestion methods found that the mildest method – using sodium acetate attack – was most effective at enhancing differences between polluted and unpolluted sediments [267]. Microwave digestion of soils and sediments was found to give better than 80% accuracy for a wide group of metals [268]. Ultrasound-assisted extraction gave two-fold better precision and better detection limits compared to microwave-assisted digestion of biological and sediment samples [269].

Real et al. [341] showed that optimising the microwave heating procedure would optimise the results obtained in sequential extraction procedures.

Torres et al. [342] found that a microwave-assisted robotic method for trace metals in soil decreased sample digestion times from two hours to three minutes.

Sturgeon et al. [343] have demonstrated that a continuous flow microwave-assisted digestion of soil samples gave an average of 90% recovery of trace elements with good precision.

An appreciable amount of work has been carried out on the application of microwave digestion techniques to the determination of heavy metals, arsenic and uranium in soils.

Lo and Fung [344] studied the recovery of heavy metals from soils during acid digestion with different acid mixtures using a block heater and by microwave heating.

Kratchvil and Mamba [345] showed that all of the zinc and copper were released from soils within seven minutes using a commercial microwave oven.

Sanchez et al. [329] achieved acceptable accuracy and precision in the determination of metals in soil using an ultrasonic bath digestion procedure.

Fortunati et al. [345] have reviewed problems associated with techniques and strategies of soil sampling.

Einax et al. [331] have used chemometric techniques to investigate the representativity of soil sampling.

Rubio and Ure [332] have discussed the risks of contaminating soil samples using inappropriate materials, containers and tools as well as possible analyte loss during sample handling. Factors affecting the realism of the collected sample have been discussed by Burton [333]. Different sampling designs are needed, depending upon whether the soil contamination is expected to be "spread" over the whole area or whether it exists in localised "hot spots".

Lame and Defize [334] showed that the fundamental sampling error for soil only affects the analytical variance when sample sizes are less than 10 g [335]. For larger samples, the variance is determined by the segregation error. A sampling board method for estimation of the segregation error was described. Skalski and Ward [335] showed that a two-way compositing strategy could be used to attribute detected contamination in composited samples directly to constituent samples without further analyses.

Meriweather et al. [277] has suggested a new type of pedologically-based soil sampling technique based on the soil horizon rather than incrementally with depth for the assessment of radionuclides in soil. He gives an example where classical sampling approaches would lead to erroneous conclusions about anthropogenic contamination.

Houba et al. [337] studied the influence of grinding procedures and demonstrated that the availability of some analytes is significantly influenced by the grinding of some soils.

Hewitt found that volatile organic compounds are readily lost from soil samples unless care is taken to limit surface area exposure and to ensure subsample isolation [338]. Volatile organic carbon losses were found to be most abundant during field collection and storage. Hewitt reported that fortified soils held in sealed glass ampoules at 4 °C, or dispersed in methanol and held at 22 °C, showed no significant losses over 20 and 98 days, respectively [339].

Hunt [340] has described a simple method of filtering soil extracts that eliminates the need for filter funnels and receivers. It therefore reduces the risk of contamination and speeds up the procedure. It also offers a convenient means of obtaining filtrates in the field for subsequent analysis. After shaking the soil suspension in the extraction bottle, a tube of filter paper folded about the centre to form a V with the open ends uppermost is inserted into the bottle. Clear filtrate collects inside the paper tube and aliquots are removed with a pipette.

References

1. Dodson A, Jennings VJ (1972) *Talanta* **19**:801.
2. Reis BF, Bergamin FH, Zegatto EAG, Krug FJ (1979) *Anal Chim Acta* **107**:309
3. Tecator Ltd. (1985) *Determination of Aluminium in Soil by Flow Injection Analysis,* Application Note ASN 78–31/85, Tecator Ltd., Bristol, UK.
4. Tecator Ltd. (1984) *Determination of Aluminium by Flow Injection Analysis,* Application Note ASTN 10/84, Tecator Ltd., Bristol, UK.
5. Zoltzer D (1882) PhD Thesis, Institut für Anorganic Chemie, University of Gottingen, West Germany.
6. Ross DS, Bartlett RJ, Magdott FR (1986) *Atom Spectrosc* **7**:158.
7. Mitrovic B, Milacic R, Pihar B (1996) *Analyst* **121**:627.
8. Kozuh N, Milacic R, Gorenc B (1996) *Ann Chim* **86**:99.
9. Gibson JAE, Willett IR (1991) *Commun Soil Sci Plant Anal* **22**:1303.
10. Garotti FV, Massaro S, Serrano SHP (1992) *Analusis* **20**:287.
11. Downard Aj, Kipton H, Powell J, Xu S (1992) *Anal Chim Acta* **256**:117.
12. Schmidt S, Koerdel W, Kloeppel H, Klein W (1989) *J Chromatogr* **470**:289.
13. Keay J, Menage PMA (1970) *Analyst* **95**:379.
14. Waughman GJ (1981) *Environ Res* **26**:529.
15. Adler JF, Gunn AM, Kirkbright GF (1977) *Anal Chim Acta* **92**:43.
16. Tecator Ltd. (1983) *Determination of Ammonia Nitrogen in Soil Samples Extractable by 2 M KCl using Flow Injection Analysis,* Application Note ASN 65-32/83, Tecator Ltd., Bristol, UK.
17. Agricultural Development & Advisory Service (ADAS) (1979) Ammonium Nitrate and Nitrite-Nitrogen, Potassium Chloride Extractable, in Moist Soil (Method 60). In: *The Analysis of Agricultural Materials,* KB 427, Second Edition, HMSO, London, UK.
18. Chikhalikar S, Sharma K, Patel KS (1995) *Commun Soil Sci Plant Anal* **26**:621.
19. Chikhalikar S, Sharma K, Patel KS (1995) *Commun Soil Sci Plant Anal* **26**:625.
20. Asami T, Kubota M, Saito S (1992) *Water Air Soil Pollut* **62**:349.
21. Woolson EA, Axley JJ, Kearney PC (1971) *Soil Sci Soc Am Proc* **35**:101.
22. Schroeder HA, Balassa JJ (1966) *J Chronic Dis* **19**:1.
23. Forehand TJ, Dupuy Jr AE, Tai H (1976) *Anal Chem* **48**:999.
24. Lopez-Garcia I, Sanchez-Merlos M, Hernandez-Cordoba M (1997) *Spectrochim Acta B* **52B**:437.
25. Ohta K, Suzuki M (1978) *Talanta* **25**:160.
26. Standing Committee of Analysts (DoE) (1987) *Selenium and Arsenic Sludges, Soils and Related Material: A Note on the Generation of Hydride Generator Kits,* HMSO, London, UK.
27. Lichte FE, Skogerboe RK (1972) *Anal Chem* **44**:1480.
28. Gurluyuk H, Tuson JF, Uden DC (2000) *Spectrochim Acta B* **55B**:935.
29. Thompson KC, Thomerson DR (1974) *Analyst* **99**:595.

30. Thompson AJ, Thoresby PA (1977) *Analyst* **102**:9.
31. Haring BJA, Van Delft W, Bou CM (1982) *Fresen Z Anal Chem* **310**:217.
32. Jiminez de Blas, Mateos NR, Sanchez AG (1980) *J AOAC Int* **79**:323.
33. Merry RH, Zarcinas BA (1980) *Analyst* **109**:998.
34. Sandhu SS (1981) *Analyst* **106**:311.
35. Thomas P, Finrie JK, Williams JG (1997) *J Anal Atom Spectrom* **12**:1367.
36. Pohaska T, Pfeffer M, Tulipan M, Stingeder G, Mentler A, Wenzel KW (1999) *Fresen J Anal Chem* **364**:467.
37. Van Laecke F, Boonan S, Moens L, Dams R (1995) *J Anal Atom Spectrom* **10**:81.
38. Hwang JD, Huxley HP, Diomiguardi JP, Vaughn WJ (1990) *J Appl Spectrosc* **44**:491.
39. Lasztity A, Krushevska A, Kotrebai M, Barnes RM, Amarasiriwardina D (1995) *J Anal Atom Spectrom* **10**:505.
40. Barra CM, Cervera ML, De la Guardia M, Santelli RE (2000) *Anal Chim Acta* **407**:155.
41. Demesmay C, Olle M (1997) *Fresen J Anal Chem* **357**:1116.
42. Agemian H, Bedak E (1980) *Anal Chim Acta* **119**:323.
43. Chappell J, Chiswell B, Olozzowy H (1995) *Talanta* **42**:323.
44. Masscheleyn PH, Delaune RD, Patrick WH (1991) *Environ Sci Tech* **25**:1414.
45. McGeehan SL, Naylor DV (1992) *J Environ Qual* **21**:68.
46. Honore Hansen G, Larsen EH, Pritzi G, Cornet C (1992) *J Anal Atom Spectrom* **7**:629.
47. Naidu R, Smith J, Mclaren RG, Stevens DP, Sumner ME, Jackson PE (2000) *Soil Sci Am J* 122.
48. Standing Committee of Analysts (DoE) (1979) *The Analysis of Agricultural Materials – Cadmium, Nitric Acid Soluble in Soil Method 11*, MAFF Technical Bulletin RB 427, HMSO, London, UK.
49. Standing Committee of Analysts (DoE) (1979) *The Analysis of Agricultural Materials – Cadmium, Extractable in Soil Method 10*, MAFF Technical Bulletin RB 427, HMSO, London, UK.
50. Bolt GH, Bruggenwerk MGM (1978) In: *Soil Chemistry, Part A*, Elsevier, Amsterdam.
51. Sperling KR (1977) *Fresen Z Anal Chem* **28**:30.
52. Berrow ML, Stein WM (1983) *Analyst* **108**:277.
53. Baucells M, Lacort G, Roura M (1985) *Analyst* **110**:1423.
54. Reddy PRK, Reddy SJ (1997) *Fresen Environ Bull* **6**:661.
55. Krishnamurti GSR, Huang PM, Van Rees KCJ, Kozak LM, Rostad HDW (1994) *Commun Soil Sci Plant Anal* **25**:615.
56. Lewin VH, Beckett OHT (1980) *Effluent Water Treat* **20**:162.
57. Christensen TH, Lun XZ (1989) *Water Res* **23**:73.
58. Turner MA, Hendricksen LL, Corey RB (1984) *Soil Sci Am J* **48**:763.
59. Roberts AHC, Turner MA, Syers JK (1976) *Analyst* **101**:574.
60. Carlosena A, Prada D, Andrade JM, Lopez P, Muniategui B (1996) *Fresen J Anal Chem* **355**:289.
61. Hirsch D, Banin A (1990) *J Anal Qual* **19**:366.
62. Feng Y, Barratt RS (1992) *Sci Total Environ* **143**:157.
63. Kim ND, Fergusson JE (1991) *Sci Total Environ* **105**:191.
64. Dubois JP (1991) *J Trace Microprobe T* **9**:149.
65. Mazzucotelli A, Soggia F, Cosma B (1991) *Appl Spectrosc* **4**:504.
66. Millward CG, Kluckner PD (1991) *J Anal Atom Spectrom* **6**:37.
67. Evans DW, Alberts JJ, Clark RA (1983) *Geochim Cosmochim Acta* **47**:1041.
68. Heong C-H, Kim C-S, Kim S-J, Park S-W (1996) *J Environ Sci Heal* **A31**:2173.
69. Fawaris BH, Johanson KJ (1995) *Sci Total Environ* **170**:221.
70. Carbol P, Skarnemark G, Skalberg M (1993) *Sci Total Environ* **130**:129.

71. Williams TM (1993) *Environ Geol* **21**:62.
72. Von Gunter HR, Benes P (1995) *Radiochim Acta* **69**:1.
73. Essington FH, Fowler EB, Polzer WI (1981) *Soil Science***132**:13.
74. Bunzl K, Schimmack W (1989) *Chemosphere* **18**:2109.
75. Wauters J, Vidal M, Elsen A, Cremers A (1996) *Appl Geochem* **11**:595.
76. Vidal M, Roig M, Rigol A, Llarado M, Rauret G, Wauters J, Elsen A, Cremers A (1995) *Analyst* **120**:1785.
77. Thiry Y, Myttenaere C (1993) *J Environ Radioactiv* **18**:247.
78. Groenewold GS, Ingram JC, McLing T, Gianotto AK, Avci R (1998) *Anal Chem* **70**:534.
79. Xing-Chu Q, Yu-Sheng Z, Yinh-Quan Z (1983) *Analyst* **108**:754.
80. Jackson YL (1958) In: *Soil Chemical Analysis*, Prentice Hall, Englewood Cliffs, NJ, USA.
81. Qi WB, Zhu LZ (1986) *Talanta* **33**:694.
82. Smith GH, Lloyd OL (1986) *Chem in Britain*, February, p. 139.
83. Sahuquillo A, Lopez-Sanchez JF, Rubio R, Rauret G, Hatje V (1995) *Fresen J Anal Chem* **119**:251.
84. Fodor P, Fischer I (1995) *Fresen Z Anal Chem* **351**:454.
85. Prokisch J, Kovacs TS, Gyori Z, Loch J (1995) *Commun Soil Sci* **26**:2051.
86. Florez-Velez LM, Gutierrez-Ruiz HE, Reyes-Salas O, Cram-Heydrich S, Baesze-Reys A (1995) *Int J Environ Anal Chem* **61**:177.
87. Chakraborty R, Das AK, Cervera ML, De la Guardia M (1995) *J Anal Atom Spectrom* **10**:353.
88. Milacic R, Stupar J, Kozuth N, Korosin J (1992) *Analyst* **117**:125.
89. Sahuquillo A, Lopez-Sanchez JF, Rubio R, Rauret G, Hatje V (1995) *Fresen J Anal Chem* **351**:197.
90. Mierzwa J, Sun YC, Yang MH (1998) *Spectrochim Acta B* **53B**:63.
91. Peterson M, Brown GE, Parks GA (1997) *Mater Res Soc Symp Proc* **432**:75.
92. Szulczewski MD, Helmke PA, Bleam WF (1997) *Environ Sci Tech* **31**:2954.
93. Kalembkiewicz J, Filar L (1997) *Chem Anal (Warsaw)* **42**:87.
94. Marques MJ, Salvador A, Morales-Rubio AE, De la Guardia M (1998) *Fresen J Anal Chem* **362**:239.
95. Standing Committee of Analysts (DoE) (1979) *The Analysis of Agricultural Materials – Cobalt, Nitric-Perchloric Acid Soluble in Soil Method 23*, MAFF Technical Bulletin RB 427, HMSO, London, UK.
96. Standing Committee of Analysts (DoE) (1979) *The Analysis of Agricultural Materials – Cobalt, Extractable in Soil Method 22*, MAFF Technical Bulletin RB 427, HMSO, London, UK.
97. Standing Committee of Analysts (DoE) (1979) *The Analysis of Agricultural Materials – Copper, Nitric-Perchloric Acid Soluble in Soil Method 27A*, MAFF Technical Bulletin RB 427, HMSO, London, UK.
98. Standing Committee of Analysts (DoE) (1979) *The Analysis of Agricultural Materials – Copper, EDTA Extractable in Soil Method 26*, MAFF Technical Bulletin RB 427, HMSO, London, UK.
99. Mesuere K, Martin RE, Fish W (1991) *J Environ Qual* **20**:114.
100. Gerringa LJA, Ver der Meer J, Cauwet G (1991) *Mar Chem* **36**:51.
101. Shapiro JB, James WD, Schweikert EA (1995) *J Radioanal Nucl Chem* **192**:275.
102. Stefanov G, Daieva L (1972) *Isotopenpraxis* **8**:146.
103. Jayman TCZ, Sivasuhramaniam S, Wijedasa MA (1975) *Analyst* **100**:716.
104. Keaton CM (1937) *Soil Sci* **43**:40.
105. Agrawal YK, Raj KPS, Desai S, Patel SG, Merh SS (1980) *Int J Environ Sci* **14**:313.
106. Chow TJ (1970) *Nature* **225**:295.

107. Motto HL, Daines RH, Chilko DM, Motto CK (1970) *Environ Sci Tech* **4**:231.
108. Low KS, Lee CK, Arshad MM (1979) *Pertanika* **2**:105.
109. Ward NI, Reeves RD, Brooks RR (1974) *Environ Pollut* **6**:149.
110. Savvin SB, Petrova K (1991) *Fresen J Anal Chem* **340**:217.
111. Standing Committee of Analysts (DoE) (1979) *The Analysis of Agricultural Materials – Lead, Nitric-Perchloric Acis Soluble in Soil Method 44*, MAFF Technical Bulletin RB 427, HMSO, London, UK.
112. Standing Committee of Analysts (DoE) (1979) *The Analysis of Agricultural Materials – Lead, Extractable in Soil Method 43*, MAFF Technical Bulletin RB 427, HMSO, London, UK.
113. Tills AR, Alloway BJ (1983) *Environ Tech Lett* **4**:529.
114. Hinds MW, Jackson KW (1989) *J Anal Atom Spectrom* **5**:199.
115. Zhang B, Tao K, Feng J (1992) *J Anal Atom Spectrom* **7**:171.
116. Hinds MW, Latimer KE, Jackson KW (1991) *J Anal Atom Spectrom* **6**:473.
117. Hinds MW, Jackson KW (1989) *J Anal Atom Spectrom* **12**:109.
118. Somer G, Aydin H (1985) *Analyst* **110**:631.
119. Fernando AR, Plambeck JA (1992) *Analyst* **117**:39
120. Sakharov AA (1967) *Pochvovedenie* **1**:107.
121. Wegrzynek D, Holynska B (1993) *Appl Radiat Isotopes* **44**:1101.
122. Rigin VI, Rigina IV (1979) *Zh Anal Khim* **34**:1121.
123. Bedrosian AJ, Skogerboe RK, Morrison GH (1968) *Anal Chem* **40**:854.
124. Xan X-P, Sperling M, Welz B (1999) *Anal Chem* **71**:4216.
125. Hinkley T (1979) *Nature* **277**:444.
126. Chen TC, Hong A (1995) *J Hazard Mater* **41**:147.
127. Lopez-Garcia L, Sanchez-Merlos M, Hernandez-Cordoba M (1996) *Anal Chim Acta* **328**:19.
128. Zaray G, Kantor T (1995) *Spectrochim Acta B* **50B**:489.
129. Manceau A, Boisset MC, Sarret G, Hazemann JL, Mench M, Cambier P, Prost R (1996) *Environ Sci Tech* **30**:1540.
130. Bermejo-Barrera P, Barciela-Alonso C, Aboal-Somoza M, Bermejo-Barrera A (1994) *J Anal Atom Spectrom* **9**:469.
131. Standing Committee of Analysts (DoE) (1979) *The Analysis of Agricultural Materials – Magnesium in Soil Method 46*, MAFF Technical Bulletin RB 427, HMSO, London, UK.
132. Standing Committee of Analysts (DoE) (1979) *The Analysis of Agricultural Materials – Manganese, Exchangeable and Easily Reducible, in Soil Method 48*, MAFF Technical Bulletin RB 427, HMSO, London, UK.
133. Alekseeva II, Davydova ZP (1971) *Zh Anal Khim* **26**:1786.
134. Kamburova M (1993) *Talanta* **40**:719.
135. Kimura Y, Miller VL (1962) *Anal Chim Acta* **27**:325.
136. Hatch WR, Ott WL (1968) *Anal Chem* **40**:2085.
137. Agemian H, Chau ASY (1976) *Analyst* **101**:91.
138. Kuwae Y, Hasegawa T, Shono T (1976) *Anal Chim Acta* **84**:185.
139. Nicholson RA (1977) *Analyst* **102**:399.
140. Hoggins FE, Brooks RR (1973) *J Assoc Off Anal Chem* **56**:1306.
141. Ure Am, Shand CA (1974) *Anal Chim Acta* **72**:63.
142. Floyd M, Sommers LE (1975) *J Environ Qual* **4**:323.
143. Head PC, Nicolson RA (1973) *Analyst* **98**:53.
144. *Methods for the Examination of Water and Associated Materials – Mercury in Waters, Effluents, Soils, Sediments, etc, Additional Methods 40453*, HMSO, London, UK, 1987.
145. Lutze RG (1979) *Analyst* **104**:979.

146. Grantham PL (1978) *Lab Practice* **27**:294.
147. Standing Committee of Analysts (DoE) (1979) *The Analysis of Agricultural Materials –
     Mercury in Soil and Plant Material, Method 86*, MAFF Technical Bulletin RB 427,
     Second Edition, HMSO, London, UK.
148. Sakamoto H, Tomlyasu T, Yonehara N (1992) *Anal Sci* **8**:35.
149. Azzaria LM, Aftabi A (1991) *Water Air Soil Pollut* **56**:203.
150. Lee HS, Jung KH, Lee DS (1989) *Talanta* **36**:999.
151. Bandyopadhyay S, Das AK (1989) *J Ind Chem Soc* **66**:427.
152. Rasemann W, Seltmann U, Hempel M (1995) *Fresen J Anal Chem* **351**:632.
153. Welz B, Sclemmeer G, Mudakavi JR (1992) *J Anal Atom Spectrom* **7**:499.
154. Baxter DC, Nichol R, Littlejohn D (1992) *Spectrochim Acta B* **47B**:1155.
155. Lexa J, Stulik K (1989) *Talanta* **36**:843.
156. Cherian L, Gupta UK (1990) *Fresen Z Anal Chem* **336**:400.
157. Saouter E, Campbell PGC, Ribeyre F, Boudon A (1993) *Int J Environ Anal Chem* **54**:57.
158. Golimowski J, Orzechowska A, Tykarska A (1995) *Fresen J Anal Chem* **351**:656.
159. Robinson L, Dyer FF, Combs DW, Wade W, Teasley NA, Carlton JE (1994) *J Radioanal
     Nucl Chem* **179**:305.
160. Carpi A, Lindberg SE (1998) *Atmos Environ* **32**:873.
161. Easterling DF, Hovassitz ES, Street KW (2000) *Anal Lett* **33**:1665.
162. Standing Committee of Analysts (DoE) (1979) *The Analysis of Agricultural Materials –
     Molybdenum, Extractable in Soil, Method 50*, MAFF Technical Bulletin RB 427, HMSO,
     London, UK.
163. Standing Committee of Analysts (DoE) (1979) *The Analysis of Agricultural Materi-
     als – Molybdenum, Total in Soil, Method 51*, MAFF Technical Bulletin RB 427, HMSO,
     London, UK.
164. Kim CH, Owens CM, Smythe LE (1974) *Talanta* **21**:445.
165. Kim CH, Alexander PW, Smythe LE (1976) *Talanta* **23**:229.
166. Ni ZM, Chin LC, Wu TH (1979) *Huan Ching K'o Hsueh* **6**:25. *Anal Abstr* (1981) **40**:5h65.
167. Maessen FJML, Balke J, de Beer JLM (1982) *Spectrochim Acta B* **37B**:517.
168. Manzoori JL (1980) *Talanta* **27**:682.
169. Milham PJ, Maksvytis A, Barkus B (1972) *Anal Chem* **44**:2102.
170. Thompson M, Zao L (1985) *Analyst* **110**:229.
171. Reisenauer MH (1965) In: *Methods of Soil Analysis, Part 2*, American Society of Agron-
     omy, Madision, WI, USA, p. 1054–57.
172. Jiao K, Jin W, Metzner H (1992) *Anal Chim Acta* **260**:35.
173. Rowbottom WH (1991) *J Anal Atom Spectrom* **6**:123.
174. Standing Committee of Analysts (DoE) (1979) textit The Analysis of Agricultural
     Materials – Nickel, Nitric–Perchloric Acid Soluble in Soil, Method 54, MAFF Technical
     Bulletin RB 427, HMSO, London, UK.
175. Standing Committee of Analysts (DoE) (1979) *The Analysis of Agricultural Materials –
     Determination of Extractable Nickel in Soil, Method 53*, MAFF Technical Bulletin RB
     427, HMSO, London, UK.
176. Sekine K, Imai T, Kasai A (1987) *Talanta* **34**:567.
177. Rubio-Montera HP, Martin Sanchez A, Crespo-Vanquez MT, Gascon Murillo JL (2000)
     Appl Radiat Isotopes **53**:259.
178. Standing Committee of Analysts (DoE) (1979) *The Analysis of Agricultural Materials –
     Potassium Extractable in Soil, Method 68*, MAFF Technical Bulletin RB 427, HMSO,
     London, UK.
179. Parker CA, Harvey LG (1962) *Analyst* **87**:558.
180. Alloway WH, Cary EE (1964) *Anal Chem* **36**:1359.

181. Watkinson JH (1960) *Anal Chem* **32**:98.
182. Watkinson JH (1966) *Anal Chem* **38**:92.
183. Ewan RC, Baumann CA, Pope AL (1968) *J Agric Food Chem* **16**:212.
184. Hall RJ, Gupta PL (1969) *Analyst* **94**:292.
185. Dagnall RM, Thompson KC, West TS (1967) *Talanta* **14**:557.
186. Azad J, Kirkbright GF, Snook RD (1979) *Analyst* **104**:232.
187. Weltz B (1976) In: *Atomic Absorption Spectrometry*, Verlag Chemie, New York, USA.
188. Pinta M (1977) In: *Atomic Absorption Spectrometry Applications to Chemical Analysis*, PWN, Warsaw.
189. Schrenk WG (1975) In: *Modern Analytical Chemistry Analytical Atomic Spectroscopy*, Plenum, New York, USA.
190. Brodie KG (1979) *Int Lab*, July/August, p. 40.
191. Laurakis V, Barry E, Golembeski T (1975) *Talanta* **22**:547.
192. Ohta K, Suzuki M (1975) *Talanta* **22**:465.
193. Thompson KC (1975) *Analyst* **100**:307.
194. Standing Committee of Analysts (DoE) (1987) *Selenium and Arsenic in Sludges, Soils and Related Materials – A Note on the Use of Hydride Generation Kits*, HMSO, London, UK.
195. Chambers JC, McClellan D (1976) *Anal Chem* **48**:2061.
196. Inhat M (1976) *Anal Chim Acta* **82**:292.
197. Henn EL (1975) *Anal Chem* **47**:428.
198. Ishizaka M (1978) *Talanta* **25**:167.
199. Montaser A, Mehrabzadeh AA (1978) *Anal Chem* **50**:1697.
200. Jiminez de Blas O, Mateos NR, Sanchez Ag (1996) *J Am Assoc Anal Chem* **79**:764.
201. Zhang YQ, Moore JN, Frankenberger Jr. WT (1999) *Environ Sci Technol* **33**:1652.
202. Martens DA, Suarez DL (1997) *Environ Sci Technol* **31**:133.
203. McCurdey EJ, Lange JD, Haygarth PM (1993) *Sci Total Environ* **135**:131.
204. Pahlavanpour B, Pullen JH, Thompson M (1980) *Analyst* **105**:274.
205. Zmijewska W, Semkow T (1978) *Chem Anal (Warsaw)* **23**:583.
206. Kronberg OJ, Steinnes E (1975) *Analyst* **100**:835.
207. Mignosin EP, Roelandts I (1975) *Chem Geol* **16**:137.
208. Van der Klugt N, Poelstra P, Zwemmer E (1977) *J Radioanal Chem* **35**:109.
209. Agemian H, Bedek E (1980) *Anal Chem* **119**:394.
210. Bem EM (1981) *Environ Health Persp* **37**:183.
211. Dong A, Rendig VV, Burau RG, Besga GS (1987) *Anal Chem* **59**:2728.
212. Rojus CJ, de Maroto SB Valenta P (1994) *Fresen J Anal Chem* **348**:775.
213. Cruvinel PE, Flocchini RG (1993) *Nucl Instrum Methods B* **75**:415.
214. Standing Committee of Analysts (DoE) (1979) *The Analysis of Agricultural Materials – Sodium, Extractable in Soil, Method 72*, MAFF Bulletin RB427, HMSO, London, UK.
215. Akcay M, Elik A, Savasci S (1989) *Analyst* **114**:1079.
216. Tagami K, Uchida S (1999) *Radioact Chem* **10**:30.
217. Tagami K, Uchida S (1993) *Radiochim Acta* **63**:69.
218. Morita S, Kim CK, Sakaku Y, Seki R, Ikeda N (1991) *Appl Radiat Isotopes* **42**:531.
219. Harvey BR, William KJ, Lovett MP, Ibbett RD (1992) *J Radioanal Nucl Chem* **158**:417.
220. De Ruck A, Vandecastell G, Dams R (1989) *Anal Lett* **22**:469.
221. Lukaszewski Z, Zembrzuski W (1992) *Talanta* **39**:221.
222. Sagar M (1992) *Mikrochim Acta* **106**:241.
223. Van Laar C, ReinkR, Simon J (1994) *Fresen J Anal Chem* **349**:692.
224. Opydo J (1989) *Mikrochim Acta* **2**:15.
225. Toole J, McKay K, Baxter M (1990) *Anal Chim Acta* **245**:83.

226. Shaw TJ, Francois R (1991) *Geochim Cosmochim Acta* **55**:2075.
227. Zbiral J, Medeki P, Kubani V, Cizmarova E, Neuiek P (2000) *Commun Soil Sci Plant Anal* **31**:2045.
228. Parsa B (1992) *J Radioanal Nucl Chem* **157**:65.
229. Mukhtar OM, Grods A, Khangi FA (1991) *Radiochim Acta* **54**:201.
230. Mann DK, Oatis T, Wong GTF (1992) *Talanta* **39**:1199.
231. Lazo EN, Roessier GS, Bervani BA (1991) *Health Phys* **6**:231.
232. Shuktomova II, Kochan IG (1989) *J Radioanal Nucl Chem* **129**:245.
233. Lazo EN (1988) *DOE Report OR/0033-T 424*, Order No.DE89010612, NTIS, Springfield, VA, USA.
234. Abbasi SA (1982) *Int J Environ Anal Chem* **11**:1.
235. Li Z, McIntosh S, Carnride GR, Slavin W (1992) *Spectrochim Acta B* **47B**:701.
236. Quinn BT, Brooke RR (1972) *Anal Chim Acta* **58**:301.
237. Prister BS, Zubach SS (1968) *Radiokhimiya* **10**:743.
238. Nass HU, Molinski VJ, Kramer HH (1972) Mater Res Stand **12**:24.
239. Premadas A, Srivastava PK (1999) *J Radioanal Nucl Chem* **242**:23.
240. Zauzow DS, Baldwin DP, Weeks SJ, Bajic SJ, D'Silva AP (1994) *Environ Sci Technol* **28**:352.
241. D'Silva AP, Bajic SJ, Zamzow D (1993) *Anal Chem* **65**:3174.
242. Boomer DW, Powell MJ *Anal Chem* **59**:2810.
243. Abkasi SA (1981) *Int J Environ Studies* **18**:51.
244. Standing Committee of Analysts (DoE) (1979) *The Analysis of Agricultural Materials – Zinc Nitric–Perchloric Acid Soluble in Soil*, MAFF Method 83, HMSO, London, UK.
245. Standing Committee of Analysts (DoE) (1979) *The Analysis of Agricultural Materials – Zinc Extractable in Soil*, MAFF Method 82, HMSO, London, UK.
246. Abdullah M, Zaman MB, Khaliquzzaman M, Khan AH (1980) *Anal Chim Acta* **118**:175.
247. Shapiro JB, James ND, Schweikert E (1995) *J Radioanal Nucl Chem* **192**:275.
248. Pickering WF (1982) *CRC Crit Rev Anal Chem* **12**:233.
249. McLaren RG, Crawford DV (1973) *J Soil Sci* **24**:172.
250. Gupta SK, Chen KY (1975) *Environ Lett* **10**:129.
251. Gatehouse S, Russell DW, Van Moort JC (1977) *J Geochem Explor* **8**:483.
252. Shuman LM (1979) *Soil Sci* **127**:10.
253. Hoffman SJ, Fletcher WK (1979) In: Watterson JR, Theobalt PR (eds) *Proc 7th Int Geochemical Exploration Symp*, Association of Exploratory Geochemistry, Rexdale, Ontario, Canada.
254. Farrah H, Pickering WF (1978) *Water Air Soil Pollut* **9**:491.
255. Beveridge A, Pickering WF (1980) *Water Air Soil Pollut* **14**:171.
256. Hatton D, Pickering WF (1980) *Water Air Soil Pollut* **14**:13.
257. Slavek J, Wold J, Pickering WF (1982) *Talanta* **29**:749.
258. Zunio H, Peirano P, Aguilera M, Schaischa E (1975) *Soil Sci* **119**:210.
259. Stevenson FJ (1977) *Soil Sci* **123**:10.
260. Lakatos B, Tibai T, Meisei T (1976) *Agrokem Taiajtan* **25**:305.
261. Martens DC (1968) *Soil Sci* **106**:23.
262. Stewart JA, Berger KC (1962) *Soil Sci* **100**:244.
263. Gibbs RJ (1973) *Science* **180**:71.
264. Crosland AR, McGrath SP, Lane PW (1993) *Int J Environ Anal Chem* **51**:153.
265. Del Castilho P, Rix I (1993) *Int J Environ Anal Chem* **51**:59.
266. Scancar J, Milacic R, Horvat M (2000) *Water Air Soil Pollut* **118**:87.
267. Adami G, Barbieri P, Reisenhofer E (1999) *Int J Environ Anal Chem* **75**:251.
268. Falciani R, Novaro E, Marchesini M, Gucciardi M (2000) *J Anal Atom Spectrom* **15**:561.

269. Lima EC, Barbosa Jr. F, Krug FJ, Silva MM, Vale MGR (2000) *J Anal Atom Spectrom* **15**:995.
270. Kingston HM, Walter PJ (1992) *J Spectrosc* **7**:20.
271. Reynolds CM (1992) In: Iskander IK, Selim HM (eds) *Engineering Aspects of Metals – Waste Management*, Lewis, Boca Raton, FL, USA, p. 49–61.
272. Sanchez J, Garcia R, Millan E (1994) *Analusis* **22**:222.
273. La Guardia M, Garrigues S (1998) *Trends Anal Chem* **17**:263.
274. Hall GEM (1997) In: Mudroch A, Azcue JM, Mudroch P (eds) *Manual of Physico-Chemical Analysis of Aquatic Sediments*, , Boca Raton, FL, USA, p. 85–145.
275. Watmough SA (1999) *Environ Pollut* **106**:391.
276. Kimbrough DE, Wakakuwa J (1994) *Analyst* **119**:383.
277. Meriweather JR, Burns SF, Thompson RH, Beck JN (1995) *Health Phys* **69**:406.
278. Jerrow M, Marr T, Cresser M, Ramsbottom DJ, Adams MJ, Sumiga J, Clarke CG, Wang W, Barnard CLR, Hartnett M, Diamond D, Kiernan L, Costello J, (1992) *Anal Proc* **29**:45.
279. Ure AM (1990) *Fresen Z Anal Chem* **387**:577.
280. Spevackova U, Kukera J (1989) *Int J Environ Anal Chem* **35**:241.
281. Sagar M (1992) *Tech Instrum Anal Che*, 12 (Hazard Met Environ) p. 133.
282. Stachel B, Elsholz O, Reincke H (1995) *Fresen J Anal Chem* **353**:21.
283. Roberts HC, Turner MA, Syers JK (1976) *Analyst* **101**:574.
284. Pederson B, Williams M, Jorgensen SS (1980) *Analyst* **105**:119.
285. Kvalheim A (ed) (1967) *Geochemical Prospecting in Fennoscandia* Interscience, NY, USA, Chap. 8, p. 99.
286. Hesse PR (1971) *A Textbook of Soil Chemical Analysis*, Murray, London, UK.
287. Gorsuch TT (1959) *Analyst* **84**:135.
288. Schroeder HA, Balassa JJ (1963) *Science* **140**:819.
289. Williams CH, David DJ (1973) *Aust J Soil Res* **11**:43.
290. Ivanova E, Stoimenova M, Gentcheva G (1994) *Fresen J Anal Chem* **348**:317.
291. Baucells M, Lacort G, Roura M (1985) *Analyst* **110**:1423.
292. Eidecker R, Jackwerth E (1987) *Fresen Z Anal Chem* **328**:469.
293. Ure AM, Ewen GJ, Mitchell MC (1980) *Anal Chim Acta* **118**:1.
294. Davis RP, Carlton Smith CH, Stark JH, Cambell JA (1988) *Environ Pollut* **49**:99.
295. Davis RP, Carlton Smith CH (1983) *Water Pollut Control* **82**:290.
296. Stupar J, Ajlec R (1982) *Analyst* **107**:144.
297. Flanagan FJ (1976) *1972 Compilation of Data on USGS Standards*, Paper No. 840, US Geological Survey, Reston, VA, USA, p. 131–183.
298. Pahlavanpour B, Thompson M, Thorne L (1980) *Analyst* **105**:756.
299. Manzoori JL (1980) *Talanta* **27**:684.
300. Link DD, Kingston HMS (2000) *Anal Chem* **72**:2908.
301. Bari-Favel G, Dalmasso J, Ardisson G (1992) *J Radioanal Nucl Chem* **156**:83.
302. Wisbrun R, Schechter I, Niessner R, Schroeder H, Kompa KL (1994) *Anal Chem* **66**:2964.
303. Que-Hee SS, Bcyle JR (1988) *Anal Chem* **60**:1033.
304. Kanda Y, Taira M (1988) *Anal Chim Acta* **207**:269.
305. Kheboian C, Bauer CF (1987) *Anal Chem* **59**:1417.
306. Duckworth DC, Barshick CM, Smith DH (1993) *J Anal Atom Spectrom* **8**:875.
307. Ure AM, Bacon JR (1978) *Analyst* **103**:807.
308. Randle K, Hartman EH (1985) *J Radioanal Nucl Chem* **90**:309.
309. Tsukada M, Yamamoto D, Sudo R, Nakahara H (1991) *J Radioanal Nucl Chem* **151**:121.
310. Muntaui H, Crossman G, Schramel P, Gallorai M, Orvini E (1987) *Fresen J Anal Chem* **326**:634.

311. Goulden PD, Anthony DHJ, Austen KD (1981) *Anal Chem* **53**:2027.
312. Edmonds TE, Guogong P, West TS (1980) *Anal Chim Acta* **120**:41.
313. Reddy SJ, Valenta P, Nurtiberg HW (1982) *Fresen Z Anal Chem* **313**:390.
314. Meyer A, de la Chevallerie-Haaf H, Henze G (1987) *Fresen Z Anal Chem* **328**:565.
315. Lee FY, Kittrick JA (1984) *Soil Sci Soc Am J* **48**:548.
316. Kim G, Burnett WC, Horwitz EP (2000) *Anal Chem* **72**:4882.
317. Horwitz EP, Chiarizia R, Diamond H, Gatrone RC, Alexandratos SD, Trochimczuk AQ, Crick EW (1993) *Solvent Extr Ion Exch* **11**:943.
318. Chiarizia R, Horwitz EP, Alexandratos SD, Gula M (1997) *J Sep Soil Technol* 32:1.
319. Maxwell Sl, Nichols ST (1999) *Report No. WSRC-MS-98-00647*, Westinghouse Savannah River Co., Aiken, SC, USA.
320. Lichtfause E, Budzinska H (1995) Analusis **23**:364.
321. Jan D, Schwedt G (1985) *Fresen Z Anal Chem* **320**:121.
322. Yamamoto KY, Cremers DA, Ferris MJ, Foster LE (1996) *Appl Spectros* **50**:222.
323. Sill CW, Duphal KW, Hindman FD (1974) *Anal Chem* **46**:1725.
324. Chao TT, Sanzolone RF (1973) *J Res US Geol Survey* **1**:681.
325. Weitz A, Fuchs G, Bachmann K (1982) *Fresen Z Anal Chem* **313**:38.
326. Department of the Environment/National Water Council Standing Technical Committee of Analysts (1983) *Methods for the Examination of Water and Associated Materials – Determination of Extractable Metals in Soil Sewage Sludge Treated Soils and Associated Materials*, 22 BC ENV, HMSO, London, UK.
327. Hinds MW, Jackson KW, Newman AP (1985) *Analyst* **110**:947.
328. Chester R, Hughes M (1967) *Chem Geol* **2**:249.
329. Fortunati GL, Banfi C, Pasturenzic M (1994) *Fresen J Anal Chem* **348**:86.
330. Kimbrough DE, Wakakuwa J (1994) *Analyst* **119**:383.
331. Einex J, Machclett B, Geiss B, Danzer K (1992) *Fresen J Anal Chem* **342**:267.
332. Rubio R, Ure A (1993) *Int J Environ Anal Chem* **51**:205.
333. Burton GA (1992) In: Burton GA (ed) *Sediment Toxicity Assessment* Lewis, Boca Raton, FL, USA, p. 37–66.
334. Lame FPJ, Defize PR (1993) *Environ Sci Technol* **27**:2035.
335. Skalski JR, Ward JQ (1994) *Environ Toxicol Chem* **13**:15.
336. Davis RE (1989) *Environ Geochem Health* **11**:137.
337. Houba VJG, Chardon WJ, Roelse K (1993) *Commun Soil Sci Plant Anal* **24**:1591.
338. Hewitt AD (1996) *Volatile Organic Compounds in the Environment*, ASTM Special Publication 1261 170, ASTM, West Conshohocken, PA, USA.
339. Hewitt AD (1996) *Volatile Organic Compounds in the Environment*, ASTM Special Publication 1261 181, ASTM, West Conshohocken, PA, USA.
340. Hunt J (1981) *Analyst* **106**:374.
341. Real C, Barreiro R, Carballeira A (1994) *Sci Total Environ* **152**:135.
342. Torres P, Ballesteros E, Luque de Castro MD (1995) *Anal Chim Acta* **308**:371.
343. Sturgeon RE, Willie SN, Methven BA, Lam JWH, Matusiewicz H (1995) *J Anal Atom Spectrom* **10**:981.
344. Lo CK, Fung YS (1992) *Int J Environ Anal Chem* **46**:277.
345. Kratchvil B, Mamba S (1990) *Can J Chem* **68**:360.
346. Rurikova D, Beno A (1992) *Chem Papers* **46**:73.
347. Wendawiak BW, Krab M (1995) *Fresen J Anal Chem* **351**:134.

3 Determination of Radioactive Elements in Soil

3.1
^{137}Caesium

This isotope has been determined in soil by low-level Compton suppression γ-counting [1].

Knizhnik et al. [2] verified a formula for evaluating the ^{134}Cs and ^{137}Cs α-radioactivity of soils using track detectors.

3.2
^{127}Iodine and ^{129}Iodine

The determination of ^{129}I in low-level radioactive waste was accomplished by radioactive instrumental neutron activation analysis [3]. A different group reported the determination of both ^{129}I and ^{127}I by neutron activation analysis and inductively coupled plasma mass spectrometry [4]. The method was very rapid – a sample could be analysed in three minutes. However, interference from ^{129}Xe resulted in limited sensitivity for ^{129}I detection.

3.3
^{63}Nickel

Scheuerer et al. [5] determined ^{63}Ni in soils. The method is applicable to analytical measurements needed during the decommissioning of nuclear power plants.

3.4
^{210}Polonium

Various workers [6,7] have reviewed methods for the determination of ^{210}polonium in soils.

3.5
^{226}Radium

Various workers have discussed methods for the determination of ^{226}radium in soils [8,9].

3.6
[89]Strontium and [90]Strontium

Martin [10] determined [89]strontium and [90]strontium in soil using total sample decomposition.

[90]Strontium is one of the most hazardous fission products because it is a bone-seeker and it has a long biological half-life. For these reasons, and the fact that it has been a principal nuclide in fallout studies for many years, the contamination of human diet with ^{90}Sr has been of great concern. Lately, short-lived ^{89}Sr also has gained significance by serving as an indicator of radiostrontium contamination of more recent origin.

The significance of these two strontium nuclides warrants reliable procedures for their analysis. Leaching methods [11,12,18] for the determination of radiostrontium in soil are applied to large-size samples for higher sensitivity but they assume that the strontium can be easily solubilised. This assumption is not needed in the procedure described, since the sample is decomposed completely. Also, the procedure can be used to analyze leached residues to check the completeness, and thus the reliability, of leaching procedures when higher sensitivities are needed.

Complete decomposition of soil samples is ensured by potassium fluoride and pyrosulfate fusions [13]. The high-sulfate system resulting from the dissolution of the fusion cake is an asset to separating strontium as a sulfate. Lead sulfate [14] was shown to be a good carrier for barium and radium, and after dissolution in ethylenediaminetetraacetic acid (EDTA), barium and radium were separated easily from lead as sulfates. Unfortunately, good recoveries of strontium on a sulfate precipitate are achieved only at lower acidities, where calcium sulfate partially precipitates. However, Eakins and Gomm [15] have shown that strontium sulfate can be precipitated from EDTA, giving a good separation from calcium and avoiding the hazardous and time-consuming fuming nitric acid method usually employed when separating strontium from calcium. They subsequently separated barium from strontium by precipitating barium chromate from a buffered EDTA solution. Montgomery [16] precipitated strontium sulfate from an EDTA–barium chromate supernate by copper displacement of strontium from an EDTA complex. Fourie and Ghijsels [17] used strontium sulfate precipitated from EDTA to gravimetrically determine strontium yields.

[90]Strontium is usually determined by separation and counting of its [90]yttrium daughter. If the strontium fraction from which yttrium is separated has been adequately purified, the yttrium can be separated as a hydroxide [15]. Otherwise the yttrium must undergo further purification such as solvent extraction. A method for the extraction of yttrium into bis(2-ethylhexyl)phosphoric acid from dilute hydrochloric acid has been reported by Butler [18]. Weaver and Kapplemann [19] reported the extraction of yttrium as well as the trivalent lanthanides into bis(2-ethylhexyl)phosphoric acid from a solution of acetic acid and sodium diethylenetriaminepentaacetate.

In the procedure described by Martin [10], strontium radionucleides are carried on strontium and lead sulfate to minimize the amount of strontium carrier. The sulfate precipitate is treated with successive portions of EDTA to preferentially dissolve lead and excess calcium, which are discarded. Subsequently, the strontium sulfate is dissolved in additional EDTA, and ferric and other insoluble hydroxides are precipitated at a pH of 12–14. The strontium is separated from residual calcium by reprecipitating strontium sulfate from EDTA at a pH of 4.0. Since the difference in stability constants between strontium and barium is twice as much for sodium diethylenetriaminepentaacetate as it is for EDTA complexes, diethylenetriaminepentaacetate rather than EDTA is used to effect a more complete separation of barium as the chromate from strontium. Strontium sulfate is metathesised to the carbonate prior to the chromate step to rid the system of interfering sulfate ion. A large excess of sodium sulfate added to the barium chromate supernate precipitates strontium sulfate without the need for copper [16] or other masking agents to displace the strontium from the strontium. Since the resultant strontium sulfate precipitate is pure enough to count for ^{89}strontium and ^{90}strontium, yttrium carrier can be precipitated from the strontium sulfate filtrate as the hydroxide, reprecipitated as the oxalate, and counted for ingrown ^{90}yttrium without further purification. When only ^{90}strontium is being determined, a barium separation from strontium is not required, and the yttrium is purified by an bis(2-ethylhexyl)phosphoric acid extraction.

The procedure described by Martin [10] has a detection limit for ^{90}strontium for a 10-gram sample and a β-counting blank of 250 counts per 100 min of 1×10^{-7} µg Ci/g. The detection limit for ^{89}strontium varies with the relative activities of ^{89}strontium and ^{90}strontium in the sample.

Lantzsch et al. [20] and Arslan et al. [21] have carried out trace determinations of ^{99}strontium and ^{89}strontium in solid samples using laser spectrometry combined with mass spectrometry [16] and accelerator mass spectrometry [17].

Grabek et al. [22] has reported a semi-automated isolation procedure and the detection of ^{90}strontium in soils. The strontium is leached from soil by employing a water suspension of Amberlite IR-120 and then separating strontium from other cations with Amberlite GC-400 or Dowex.

3.7
^{99}Technetium

Several groups have studied the determination of ^{99}technetium in environmental samples including soil. High-resolution inductively coupled plasma mass spectrometry [23, 24] was used in these studies. In one study ^{99}technetium was eluted from the soil by nitric acid and the analyte was separated by three solvents using 30% trioctylamine, methylethylketone and cyclohexanone [23]. Purification of the ^{99}techretium extract was performed by using an anion exchange column to reduce the dissolved solids content. In another investigation [24], samples of soil were fused with sodium hydroxide, extracted in a column containing methyltrioctylammonium chloride, and extracted by solvent with N-benzoyl-N-phenylhydroxylamine.

In a procedure reported by Jordan [25], the soil extract is enriched on an anion exchange column, ashed, fused with sodium carbonate and potassium carbonate, and then [99]technetium is determined by conventional liquid scintillation counting.

3.8
Transuranic Elements

3.8.1
Americium

Sill et al. [26] have discussed a spectrometric method for the determination of americium and other alpha-emitting nuclides, including curium and californium, in potassium fluoride–pyrosulfate extracts of soils. Sekine [27] used α-spectrometry to determine americium in soils with a chemical recovery of 60–70%. Joshi [28] and Livens et al. [29] have discussed methods for the determination of [241]americium in soils.

3.8.2
Californium

See under americium in Sect. 3.8.1 [26].

3.8.3
Neptunium

Kim et al. [30] have demonstrated good agreement between methods of determining [237]neptunium in soils, based on inductively coupled plasma mass spectrometry, neutron activation analysis and α-spectrometry. Kim et al. [30] determined the [240]plutonium to [239]plutonium ratio in soils using the fission track method and inductively coupled plasma mass spectrometry (ICPMS).

3.8.4
Plutonium

Talvitie [31] has described a radiochemical method for the determination of plutonium in soil based on chromatography on an anion exchange resin of a 9 M hydrochloric acid extract of the sample. Following clean-up, plutonium is desorbed by reductive elution with 1.2 M hydrochloric acid, 30% hydrogen peroxide (50:1) at pH 2, and then α particle counting. The lowest detectable activity for 1000 m of counts was 0.02 pCi for ^{239}Pu, which is sufficient to detect global nuclear contamination in one gram of soil.

Various workers [32–34] have discussed mass spectrometric and other methods for the determination of plutonium in soils. Plutonium in soils has been quantified using ^{238}plutonium as a yield tracer. Hollenbach et al. [36] used flow injection preconcentration for the determination of ^{230}Th, ^{234}U, ^{239}Pu and ^{240}Pu in soils. Detection limits were improved by a factor of about 20, and greater freedom from interference was observed with the flow injection system compared to direct aspiration.

Packed column gas chromatography has been used to determine various plutonium isotopes in soils [37].

Dienstbach and Bachmann [38] have determined plutonium in amounts down to 20 fCiP/ug soil in sandy soils by an automated method based on gas chromatographic separation and α-spectrometry. In this procedure, the sample is decomposed completely by hydrogen fluoride. The hydrogen fluoride is evaporated and the residue is chlorinated. Plutonium is separated from the sample by volatilisation and separation of the chlorides in the gas phase. The plutonium is deposited on a glass disk by condensation of volatilised plutonium chloride. The concentration of plutonium is then determined by α spectroscopy.

Sekine et al. [27] used α-spectrometry to determine plutonium (and americium) in soil. The chemical recovery of plutonium was 51–99% and averaged 81%, while for americium the recovery was 60–70%. The method is coupled with the liquid–liquid extraction stage, taking about two days less than the ion exchange method; a complete analysis takes about one week.

The analytical separation of ^{239}plutonium and ^{240}plutonium in environmental samples is usually not possible using α-particle spectrometry because of the small difference in α energies, but this problem was solved in one study by additional measurement of the L X-rays emitted from the same sources [39].

Plutonium oxide becomes extremely refractory and difficult to dissolve when it is heated strongly [40, 41], such as occurs during a fire, a reactor incident, or ignition in the laboratory to burn off organic material prior to analysis. Many laboratories analysing soils for this toxic element ignore these facts and attempt to leach the compound selectively from the insoluble matrix with hydrochloric and/or nitric acids [42–47]. The results are frequently low and erratic. Occasionally reproducible results are obtained, but the results can still be shown to be low in many cases by analysing the insoluble residue remaining after leaching.

Hydrofluoric acid is best known for its ability to dissolve silica, but even small quantities exert a marked catalytic effect on the nitric acid dissolution of many refractory oxides, including ignited plutonium oxide, due to the high stability of the fluoride complexes. Consequently, use of hydrofluoric acid in leaching procedures generally gives more accurate results. However, erratic results are frequently obtained on larger samples due either to incomplete dissolution of plutonium oxide and/or the siliceous matrix, or to incomplete elimination of fluoride with consequent deleterious effects on the subsequent separations.

The accuracy of any analytical procedure is generally difficult to determine when starting from actual samples whose compositions and previous histories are unknown rather than from pure solutions using water-soluble tracers. An accurate and reliable analytical procedure should give complete decomposition of the siliceous matrix with complete elimination of insoluble material, but must also guarantee complete conversion of the plutonium to a soluble monomeric ionic form before any chemical separations are attempted. Existing total decomposition procedures employ either sodium carbonate or hydroxide fusions, neither of which can guarantee complete dissolution of refractory components. Subsequent steps generally involve so many precipitations, leaching, repeated ion exchange separations, etc., that the accuracy and reliability of the procedure remains in doubt. Yields are frequently only 50% and sometimes as low as 2% [47]. While many of these chemical inadequacies can be corrected for by using ^{236}Pu tracer, such a technique cannot help to correct for the plutonium remaining undissolved or polymerised because heterogeneous exchange does not occur to any significant extent [48]. Also, when the time required for complete sample decomposition is prolonged, losses of tracer added in water-soluble form might exceed losses of insoluble sample, and high results can be obtained.

3.8.5
Uranium

Parsa [49] has described a sequential radiochemical method for the determination of uranium (and thorium) in soils.

Methods involving solutions in hydrogen chloride gas and microwave dissolution have been compared for the dissolution of uranium in soil [50].

Boulyga et al. [51, 52] characterised different inductively coupled plasma mass spectrometric configurations for the determination of the ^{236}U/^{238}U ratio and ^{240}Pu/^{239}Pu ratio in soil.

Pavetko et al. [53] distinguished natural ^{230}U + ^{240}Pu and nonfinite α-emitters in contaminated soil profiles using an autoradiography approach.

Miscellaneous

Sill et al. [26] have described a procedure by which virtually all alpha-emitting nuclides of radium through californium can be determined in soil, singly or in any combination in a single sample.

The main objective of this work was to develop an accurate and reliable method for plutonium in soils in which complete dissolution of both the siliceous matrix and the most insoluble forms of plutonium oxide can be guaranteed routinely in a short period of time. It was also desired that the procedure be specifically applicable to the insoluble residues remaining from leaching procedures and to all other analytical fractions, such as ion exchange resins, so that complete material balances can be obtained for the analytical procedure. Analysis of residues is a much more sensitive, direct and unequivocal way of determining the adequacy of selective leaching than direct comparison of absolute numerical values obtained by different procedures or by different laboratories. Obviously, if complete exchange with the tracer does not occur and plutonium is present in the residue, the quantity found in the acid leach cannot accurately reflect the true concentration in the sample regardless of the precision with which the value can be reproduced. Furthermore, the fact that plutonium can be leached completely from the insoluble matrix is in itself no guarantee that it has been converted to a chemically reactive form.

Fusion with anhydrous potassium fluoride in a platinum dish is undoubtedly the simplest, most effective and reliable method available for the complete dissolution of a wide variety of siliceous materials. The potassium fluoride cake can then be transposed in the same container to a pyrosulfate fusion with rapid and complete volatilisation of both hydrogen fluoride and silicon tetrafluoride [54]. Except for a small quantity of barium sulfate, the pyrosulfate cake will dissolve completely in dilute hydrochloric acid. The resulting pyrosulfate fusion is one of the simplest and most effective methods available for rapid, complete and dependable dissolution of nonsiliceous materials, particularly high-fired oxides. This fusion has the distinct advantage that the flux can be obtained by simply adding easily purified alkali metal sulfates to sulfuric acid, and the fusion can be carried out in either borosilicate flasks or platinum vessels with very little contamination from either reagents or containers.

There are several other alpha emitters besides plutonium whose identification and determination in the environment is important, such as the isotopes of uranium, thorium and the other transuranium elements, particularly [241]americium. Because determinations of these other radionucleides have problems almost identical to those encountered with plutonium, Sill et al. [26] aimed to choose separations that would permit recovery of all alpha emitters from radium though californium. Previous work [55, 56] has demonstrated that virtually all alpha emitters can be carried on barium sulfate with extremely high efficiency from acid solutions containing high concentrations of potassium sulfate. This separation is particularly appropriate for application to the high potassium sulfate solutions resulting from the total sample decomposition employed. In the separations used by Sill et al. [26], protactinium and neptunium are carried quantitatively in the plutonium fraction. Uranium can be included either in the plutonium fraction or in one by itself. Thorium isotopes and the tervalent actinides are also obtained quantitatively in separated fractions. Because of the high resolution of α-spectrometers and the large differences in energies of the various nuclides, little mutual interference is encountered because of the presence of several nuclides in the same fraction. The Sill et al. [26] procedure was checked on soils spiked with known quantities of various α-emitting nuclides. Detailed material balance studies (sum of all measured fractions) carried out on [230]thorium, [241]palladium, [238]uranium, [237]neptunium, [239]plutonium, [241]americium, [244]curium and [252]californium gave values ranging from 98.4% for [244]curium to 101% for [237]neptunium.

The National Bureau of Standards [57] has issued standard reference materials for α-emitting radionuclides in soils. These include [60]cobalt, [90]strontium, [90]yttrium, tritium, and [106]ruthenium.

Zhu [58] described the use of liquid scintillation analysis for the monitoring of α-emitting and transuranium nuclides in environmental samples.

Wu and Landsberger [59] found that various conditions were required for the determination of medium-lived radionuclides in soil by neutron activation methods.

Several reviews on the measurement of environmental radioactivity have been published [60–63].

References

1. Iskander FY, Wuberger S, Warren SD (2000) *J Radioanal Nucl Chem* **244**:159.
2. Knizhnik EI, Prokopenko VS, Stolyarov SV, Tokarevskii VV (1994) *Atom Energy* **76**:113.
3. Burns KI, Ryan MR (1995) *J Radioanal Nucl Chem* **194**:15.
4. Muramatsu Y, Yoshida S (1995) *J Radioanal Nucl Chem* **197**:149.
5. Scheuerer E, Schupfer R, Schuettelkopf J (1995) *J Radioanal Nucl Chem* **193**:127.
6. Nakamisha T, SatahM, Takei M, Ishikawa A, Murato A, Daryah M, Higuchi S (1990) *J Radioanal Nucl Chem* **138**:321.
7. El Daoushy F, Garcia-Tenorio R (1990) *J Radioanal Nucl Chem* **138**:5.
8. Williams LR, Leggett RW, Espearen ML, Little CA (1989) *Environ Monit Assess* **12**:83.
9. Hafez AM, Morraram BM, El Khatib AM, Adel Naby A (1991) *Isotopraxis* **27**:185.
10. Martin DB (1979) *Anal Chem* **51**:1968.
11. HASL (1972) Radiochemical Determination of Strontium-90. In: *Procedures Manual HASL-300*, Health and Safety Laboratory (now Environmental Measurements Laboratory), US Department of Energy, New York, USA.
12. Gregory LP (1972) *Anal Chem* **44**:2113.
13. Sill CW, Puphal KW, Hindman FD (1974) *Anal Chem* **46**:1725.
14. Percival DR, Martin DB (1974) *Anal Chem* **46**:1742.
15. Eakins JD Gomm PJ (1966) *Health Phys* **12**:1557.
16. Montgomery HAC (1960) *Analyst* **85**:524.
17. Fourie HO, Ghijsels JP (1969) *Health Phys* **17**:685.
18. Butler FE (1962) *Health Phys* **8**:273.
19. Weaver B, Kappelmann J (1968) *J Inorg Nucl Chem* **30**:263.
20. Lantzsch J, Bushaw BA, Herrmann G, Kluge H-J, Monz L, Niess S, Otten EW, Schwalbach R, Schwartz M, Stenner J, Trautmann N, Walter K, Wendt K, Zimmer K (1995) *Angew Chem Int Ed Engl* **34**:181.

21. Arslan F, Behrendt M, Ernst W, Finckh E, Greb G, Gumbmann F, Haller M, Hofmann S, Karschnick R, Klein M, Kretschmer W, Mackiol J, Morgenroth G, Pagels C, Schleicher M (1995) *Angew Chem Int Ed Engl* **34**:183.
22. Grabek Z, Eskinja I, Kosutik K, Lulic S, Kuastek KJ (1999) *J Radioanal Nucl Chem* **241**:617.
23. Yamamoto M, Syarbaini, Kofuji K, Tsumura A, Komura K, Ueno K, Assinder DJ (1995) *J Radioanal Nucl Chem* **197**:185.
24. Hepiegne P, Dall'ava D, Clement R, Degros JP (1995) *Talanta* **42**:803.
25. Jordan D, Schupfner R, Schuettelkopf H (1995) *J Radioanal Nucl Chem* **193**:113.
26. Sill CW, Puphal KW, Hindman FD (1974) *Anal Chem* **46**:1726.
27. Sekine K, Qmai T, Kasai A (1987) *Talanta* **34**:567.
28. Joshi SR (1989) *Appl Radiat Isotpoes* **40**:691.
29. Livens FR, Singleton DL (1989) *Analyst* **114**:1097.
30. Kim CK, Takaku A, Yamamoto M, Kawawamura H, Shiraishi K, Igarashi Y, Igarashi S, Takayama H, Ikeda N (1989) *J Radioanal Nucl Chem* **132**:131.
31. Talvitie NA (1971) *Anal Chem* **43**:1827.
32. Green LW, Miller FC, Sparling JA, Joshi SR (1991) *J Am Soc Mass Spec* **2**:240.
33. Halgye Z (1991) *J Radioanal Nucl Chem* **149**:275.
34. Barci-Funel G, Dalmasso J Ardisson G (1992) *J Radioanal Nucl Chem* **156**:83.
35. Zhu HM, Tang XZ (1989) *J Radioanal Nucl Chem* **130**:443.
36. Hallenbach M, Grohs J, Kraft M, Mamich S (1995) *Applications of Inductively Coupled Plasma Mass Spectrometry to Radionucleide Determination* (ASTM STP 1129), ASTM, West Conshohocken, PA, USA.
37. Jia G, Testa C, Desideri D, Meli MA (1989) *Anal Chim Acta* **220**:103.
38. Dienstback E, Bächmann K (1980) *Anal Chem* **52**:620.
39. Arnold D, Kolb W (1995) *Appl Radiat Isotopes* **46**:151.
40. Wick OJ (ed) (1967) *Plutonium Handbook*, Gordon and Breach, New York, USA.
41. Cleveland JM (1964) *J Inorg Nucl Chem* **26**:1470.
42. Chu N (1971) *Anal Chem* **43**:449.
43. Harley JH (1972) *US Atomic Energy Commission Document, HASL 300*, Health and Safety Laboratory (now Environmental Measurements Laboratory), US Department of Energy, New York, USA.
44. Coleman GH (1965) *The Radiochemistry of Plutonium*, Nuclear Science Series NAS-NS 3058, National Academy of Sciences, Washington, DC, USA.
45. Benck RF, Smith RJ, Bogdan J, Lada H, Mortko H (1964) *NDL-SP-7*, US Army Nuclear Defense Laboratory, Edgewood Arsenal, MD, USA.
46. Bortoli MC (1967) *Anal Chem* **39**:375.
47. Inoue Y, Sakancue M (1970) *J Radiat Res* **11**:98.
48. Sill CW (1974) *Some Problems in Measuring Plutonium in the Environment*, Presented at Life Sciences Symp, Los Alamos Scientific Laboratory, Los Alamos, NM, USA.
49. Parsa B (1992) *J Radioanal Nucl Chem* **157**:65.
50. D'Silva AP, Bajic SJ, Zamzow D (1993) *Anal Chem* **65**:3174.
51. Boulyga SF, Matusevich JL, Mironov VP, Kudrjashov VP, Halicz L, Segal I, McLean JA, Montaser A, Becker JS (2002) *J Anal Atom Spectrom* **17**:958.
52. Boulyga SF, Testa C, Desideri D, Becker JS (2001) *J Anal Atom Spectrom* **16**:1283.
53. Pavetko OG, Higley KA (2001) *J Radioanal Nucl Chem* **248**:561.
54. Sill CW (1961) *Anal Chem* **33**:1684.
55. Sill CW (1969) *Health Phys* **17**:89.
56. Sill CW, Williams RL (1969) *Anal Chem* **41**:1624.
57. Inn KGW (1987) *J Radioanal Nucl Chem* **115**:91.

58. Zhu Y, Yang DZ (1995) *J Radioanal Nucl Chem* **194**:173.
59. Wu D, Landsberger S (1994) *J Radioanal Nucl Chem* **179**:155.
60. Bowlt C (1994) *Contemp Phys* **35**:385.
61. Al-Masari MS, Blackburn R (1996) *Radiat Phys Chem* **47**:171.
62. Crain JS (1996) *Spectroscopy* **11**:30.
63. Bandong BB, Guthrie EB, Kreek SA, Zaka FA, Ruth MA, Dupzyk IA, Edwards WL, Hall HL, Marsh KV, Bazan JM (1996) *Report GRI-94/0323*, Order No.PB96-178496GAR, NTIS, Springfield, VA, USA.

4 Determination of Organic Compounds in Soils

Sampling procedures and, where applicable, methods of preliminary extraction of the analyte from the soil sample are reviewed in Chap. 1. Analytical methods are discussed in an order that is as logical as possible over the next sections.

4.1
Aliphatic Hydrocarbons

Commonly used methods for the determination of petroleum hydrocarbons in soil are modifications of the EPA method 418.1, which uses sonication or Soxhlet extraction to separate the hydrocarbons from the soil prior to either infrared spectroscopy [1] or gas chromatography with flame ionisation detection [2, 3].

Regardless of the analytical method used following the extraction, both modifications use Freon-113, which has been implicated as a cause of ozone depletion. Therefore, alternative methods are being sought for the determination of hydrocarbon contamination in environmental samples that reduce the need for this halogenated solvent.

Supercritical fluid extraction with carbon dioxide has been shown to be an excellent alternative to conventional solvent extraction for the removal of hydrocarbon pollutants from solid samples [4–7]. It is fast (~30 minutes), nonpolluting, and relatively simple to implement. Additionally, recent work has shown that (supercritical fluid extraction using carbon dioxide is generally applicable to soil samples that have been contaminated with petroleum hydrocarbons ranging from those found in gasoline to those in medium crude oil (i.e., $\lesssim C_{30}$) hydrocarbons) [8–10].

In general, supercritical fluid extractions can be performed in either an on-line extraction mode or an off-line extraction mode. Off-line supercritical fluid extraction is the most common mode and involves extracting the analytes from the matrix and collecting them in either a sorbent trap or a collection solvent [11]. Following the collection step, the analytes are determined on a separate instrument (for example, on a chromatograph or an infrared spectrometer). In the on-line supercritical fluid extraction experiment, the outlet of the supercritical fluid extraction system is connected to a second analytical instrument [12–17]. This direct interface eliminated the need to collect the extracted analytes in either a sorbent trap or a collection solvent. Consequently, analyte loss is avoided and, more importantly, organic solvent use is eliminated.

Laing and Tilotta [5] have described the use of supercritical argon for the extraction of petroleum hydrocarbons from soil samples. Argon is an attractive solvent because it is inexpensive and inert. Additionally, it has a clear spectral window in the infrared region which makes it useful for on-line (i.e., directly coupled) experiments. Spiking studies conducted with gasoline, No. 1 fuel oil, and No. 5 fuel oil on sand, loam, and clay show that component recovery rates for argon supercritical fluid extraction generally increase with increasing pressure and/or temperature. The highest recovery rates (and recoveries) were obtained for argon supercritical fluid extraction at 500 atm and 150 °C. Under these conditions, the components of the gasoline and No. 1 fuel oil spikes could be recovered in as little as 12 minutes. However, the No. 5 fuel oil components could not be quantitatively removed from the loam and clay matrixes, even for extraction times as long as 100 minutes.

Yang et al. [11] compared sorbent trapping with solvent trapping after the supercritical fluid extraction of volatile petroleum hydrocarbons in soil. Sorbent trapping yielded quantitative collections of n-alkanes as volatile as n-hexane, while solvent extraction trapping effectively collected n-alkanes as volatile as n-octane.

Various other extraction techniques have been used to recover hydrocarbons from soil including microwave-assisted extraction [19] and supercritical fluid extraction coupled with on-line infrared spectroscopy detection [20, 21]. The on-line SFA infrared procedure produced results similar to those obtained by Soxhlet extraction.

Spectrofluorimetry

Morel et al. [22] compared several methods for the determination of hydrocarbons in soil and found that molecular spectrofluorimety in a Shpoliskii matrix was rapid, accurate and could be automated.

Fourier Transformed Infrared Spectroscopy

Progress has been reported in the use of multivariate analysis of infrared spectra of hydrocarbon-contaminated wet soil for real time in situ underground measurements [23].

Miscellaneous

Headspace analysis, purge and trap analysis and gas chromatography coupled to mass spectrometry have all been employed in determinations of gasoline hydrocarbons in soil, yielding detection limits as low as 5 μg/g [24, 25].

Thermoanalysis methods such as pyrolysis–gas chromatography–mass spectrometry [GC–MS] and thermogravimetry mass spectrometry have been used to characterise hydrocarbon sludges from petrochemical plants and polluted soils [26,27]. In combination with conventional extraction and supercritical fluid extraction followed by [GC–MS], over 100 constituents were identified in samples. White et al. [28] also applied pyrolysis–[GC–MS] to the determination of hydrocarbons and showed that the analysis can be complicated by the presence of natural organic matter. White [28] inferred the presence of biogenic compounds in Alaskan soil.

Peuron and Daugherty [29] have described a method of distinguishing between liquid and dissolved-phase hydrocarbons and assessing the levels of nonaqueous-phase liquids in gasoline-polluted soils.

Ostendorf et al. [30] have described two different methods of field sampling for residual gasoline in sandy soil. Both methods gave precise estimates of the vertically integrated mass of aviation gasoline in a given location.

Robbins et al. [31] assessed gasoline contamination of soil using a reclosable polyethylene bag and a total organic vapour detector.

Karasek et al. [32] determined hydrocarbons in benzene water extracts (pH 7) of soil and in incinerator or fly ash by a variety of techniques, including gas chromatography with flame ionisation, electron capture and mass spectrometric detection. Benzene water extractants were adjusted to pH 4, 7 and 10 before the extraction in order to selectively extract various types of acidic and basic organic compounds in addition to hydrocarbons.

Thermal desorption mass spectrometry is a rapid technique for determining oil in soils and sediments [33]. This method exhibited lower analytical variance compared to Soxhlet extraction, i.e., followed by conventional analysis. The analysis time for wet soil samples was about 20 minutes.

4.2
Aromatic Hydrocarbons

Kester [34] discussed the application of purge and trap GC to the determination of aromatic hydrocarbons such as benzene, ethyl benzene, toluenes and xylene in soil. In this method, a 4 g portion of the soil is dispersed in 9 ml methanol and 1 ml of a methanoic surrogate spike containing deuterated compounds for mass spectrometric recovery analysis in a 15 ml screw-capped vial. Volatile compounds are dissolved in the solvent by shaking for one minute or by sonicating for 30 minutes. The slurry is allowed to settle, centrifuging if necessary, and 10 to 100 μl aliquot of the extract is added to organic-free water and dry-pumped at ambient temperature into a Tenax trap. The purge gases are analysed by GC using a photoionisation detector or can be analysed by mass spectrometry.

De Leeuw et al. [35] have described a method based on Curie Point flash evaporation–pyrolysis GC–MS for the fast screening of anthropogenic aliphatic hydrocarbons in soils. The detection limit is in the low µg/kg range. Polycyclic aromatic hydrocarbons, heteroaromatic hydrocarbons, haloorganic compounds and pyrolysis products of polymers can also be screened by this method.

Various workers [36–39] have discussed various aspects of the determination of total petroleum hydrocarbons and benzene, toluene, ethyl benzene and xylene in soils.

Greco [40] determined optimal extraction conditions for the recovery of nitrogen-containing aromatic compounds from soil.

Solvent extraction with methanolic hydrolysis of the soil has been used to extract aromatic hydrocarbons. Significantly higher quantities of organics were recovered compared to the use of only an organic solvent extraction [41].

4.3
Polycyclic Aromatic Hydrocarbons [PAH]

The interest in determining the concentration of polycyclic hydrocarbons in soil is evidenced by the vast number of publications on this subject over the past decade.

Polycyclic aromatic hydrocarbons represent a class of compounds of great environmental concern due to their suspected mutagenic and carcinogenic properties [42–47]. Unease over the potential adverse health effects of polycyclic aromatic hydrocarbons is evident in the recent inclusion of P6 polyaromatic hydrocarbons in the Environmental Protection Agency's priority contaminates list. Polycyclic aromatic hydrocarbon contaminates pose several potential health risks due to the persistence of these compounds in the environment [48, 49], their tendency to strongly bind to soil surfaces [50–52], and their presence in a wide variety of common media (air, dust, soil and food) [53]. Possible risks are associated with skin contact, inhalation or ingestion of contaminated dust, soil, or air, and ingestion of contaminated food.

Environmental polycyclic aromatic hydrocarbon contamination has many different sources. Petroleum-based fuels and oils are known polycyclic aromatic hydrocarbon sources, with total polycyclic aromatic hydrocarbon contents as high as 4 wt% for diesel fuel and 5 wt% for gasoline [54]. Pipeline ruptures, tanker failures, underground and aboveground storage tank leaks, and various other production and transportation accidents frequently produce hydrocarbon-contaminated soil and groundwater on enormous scales. Therefore, hydrocarbon spills represent a large and widespread cause of soil and groundwater polycyclic aromatic hydrocarbon contamination. In addition, fossil fuel combustion produces airborne particulate matter containing polycyclic aromatic hydrocarbons. Petroleum hydrocarbon contamination may also occur naturally though seepage from underground oil and natural gas reserves. However, the transportation and storage needs for petroleum-based fuels in highly industrialised nations are the main source of bulk fuel contamination problems.

Extraction Methods for Polycyclic Aromatic Hydrocarbons in Soil

(See also Sect. 1.1.4).

Soxhlet Extraction

Ethyl acetate [55] and other solvents [56] have been employed.

Pressurised Liquid Extraction

Pressurised hot water extraction has been employed [57–59]. It is rapid and typically needs only 100 µl of extractant.

Accelerated Solvent Extraction

Saim et al. [60] investigated the dependence of accelerated solvent extraction operating variables (pressure, temperature, extraction time) on the recovery of 16 polycyclic aromatic hydrocarbons from native, contaminated soil. At the 95% confidence interval, no significance in terms of the three operating parameters was found regarding the total polycyclic aromatic hydrocarbon recovery. However, when individual polycyclic aromatic hydrocarbons were considered, some compounds were found to be dependent on operating variables. The most significant operating variable was extraction temperature. Low extraction temperature (40 °C) was found to be significant for naphthalene, chrysene, and benzo[b]fluoranthene. Using constant operating conditions (100 °C, 14 MPa and an extraction time of five minutes plus five minutes of equilibration time), the influence of extraction solvent was evaluated. No dependence on recovery was found when polar organic solvents, i.e., a dipole moment of > 1.89, were used.

Subcritical Water Extraction

Hawthorne et al. [61] coupled subcritical water extraction of polycyclic aromatic hydrocarbons with extraction using styrene–divinyl benzene extraction discs. The discs can be stored in autosampler vials without loss of polycyclic aromatic hydrocarbons.

Supercritical Fluid Extraction

Lagenfeld et al. [62] studied the effect of temperature and pressure on the supercritical fluid extraction efficiencies of polyaromatic hydrocarbons and polychlorobiphenyls in soils. At 50 °C, raising the pressure from 350 to 650 atm had no effect on recoveries.

Reindt and Hoffler [63] optimised parameters in the supercritical fluid extraction of polyaromatic hydrocarbons from soil. These workers used carbon dioxide–8% methanol for extraction and obtained 88–101% recovery of polyaromatic hydrocarbons in the final HPLC.

Barnabas et al. [64] has discussed an experimental design approach for the extraction of polyaromatic hydrocarbons from soil using supercritical carbon dioxide. They studied 16 different polyaromatic hydrocarbons using pure carbon dioxide and methanol-modified carbon dioxide. The technique is capable of determining down to 100 mg/kg polyaromatic hydrocarbons in soils.

Tena et al. [65] carried out a screening of polyaromatic hydrocarbon types in soil by on-line fibre optic interfaced supercritical fluid extraction spectrofluorimetry. Measurements can be carried out with a relative standard deviation of less than 5%.

Accelerated solvent extraction is a new technique for the extraction of a range of organic pollutants from soils and related material. The technique is based on the use of a solvent or combination of solvents to extract organic pollutants at elevated pressure and temperature from a solid matrix. The range of organic pollutants for which the technique is proposed includes semi volatile compounds, organochlorine pesticides, organophosphorous pesticides, chlorinated herbicides, polychlorinated biphenyls and polycyclic aromatic hydrocarbons [66–69].

Various aspects of supercritical fluid extraction have been discussed, including the use of liquid/solid traps [70], and the use of methylene and dichloride and methanol as static modifier [71–73].

It has been shown [73] that carbon dioxide is less efficient as an extractant for the heavier polycyclic aromatic hydrocarbons than nitrous oxide and Freon-22. This deficiency can be remedied by using a mixture of water, methanol and methylene dichloride [73].

Guo et al. [74] has pointed out that whilst supercritical carbon dioxide is effective at extracting nonpolar and slightly polar chemicals from soil, it is unsatisfactory for recovering polar chemicals in soils. They developed a supercritical fluid extraction procedure to quantitatively recover polar and nonpolar chemicals from soils. The nonpolar chemicals and slightly polar chemicals used as model compounds were common pesticides and environmental pollutants such as polycyclic aromatic hydrocarbons. The procedure required pretreatment of the samples with 15% water (g/g), 5% (ethylenedinitrilo)-tetraacetic acid tetrasodium salt (g/g) and 50% methanol polycyclic aromatic hydrocarbons (ml/g) prior to extraction using supercritical carbon dioxide at 60 °C and 34.5 MPa. Recoveries ranged from 90 to 106% for the aromatic acids using the tetrasodium EDTA-assisted supercritical fluid chromatography, compared with only 7–63% recoveries for the corresponding chemicals when no tetrasodium EDTA was used. The method quantitatively extracted 2,4-D and its close analogues that had been aged in the soil for 2–30 days. The tetrasodium EDTA-assisted supercritical fluid chromatography was also adequate foe extracting phenolic analytes, including picric acid and pentachlorophenol, with recoveries of 85 to 104%. Tetrasodium EDTA was a good enhancer in the extraction of 29 analytes representing a wide range of polarities from soil using supercritical carbon dioxide. The method is valuable for the analysis of parent pollutants and transformed products, particularly oxygen-borne metabolites in the environment.

The various instrumental techniques that have been used to determine polycyclic aromatic hydrocarbons in soil are now reviewed.

High-Performance Liquid Chromatography HPLC

This technique, with fluorescence detection, has been applied by Reindl and Hoeffer [75], Lopez et al. [76], Fenandez Perez et al. [80], and Krahn et al. [81]. Liquid chromatography has also been employed [77–79].

HPLC with fluorescence detection gave results for polycyclic aromatic hydrocarbons which were comparable to those obtained by gas chromatography with mass spectrometric detection.

Gas Chromatography–Mass Spectrometry (GC–MS)

This method has been applied to hexane extracts of soils [82, 83] using deuterated internal standards.

Mass Spectrometry

Dale et al. [84] has described a method for the determination of polycyclic aromatic hydrocarbons in contaminated soils which involves use of a laser desorption, laser photoionisation time-of-flight mass spectrometer. This method can be applied directly to soils without extraction and cleaning procedures, and consequently it has great potential as an on-site screening tool.

Rodgers et al. [85] identified soil surface-bound polycyclic aromatic hydrocarbons through the use of real-time aerosol mass spectrometry in two NIST standard research material soils (Montana SRM 2710 and Peruvian SRM 4355), each contaminated separately with three common petroleum hydrocarbons (diesel fuel, gasoline and kerosene). This method required no sample preparation. Direct laser desorption/ionisation mass spectrometric analysis of individual soil particles contaminated with each of the petroleum hydrocarbons at three different contamination levels (0.8, 8, and 80 ppth (wt/wt)) yielded detectable polycyclic aromatic hydrocarbon cation distributions that ranged from m/z 128 to 234, depending on the fuel contaminant. The same analysis performed on uncontaminated standard research material soils revealed very little (Peruvian) to no (Montana) detectable polycyclic aromatic hydrocarbons species. Size analysis showed that most of the individual soil particles analysed were between 1 and 5 μm in diameter. Tandem mass spectrometry experiments identified alkyl-substituted two- and three-ringed polycyclic aromatic hydrocarbons in all three petroleum hydrocarbon-contaminated soils. However, due to similarities in fragmentation patterns, tandem mass spectrometry of higher molecular weight species ($m/z > 200$) was unable to distinguish between the possibility of higher alkyl-substituted three-ringed polycyclic aromatic hydrocarbons and four-ringed polycyclic aromatic hydrocarbons. The technique offers the direct, rapid determination and characterisation of surface-bound polycyclic aromatic hydrocarbons in petroleum-contaminated soils at part per million levels without extraction, separation or other sample preparation methods.

Other Techniques

Other techniques that have been used to determine polycyclic aromatic hydrocarbons in soil extracts include ELISA field screening [86], micellar electrokinetic capillary chromatography [87], supersonic jet laser-induced fluorescence [88,89], fluorescence quenching [90], thermal desorption gas chromatography–mass spectrometry [81, 90, 100], microwave-assisted extraction [91], thermal desorption [92], immunochemical methods [93, 94], electrophoresis [96], thin layer chromatography [95], and pyrolysis gas chromatography [35].

   The addition of methanolic alkali or boron trifluoride methanol [97, 99] as a modifier has been recommended for the extraction of polycyclic aromatic hydrocarbons from soils containing a high proportion of humic acids.

   Lichtfouse et al. [98] and others [102, 103] demonstrated that polycyclic aromatic hydrocarbons in soil are mainly pyrolytic in origin.

4.4
Heteropolycyclic Hydrocarbonyl [NSO]

Niederer [100] used ion trap mass spectrometry and negative ion chemical ionisation to determine nitro- and oxypolyaromatic hydrocarbons in soils. Meyer et al. [101] have described a simple and reproducible method which provides the simultaneous determination of polycyclic aromatic hydrocarbons and heteropolycyclic aromatic hydrocarbons (N, S, O) and their metabolites in contaminated soils. Contaminants extracted from the soil sample were separated by polarity and acid–base characteristics using solid-phase extraction on silica gel and a strong basic anion exchange material. A subfraction containing PANHs and neutral metabolites was subsequently fractionated into neutral and basic compounds using a strong acidic cation exchange material. The identification and quantification was performed using different gas chromatographic and high-performance liquid chromatographic methods. A method validation was carried out for 21 polycyclic aromatic hydrocarbons, 22 heteropolycyclic aromatic hydrocarbons, and 19 metabolites in a five-level matrix calibration. The method showed good linearity (coefficient of correlation) and high precision (coefficient of variation) for major and minor compounds over a wide range of concentrations as well as high sensitivity (limit of detection). The method was successfully applied to different authentic tar oil-contaminated sites. Aside from the typical tar oil polycyclic aromatic hydrocarbons, heteropolycyclic aromatic hydrocarbons and metabolites predominantly with ketonic or quinonic structures were identified at a mg/kg concentration level.

4.5
Oils and Greases

Freon or dichloromethane extraction gives precise and accurate estimates of oil and grease contents of soil [104]. Thermal desorption mass spectrometry has been used as a rapid method for the determination of oil in soil. The analysis takes only 20 minutes [33].

4.6
Polystyrene

A Curie Point flash evaporation–pyrolysis gas chromatography–mass spectrometric method [35] has been applied to the determination of polystyrenes in soil via identification and determination of their unzipping pyrolysis products, such as styrene monomer, α-methylstyrene, 3-methylstyrene, 4-methylstyrene, α-3-dimethylstyrene, 3-ethylstyrene, α-4-dimethylstyrene, 3,5-dimethylstyrene, α-2- or 2,5- or 2,4-dimethylstyrene, as well as various phenyl ethers.

4.7
Phenols

Phenols in soil come from natural sources (e.g., degradation of lignin and humic acids present in soils) and anthropogenic sources (e.g., degradation of pesticides, herbicides, and fungicides containing phenols; direct emissions from chemical factories). Phenols in the environment (especially in soils) are usually present as pollutants. Because of their toxicological potential, some have been included on lists of priority pollutants from the European Community (EC) and the US Environmental Protection Agency (EPA) [120].

Traditional Soxhlet extraction is gradually being superseded by new alternative approaches, including supercritical carbon dioxide [104–106] and supercritical water extraction [107], accelerated solvent extraction [108, 109], microwave-assisted extraction [91,108], and solid–liquid extraction [111,112]. Supercritical fluid extraction is increasingly being used to extract organic pollutants from environmental solids. However, obtaining adequate supercritical fluid recoveries of phenols entails using a high temperature [110], an organic modifier to increase the polarity of the carbon dioxide, or supercritical water as extractant [110].

To overcome some of the difficulties, Crespin et al. [113] developed a semi-automatic module for the direct continuous extraction and preconcentration of phenols from soils.

The extraction fluid flows were controlled by multiport rotary valves and were driven by peristaltic pumps and by compressed nitrogen. The interface plays a crucial role in changing the pH and homogenising the sample before preconcentration; in addition the sample volume was controlled at the interface. Spiked uncontaminated soils were prepared two months before treatment in order to simulate weathering and allow for the occurrence of analyte–matrix interactions. Soil extractions were done with alkaline aqueous solutions and solid-sorbent preconcentration in an acid medium, using XAD-2 as sorbent. Soil samples (0.1 – 10 g) containing 50 – 5000 ng/g phenols were analysed by gas chromatography with a high precision (4 – 7%). Average recoveries of 60 – 80% were achieved, except for 2-*tert*-butyl-4-methylphenol, the mean recovery of which was rather poor (9%). The adsorption–desorption of phenols in agricultural soils was evaluated, and their clay mineral and organic matter contents were found to affect the recovery of alkylphenols to a different extent than those of chlorophenols and nitrophenols.

Llopart-Vizoso et al. [114] determined phenols and cresols in soil by direct acetylation followed by gas chromatography headspace analysis. Danis and Albanis [115] also used a technique based on acetylation–gas chromatography. Three gas chromatographic detectors were employed: flame ionisation, electron capture and mass spectrometric.

Talsky [116] has described a higher order derivative spectrometric method for the determination of phenols in soils.

Karasek et al. [32] determined phenols in soils by extraction with a mixture of benzene and water modified to pH 10 by the addition of 2-methoxyethyl-amine. The phenol in the extract was identified and determined by gas chromatography using a variety of detectors, including flame ionisation, electron capture and mass spectrometry.

4.8
Alcohols, Ketones, Aldehydes and Organic Acids

This technique has been applied [34] to the determination of ethanol, methyl ethyl ketone, paraldehyde and acrolein in soils. Following extraction of the soil with methanol and gas purging, the purge gas is trapped on a Tenax column. The purgate obtained by heating the Tenax column is analysed by gas chromatography and/or mass spectrometry.

Langbehn et al. [117] used supercritical fluid chromatography to determine organic acids and ketones in soil.

4.9
Volatile Organic Compounds

Contamination of the environment with volatile organic compounds has become an important issue over the past few decades, since many volatile organic compounds are toxic and may pose serious health risks. The main emission sources of volatile organic compounds are industry, traffic and energy production. Serious local contamination problems often follow from accidents, leakage of petrol and diesel fuel from underground storage tanks, and improper waste treatment. Volatile organic compounds in soil easily diffuse from the point of emission over wide areas, finding their ways into groundwater [118] and, through construction or sewerage, into households [119, 120]. The contamination of groundwater is becoming one of the most serious environmental problems. Several water catchments have already had to be closed because of high concentrations of volatile organic compounds or other organic contaminants. Intensified research and reconditioning of contaminated areas are needed, as well as continuous monitoring of the quality of drinking water.

Ultrasonic extraction, methanol extraction [147] and supercritical fluid extraction have all been applied to the extraction of or the determination of volatile organic compounds [121, 122] in soils. However, methods based on headspace analysis or on mass spectrometry are now the methods of choice.

Kawata et al. [128] have described the effects of headspace conditions on recoveries of volatile organic compounds from sediments and soils. Hewitt [129] compared three vapour partitioning headspace and three solvent extraction methods for the preparation of soil samples for volatile organic carbon determination in soils. Methanol extraction was the most efficient method of spiked volatile organic carbon recovery, which depended on the soil organic carbon content, the octanol–water partitioning coefficients of analytes and the extraction time.

Papaefstathion and Luque de Castro [130] used pervaporation as an alternative to headspace analysis for the analysis of down to 1 ng/g of volatile organic compounds in soils.

James and Stack [131] found that solid-phase microextraction is an effective technique for determining volatile organic compounds in landfill sites. The headspace above the sample was sampled.

Headspace Analysis

Headspace analysis is the method of choice for determining volatile organics [123–131] in soil. A limitation of this method is incomplete desorption of the contaminants in soil/water mixtures, but this problem can be overcome by the addition of methanol to the sample [126, 129].

Stuart et al. [127] studied the analysis of volatile organic compounds using an automated static headspace method. Recoveries decreased in the following order: water, pure sand, sandy soil, clay and topsoil. A full evaporation technique that uses little or no aqueous phase and higher equilibration temperature gave the most reproducible analyte recoveries.

Askari [132] compared purge and trap methanol immersion and hot solvent extraction for the determination of volatile organic compounds in aged soils, and found the hot solvent extraction method to be much more effective than the EPA approval purge and trap technique.

Mass Spectrometry

Bianchi et al. [134] and Yokouchi and Sano [135] obtained good recoveries of volatile organic compounds in soils employing thermal vaporisation followed by trapping on Tenax GC and analysis by gas chromatography–mass spectrometry.

Krock and Wilkins [136] used multidimensional gas chromatography with infrared and mass spectrometric detection to determine organic compounds in soil.

Barrio et al. [137] used pyrolysis–gas chromatography to study organic matter evolution in sewage sludge-amended soils. Nitrogen–phosphorus specific flame ionisation and mass spectrometric detectors were used.

Other workers [138–140] have discussed gas chromatography–mass spectrometry. Kostiainen et al. [141] have described a new method, purge and membrane mass spectrometry, for the analysis of volatile organic compounds in water and soil samples. In this method, volatile organic compounds are purged from water or soil samples with an inert gas and the stream is directed through a sheet membrane module. The volatile organic compounds pervaporate through the membrane directly into the ion source of a mass spectrometer. The limits of detection for nonpolar volatile organic compounds such as halogenated hydrocarbons, benzene, toluene and xylenes were at low micrograms per kilogram levels in soil samples. The correlation coefficients measured for the compounds studied were typically better than 0.9999 and 0.9975 in water and soil samples, respectively. The relative standard deviations were between 0.5 and 2.0% for water samples and between 4.8 and 14.0% for soil samples. These results demonstrate excellent linearity and repeatability. Purge and membrane mass spectrometry thus provides a highly sensitive, selective, accurate, solvent-free and rapid analytical method. Tens of samples can be analysed within an hour.

Infrared Spectroscopy

Budzinski et al. [142] have reported an infrared spectroscopic method for the determination of down to 200 – 300 ppb of semivolatiles such as polycyclic aromatic hydrocarbons and polychlorobiphenyls in soil.

Yang and Her [143] have described a method for the determination of semivolatiles in soil based on coupling solid-phase microextraction with attenuated total reflectance Fourier transform infrared spectroscopy. A trapezoidal internal reflection element was mounted horizontally in a flow cell with the inlet port connected to a temperature-controlled glass extraction chamber. Soil samples were placed inside the glass tube and heated to the desired temperature. Vaporised, semivolatile compounds were carried by a stream of nitrogen gas to the attenuated total reflection–infrared flow cell. To increase the trapping efficiency, the attenuated total reflection crystal was coated with a hydrophobic polyisobutylene polymer that acted as the solid-phase microextraction phase. The method proved to be very sensitive in the detection of semivolatile compounds in soils. The relationship between various parameters affecting chemical quantitation, such as the film thickness, gas flow rate and water contents, was also studied. Three different compounds, 1-chloronaphthalene, nitrobenzene and 2-chlorotoluene, were used to investigate the feasibility of this method in the analysis of organic compounds in sand and soil. Results indicated that a linear relationship between concentration and infrared signals can be obtained for the three analytes. The detection limit of this method was in the range 200 – 300 ppb.

Miscellaneous

Hewitt found that volatile organic compounds are readily lost from soil samples unless care is taken to limit surface area exposure and to ensure subsample isolation [144]. Volatile organic compound losses were found to be most abundant during field collection and storage. In a separate investigation, Hewitt reported that fortified soils held in seated glass ampoules at 4 °C, or dispersed in methanol and held at 22 °C, showed no significant losses over 20 and 98 days, respectively [145].

Liikala et al. [146] compared different sampling methods in the determination of volatile organic compounds in soil. Large negative biases occurred for aromatic compounds when conventional sampling was used compared to placing the soil aliquot in methanol.

4.10
Chloroaliphatic Hydrocarbons

Deetman et al. [148] have devised an electron capture gas chromatographic technique, applicable to mud samples, for the determination of down to

1 ng/l of 1,1,1-trichloroethane, trichloroethylene, perchloroethylene, 1,1,1,2-tetrachloroethane, 1,1,2,2-tetrachloroethane, pentachloroethane, hexachloroethane, pentachlorobutadiene, hexachlorobutadiene, chloroform and carbon tetrachloride. These workers used extraction of the samples with n-pentane as a means of isolating the chlorinated compounds from the sample. Recoveries of 95% were obtained.

Neumayr [149] carried out soil atmosphere studies using capillary gas chromatography and electron capture and flame ionisation sequential detection and used this as a means of pinpointing zones of soil and groundwater contamination. Methods have been described for determining chlorinated aliphatic hydrocarbons in soil and chemical waste disposal site samples. The latter method involves a simple hexane extraction and temperature-programmed gas chromatographic analysis using electron capture detection and high-resolution glass capillary columns. Combined gas chromatography–mass spectrometry was used to to confirm the presence of the hydrocarbons in the samples [150].

Kester [34] has discussed the application of the purge and trap gas chromatographic method to the determination of aliphatic chlorocompounds in soil. Following methanol extraction of the soil, the extract is gas-purged and the purge gases trapped on a Tenax silica gel/charcoal trap, followed by thermal desorption from the trap and examination by gas chromatography and mass spectrometry.

Kerfoot [151] examined the performance of a grab sampling technique for soil gas measurement analyses at a site with groundwater known to be contaminated with chloroform. The study assessed the correlation between soil gas and groundwater analyses with chloroform as a model volatile organic compound. Chloroform concentration in soil gas increased linearly with depth in the unsaturated zone.

A study of the vertical profile of chlorinated solvents in the soil enables the source of contamination to be distinguished; for atmospheric inputs a peak occurred a short distance below ground, whereas for inputs from groundwater the concentration increased progressively as the water table was approached.

Mehran et al. [152] determined the distribution coefficient of trichloroethylene in soil water systems. The distribution coefficient of trichloroethylene could be used to define the retardation factor, which expressed the velocity of trichloroethylene migration relative to an advancing water front. The two methods used to obtain the distribution coefficient were field measurements based on trichloroethylene concentrations in soil at various depths, and theoretical methods based on the total organic carbon content of the soil and the octanol–water partition coefficient for trichloroethylene. The average distribution coefficient was 0.18 ml/g, and the average retardation factor was 2.48 (19 field samples). Theoretical methods were valid for soils with greater than 1% organic carbon. Reasonable estimates for actual migration rates could be provided for soils low in organic carbon. Field methods were still preferred, as the effects of various factors on the partitioning of trichloroethylene were integrated.

4.11
Chloroaromatic Hydrocarbons

Chlorinated aromatic compounds are commonly found as contaminants in environmental soil samples. For example, chlorobenzenes have been listed as priority pollutants and can be found in various matrixes such as water, soils, sediments and sewage sludges. Polychlorinated biphenyls are probable human carcinogens but have been applied in large doses in various industrial products. Analysis of these compounds in solid matrixes, such as soils and sediments, requires several steps.

Supercritical fluid extraction [153, 154], accelerated solvent extraction [68] and subcritical fluid extraction [107, 155] have been studied. To reduce the equipment cost and the analysis time in the extraction process and sample preconcentration, a solid-phase microextraction method was proposed by Pawliszyn and coworkers [156–158].

Young and Her [159] used the principle of solid-phase microextraction combined with Fourier transform infrared sensing to determine chlorinated organic compounds in soil.

The sensing device of this method was based on an infrared hollow waveguide, the inner surface of which was coated with a hydrophobic film. Vaporised chlorinated aromatic compounds from soils were trapped onto the hydrophobic film of the hollow waveguide sampler following detection by Fourier transform infrared spectrometry. The extraction process in this method was similar to the headspace solid-phase microextraction in principle. Means of increasing the speed of transfer of the vaporised organic species to the sampler were also studied. Results indicated that, with a negative pressure on the end of the sampler, the speed of transfer increased significantly. Vapour pressures of the analytes were used as an indication in order to test the limitations of this method in the analysis of organic compounds in soils. Results showed that analytes with vapour pressures lower than 1600 Pa could be detected quantitatively. Typical R^2 values for the regression on the concentration and infrared signals were around 0.99, and typical detection limits were in the range of hundreds of parts per billion.

Efforts are being made to reduce both the use of organic solvents and the time-consuming clean-up and preconcentration steps. One such approach is to apply supercritical fluid extraction (SFE) [164] with carbon dioxide as the extraction medium for sample preparation.

Wennrich [167] optimised important accelerated solvent extraction parameters, such as extraction temperature and time, using a spiked wetland soil. The effect of small amounts of organic modifiers on the extraction yields was studied. An extraction temperature of 125 °C and ten-minute extractions performed three times proved optimal. Two accelerated solvent extraction–solid-phase microextraction procedures without and with an organic modifier (5% acetonitrile) were evaluated with respect to precision and detection limits.

The reproducibility of replicate water extractions/solid-phase microextraction determinations ($n = 6$) was in the range of 7–20% relative standard deviation for the nine chlorophenols investigated. Limit of detection values in the low ppb range were achieved for all chlorophenols.

4.12
Polychlorophenols

The higher order derivative spectrophotometric method described by Talsky [116] for the determination of phenols in soils has been used to determine pentachlorophenol.

Stark [168] has described a gas chromatographic method for the determination of pentachlorophenol as the trimethylsilylether in amounts down to 0.5 mg/kg in soil.

Renberg [169] used an ion exchange technique for the determination of chlorophenols and phenoxyacetic acid herbicides in soil. In this method, the soil extracts are mixed with Sephadex QAE A-25 anion exchanger and the adsorbed materials are then eluted with a suitable solvent. The chlorinated phenols are converted into their methyl ethers and the chlorinated phenoxy acids into their methyl or 2-chloroethyl esters for gas chromatography.

Stable-isotope dilution analysis is an analytical technique in which a known quantity of a stable-labelled isotope is added to a sample prior to extraction, in order to quantitate a particular compound. The ratio of the naturally abundant and the stable-labelled isotope is a measure of the naturally abundant compound and can be determined only by gas chromatography–mass spectrometry, since the naturally abundant and the stable-labelled isotope cannot be completely separated gas chromatographically.

Lopez-Avila et al. [170] used a stable isotope dilution gas chromatography–mass spectrometry technique to determine down to 0.1 ppb of pentachlorophenol (also atrazine, diazinon and lindane) in soil. Soil samples are extracted with acetone and hexane. Analysis is performed by high-resolution gas chromatography–mass spectrometry with the mass spectrometer operated in the selected ion monitoring mode. An accuracy greater than 86% and a precision of better than 8% were demonstrated by use of spiked samples.

Chlorinated phenols are among the most important contaminants in the environment (aqueous systems and soils) due to their widespread use in industry and agriculture and for domestic purposes for over 50 years. It is well-known that chlorophenols are toxic at low levels. The more highly chlorinated phenols such as trichlorophenols and pentachlorophenol are also persistent. Five of the chlorophenols (2-chlorophenol, 2,4-dichlorophenol, 4-chloro-3-methylphenol, 2,4,6-trichlorophenol and pentachlorophenol) have been classified as priority pollutants by the US Environmental Protection Agency (EPA).

Determining chlorophenols in soil samples usually involves various methods of liquid–solid extractions, e.g., Soxhlet [160, 161], microwave- or ultrasonically assisted [115, 162], and accelerated solvent (ASE) [163] extraction with an organic solvent or solvent mixture followed by both clean-up and preconcentration procedures and subcritical hot water extraction [164–166].

Wennrich et al. [167] investigated the capabilities of coupling accelerated solvent extraction with water as the extraction solvent and solid-phase microextraction to determine chlorophenols in polluted soils. Subcritical water extraction was performed using a commercially available accelerated solvent extractor. This system solves the problem of the analytes partitioning back to the soil matrix, which can occur in straightforward subcritical water extraction because in the Wennrich et al. method [167] the aqueous phase and the soil are separated under the extraction conditions.

4.13
Polychlorobiphenyls

A wide variety of techniques have been employed to extract polychlorobiphenyls from soil prior to analytical determination (Table 4.1).

Table 4.1. Extraction methods for the isolation of polychlorobiphenyls from soil (from author's own files)

Subtraction method	Comment	Reference
Solvent extraction	Study of polar extraction solvents; comparison of extraction solvents	[171, 172]
Microwave extraction	Comparison of gas chromatography to ELISA after microwave extraction	[173]
Solid phase microextraction	With gas chromatography–mass spectrometry to finish	[174]
Subcritical water extraction	Complete extraction at $250-350\,^{\circ}\mathrm{C}$ and 50 atm pressure	[175]
Subcritical water extraction	Combination of static subcritical water extraction and solid-phase microextraction	[155]
Supercritical fluid extraction	Comparison of $CHClF_2$, N_2O and CO_2 extractants. $CHClF_2$ gave highest recovery, methanol-modified CO_2 gave 90% recovery	[78]
Supercritical fluid extraction	Combination of supercritical fluid extraction with off-line Fourier transform infrared spectroscopy	[177]
Supercritical fluid extraction	PCB phase diagrams with carbon dioxide	[178]
Supercritical fluid extraction	Effect of temperature and pressure on extraction of PCB and polyaromatic hydrocarbons	[62]
Supercritical fluid extraction	Combination of solid-phase carbon trap with supercritical fluid chromatography for PCB, pesticides, polychlorodibenzo-p-dioxins and polychlorofurans	[179]

ELISA: enzyme-linked immunosorbent assay

Methods for the determination of polychlorobiphenyls seem to be limited mainly to those based on gas chromatography and enzyme-linked immunosorbant assay (ELISA).

Gas Chromatography

Thermal desorption gas chromatography, mass spectrometry [180–182] and multidimensional gas chromatography have been used [141, 142].

Benicka et al. [183] used multidimensional gas chromatography to separate the atropisomers of polychlorobiphenyl congeners in extracts of soil. The correct enanatiomeric ratio was determined from the peak areas obtained by deconvolution of the chromatograms.

Jensen et al. [184, 185] pointed out that bottom soils in rivers contain elementary sulfur, which greatly interferes with gas chromatographic methods for the determination of polychlorobiphenyls and chlorinated insecticides. They discuss methods of overcoming such interferences. Chiarenzelli [186] found that air-drying soils and sediments for 24 h at ambient conditions resulted in validation losses of 14–23%, with most occurring within the first eight hours. Polychlorobiphenyl loss was strongly correlated with water loss. Microwave-assisted extraction with electron capture gas chromatography has been compared to ELISA for the field determination of polychlorobiphenyls in soils and sediments. Both techniques were found to be amenable to field screening and monitoring applications.

Teichman et al. [187] separated polychlorobiphenyls from chlorinated insecticides in soil samples using gas chromatography coupled to mass spectrometry. Polychlorobiphenyls were separated from DDT and its analogues and from the other common chlorinated insecticides by adsorption chromatography on columns of alumina and charcoal. Elution from alumina columns with increasing fractional amounts of hexane first isolated dieldrin and heptachlor from a mixture of chlorinated insecticides and polychlorobiphenyls. The remaining fraction, when added to a charcoal column, could be separated into two fractions, one containing the chlorinated insecticides and the other containing the polychlorobiphenyls, by eluting with acetone–dimethyl ether (25:75) and benzene, respectively. The polychlorobiphenyls and insecticides were then determined by gas chromatography on the separate column eluates without cross-interference.

Teichman et al. [187] used a gas chromatograph (Aerograph 1200) containing a glass column (180 cm × 3 mm) packed with 4% SE-30, 6% SP-4201 on Chromosorb W (100–120 mesh). They also used an Aerograph 204 gas chromatograph containing a glass column (180 cm × 3 mm) with 4% SE-30, 6% QF-1 on Chromosorb W (80–100 mesh). Both instruments contained an electron capture detector with a tritium foil source. For gas chromatography-mass spectrometry, a Varian 1400 gas chromatograph coupled to a Finnegan 3000 mass spectrometer was used. The 1400 was equipped with a glass column (180 cm × 2 mm id) packed with 4% SE-30, 6% SP-4201 on Supelcoport (100–120 mesh). The operating conditions were: column temperature: 210 °C; transfer line temperature: 250 °C, gas jet separator temperature: 255 °C, flow rate of helium gas: 12 ml/min; sensitivity: 10^{-7} A/V; electron multiplier voltage: 2.25 kV; electron ionisation current: 6.95 eV.

The recovery of polychlorobiphenyls from soil samples obtained in spiking experiments was 100%, while that of chlorinated insecticides ranged from 81.5% (heptachlor) to 96.3% (dieldrin). A limit of detection of 6.5 ppb was obtained from Aroclors 1254 and 1260.

Lopshire [188] explored the exchange reaction of chlorine by oxygen with polychlorobiphenyl anions as a method of compound-selective polychloro-biphenyl congener detection in a gas chromatography–mass spectrometric system. Multiple reaction monitoring allowed separate chromatograms to be detected for each different polychlorobiphenyl composition from tetra-through nonachloro.

Polychlorobiphenyls were recovered from sediments in another investigation by steam distillation/solvent extraction, followed by enantioselective analysis using multidimensional electron capture gas chromatography [189].

Enzyme-Based Immunoassay

Johnson and Van Emon [190] described a quantitative enzyme-based immunoassay procedure for the determination of polychlorinated biphenyls in soils and sediments and compared the results with those obtained by a gas chromatographic method. The soil is extracted with methanol or Soxhlet-extracted or extracted with supercritical fluid. In the case of the latter two extractants, good agreement was obtained between immunoassay and gas chromatographic methods. Spiking recoveries from soil ranged from 104% (Aroclor 1248) to 107% (Aroclor 1242). Detection limits were 9 μg/kg (Aroclor 1245) and 10.5 μg/kg (Aroclor 1242). Chlorinated anisoles, benzenes or phenols did not interfere.

Schuetz [191] has developed an enzyme immunoassay method for the determination of all polychlorobiphenyls up to biphenyl in soils before determination by ELISA.

Pullen et al. [192] have described a polychlorobiphenyl immunochemical test kit for soil analysis and showed that results corresponded well with those obtained from gas chromatographic electron capture detection.

Miscellaneous

Methods based on photoactivated luminescence [193] and room-temperature phosphorescence [194] have been used to determine polychlorobiphenyls in soil.

Klingston et al. [195] have described a vacuum-drying procedure for the preparation of samples for polychlorobiphenyl analysis.

Alcock et al. [196] demonstrated the contamination of soil samples in laboratory air with polychlorobiphenyls. The calculated average net dry deposition from laboratory air to soil was calculated at 5 ng total polychlorobiphenyl/m²/day.

Hellman [197] studied the adsorption and desorption of polychlorobiphenyls and hexachlorobenzene from clays in contact with water. Appreciable adsorption of these compounds onto clays occurred.

Yu Ma and Bayne [198] differentiated different aroclors in soil using linear discrimination and analyses by electron-capture negative-ion chemical ionisation mass spectrometry.

4.14
Polychlorodibenzo-*p*-dioxins and Polychlorodibenzofurans

Polychlorinated dibenzo-*p*-dioxins, polychlorinated dibenzofurans and *ortho*-unsubstituted polychlorinated biphenyls (non-ortho polychlorobiphenyls) are three structurally and toxicologically related families of anthropogenic chemical compounds that have in recent years been shown to have the potential to cause serious environmental contamination due to their extreme toxicities [199–204]. These substances are trace-level components or byproducts in several large-volume and widely used synthetic chemicals, principally polychlorobiphenyls and chlorinated phenols [205, 206], and can also be produced during combustion processes [206–209] and by photolysis [210, 211]. In general, polychlorodibenzo-*p*-dioxins and dibenzofurans and non-ortho polychlorobiphenyls are classified as highly toxic substances [212], although the toxicities are dramatically dependent on the number and positions of the chlorine substituents [213]. About ten individual members of a total of 216 polychlorodibenzo-*p*-dioxins and dibenzofurans and non-ortho polychlorobiphenyls are among the most toxic manmade or natural substances to a variety of animal species [199–202]. The toxic hazards posed by these chemicals are exacerbated by their propensity to persist in the environment [214–216] and to readily bioaccumulate [217–219], and although the rate of metabolism and elimination is strongly species-dependent [211], certain highly toxic isomers have been observed to persist in the human body for more than ten years [220].

The majority of scientific and governmental concerns for the hazards of these compounds have been directed towards analytical methodologies, toxicology, epidemiology and determination of the disposition in the environment for the single most toxic isomer, 2,3,7,8-tetrachlorodibenzo-*p*-dioxin [199–204, 206].

Extraction Methods

Von Bavel et al. [179] have developed a solid-phase carbon trap (PX-21 active carbon) for the simultaneous determination of polychlorodibenzo-*p*-dioxins and polychlorodibenzofurans as well as polychlorobiphenyls and chlorinated insecticides in soils using superfluid extraction liquid chromatography for the final determination.

Walters and Guiseppe-Elle [221] studied the sorption of 2,3,7,8-tetrachloro-dibenzo-p-dioxin to soils from aqueous methanol mixtures and evaluated the applicability of the cosolvent theory to such sorption. Sorption kinetics were influenced by the fraction of methanol in the liquid phase and the soil type.

Onuska and Terry [222] used supercritical fluid chromatography to extract dioxins from soil.

Richter et al. [223] showed that accelerated solvent extraction gave essentially equivalent recoveries of chlorinated dibenzo-p-dioxins and dibenzofurans from soil compared to Soxhlet extraction, but in less time and using much less solvent.

No significant difference was found in the determination of chlorodibenzo-p-dioxins and dibenzofurans in soil between Soxhlet extractions using toluene or methylene chloride–acetone as extractants [224].

Hengstmann et al. [225] studied five different extraction methods for the determination of chlorodioxins in soil and concluded that nearly quantitative extractions could be achieved in two hours using "supersonic" Soxhlet extraction.

Gas Chromatography–Mass Spectrometry

Gas chromatography–mass spectrometry has been used to determine 2,3,7,8-tetrachloro-p-benzo dioxin in soil [226].

Tong et al. [227] have described a high-resolution gas chromatographic mass spectrometric method for the determination of monobromopoly-chloro-dibenzo-p-dioxin in soils and incinerator wastes.

A good example of the application of gas chromatography–mass spectrometry to the determination of polychlorodibenzo-p-dioxin and dibenzofurans up to the octochlorocongeners in soils and sediments is that of Smith et al. [228], which, it is claimed, is sufficiently sensitive to determine down to 1–5 parts per trillion of these substances.

Di Domenico et al. [229] have discussed analytical techniques used for the determination of 2,3,7,8-tetrachlorodibenzo-p-dioxin in environmental samples taken after the industrial accident at Sevesco, Italy. Detection thresholds of 2–50 ppt were achieved for agricultural soil samples.

4.15
Trifluoroacetic Acid

In the early 1990s, chlorofluorocarbon refrigerants and propellants were largely replaced by hydrofluorocarbons and hydrochlorofluorocarbons in order to reduce stratospheric ozone depletion. The hydrochlorofluorocarbons and hydrofluorocarbons, unlike the older chlorofluorocarbons, are unstable in the troposphere and can degrade to trifluoroacetic acid as a byproduct [230]. As a result of its high water solubility and low Henry's constant, trifluoroacetic acid is removed from the atmosphere primarily through wet deposition [231, 232], where it tends to accumulate in aquatic ecosystems with little outflow or seepage and high evaporation rates [233, 234].

Trifluoroacetic acid has two characteristics that are cause for environmental concern: exceptional stability and phytotoxicity. Its stability arises from its effective immunity to environmental oxidation and reductive dehalogenation [235, 236]. As a result of its stability, it has been demonstrated that trifluoroacetic acid accumulates in terminal water bodies [233, 237, 238].

Currently, typical trifluoroacetic acid concentrations in surface water range from approximately 100 to ~ 500 ng/l, although trifluoroacetic acid concentrations as great as 6.4 μg/l have been found in certain terminal lakes [237, 239].

Trifluoroacetic acid has been shown to impact sensitive species of algae, such as *Raphidocelis subcapitata*, with a lowest observed effective dose of 360 μg/l [236]. Most plant species tested were impacted by trifluoroacetic acid at concentrations in the mg/l range [236, 240]. Although these toxic concentrations are 2–3 orders of magnitude higher than typical current environmental levels [237–239], there is concern that trifluoroacetic acid may accumulate over many years in terminal water bodies to the point that concentrations toxic to algae and some higher plants may be achieved.

Cahill et al. [241] have developed a simple and sensitive analytical procedure for determining the concentration of trifluoroacetic acid in plant, soil, and water samples. The analysis involves extraction of trifluoroacetic acid by sulfuric acid and methanol followed by derivatisation to the methyl ester of trifluoroacetic acid. This is accomplished within a single vial without complex extraction procedures. The highly volatile methyl ester is then analysed using headspace gas chromatography. The spike recovery trials from all media ranged from a low of 86.7% to a high of 121.4%. The relative standard deviations were typically below 10%. The minimum detectable limit for the method was 34 ng/g for dry plant material, 0.20 ng/g for soil and 6.5 ng/l for water.

4.16
Compounds Containing Nitrogen, Sulfur and Phosphorus

4.16.1
Nitrogen Compounds

4.16.1.1
Aromatic Amines

Supercritical fluid chromatography has been used to determine aromatic amines in soil [242–246].

Talsky [116] has used higher order derivative spectrophotometry to determine aniline in soil.

Kester [34] has discussed the application of purge and trap chromatography to the determination of acrylonitrile in soil. In this method, the soil sample is heated for 30 minutes to 85 °C and dry purged with dry helium and the volatiles collected in a Tenax trap. Subsequent release of acrylonitrile and acetonitrile by heating the Tenax trap to 100 to 180 °C is followed by collection of the volatiles and analysis by gas chromatography using a Chromasorb 101 column programmed from 80 to 150 °C and a flame ionisation detector.

Kido et al. [245] determined basic organic compounds such as quinoline, acridine, aza-fluorene and their N-oxides in marine sediments found in an industrial area. The sediments were extracted with benzene using a continuous extractor for 12 hours. Hydrochloric acid solution (1 N) was added to the benzene extracts and the mixture was shaken for five minutes; the acid layer separated from the benzene layer was made alkaline by the addition of sodium hydroxide, and the alkaline aqueous solution was extracted with diethyl ether; the ether extracts were then dehydrated with anhydrous sodium sulfate and concentrated with a Kuderna-Danish evaporator. The concentrations were separated and analysed by gas chromatography–mass spectrometry and gas chromatography–high-resolution mass spectrometry.

Krone et al. [246] used capillary column gas chromatography with nitrogen-specific detection and gas chromatography–mass spectrometry to determine nitrogen-containing aromatics originating from creosote oil in solvent extracts of sediments taken in Eagle Harbour, Puget Sound and in uncontaminated areas. Organic sediment extracts and the commercial creosote oil were fractionated by silica alumina column chromatography. No nitrogen-containing aromatics were detected in sediments from a pristine reference area.

Ethylenediamine Tetraacetic Acid

Nowack et al. [263] determined adsorbed iron III– and nickel–EDTA species in soil by reverse-phase ion-pair high-performance liquid chromatography. Iron III–EDTA was found to be the main species present, occurring at 30–70%, while nickel–EDTA species were present in considerably lower amounts (<10%). The adsorbed metal–EDTA species were detected in lake sediment and soil cores.

Tricyclazole and tetracycline have been determined by gas chromatography – mass spectrometry [247]. Persistent tetracycline residues have been determined in soil by high-performance liquid chromatography with electrospray ionisation tandem mass spectrometry [248].

Nitrogen-containing explosives [249] and trinitrotoluene [250] have been determined in soil by gas chromatography with thermionic NP detection and reverse-phase high-performance liquid chromatography. Warmont et al. [251] used tunable infrared laser detection to study the pyrolysis products of explosives in soil.

Surface-enhanced Raman spectroscopy [252], a method based on an array of sensory polymers attached to fibre optics [253], has been used to determine down to 5 ppb and 100 ppb of 2,4-dinitrotoluene in soil.

Emery et al. [254] studied the binding of trinitrotoluene to soil using ^{2}H magic angle spinning NMR.

Trinitrotoluene and RDX have been determined in soil using a field-portable continuous-flow immunosensor. Results agreed with those obtained by high-performance liquid chromatography [255, 256].

Grant [257] found nitramine and nitroaromatic explosive residues in real field soil samples. The samples were stable under refrigeration, but the nitroaromatics used to fortify the samples degraded rapidly, even when the samples were refrigerated. Therefore fortified soils can lead to significant errors.

4.16.2
Sulfur Compounds

Garlucci et al. [258] discuss a method for determining tetrahydrothiophen contaminant in soil using headspace high-resolution gas chromatography together with mass spectrometry. Down to 10 ng of this substance could be determined.

4.16.3
Phosphorus Compounds

David and Seiber [259] compared the efficiencies of various extraction techniques, including supercritical fluid [260], high-pressure solvent and Soxhlet extraction, for the removal of organophosphorus hydraulic fluids from soil. High-pressure solvent extraction at temperatures up to 200 °C and pressures up to 17 MPa was the favoured technique. Extraction efficiencies were similar in all three methods, but the favoured method was more rapid and less expensive to operate.

Ingram et al. [259] applied static secondary ion mass spectrometry to determine down to 70 pg/m^2 of tributyl phosphate in soil surfaces.

4.17
Miscellaeous Organic Compounds

4.17.1
Flame Retardants

A new potential source of environmental contamination is the use of flame retardants composed of brominated aromatics, many of which have close structural relationships to polychlorobiphenyls and other known persistent organic pollutants. These compounds have recently received much publicity after accidental contamination of livestock rations, resulting in the forced destruction of 23 000 cattle, 4000 swine, 1.5 million chickens, and tons of eggs, milk, butter, cheese and feed. The flame retardant, FireMaster PB6, is a mixture of at least 18 different compounds, with penta-, hexa-, and heptabromobiphenyls as major components [261, 262].

Incubation in soils showed that polybrominated biphenyls were resistant to degradation, but were apparently not taken up by plants or leached into groundwater [261]. Commercial formulations of brominated aromatic flame retardants had variable composition; some contained highly brominated phenols, but no evidence of contamination with dibenzodioxins and dibenzofurans was found [198].

4.17.2
Humic and Fulvic Acids

Saar and Weber [264] compared methods based on spectrofluorimetry and ion-selective electrode potentiometry for determining the complexes formed between fulvic acid and heavy metal ions.

The fluorescence properties of two fulvic acids, one derived from the soil and the other from river water, were studied. The maximum emission intensity occurred at 445–450 nm upon excitation at 350 nm, and the intensity varied with pH, reaching a maximum at pH 5.0 and decreasing rapidly as the pH dropped below 4. Neither oxygen nor electrolyte concentration affected the fluorescence of the fulvic acid derived from the soil. Complexes of fulvic acid with copper, lead, cobalt, nickel and manganese were examined and it was found that bound copper II ions quench fulvic acid fluorescence. Ion-selective electrode potentiometry was used to demonstrate the close relationship between fluorescence quenching and fulvic acid complexation of cupric ions. It is suggested that fluorescence and ion-selective electrode analysis may not be measuring the same complexation phenomenon in the cases of nickel and cobalt complexes with fulvic acid.

Wilson et al. [265] carried out a compositional and solid-state nuclear magnetic resonance (NMR) spectroscopic study of humic and fulvic acid and fractions present in soil organic matter. The ^{13}C NMR study utilised cross polarisation–magic angle spinning (CP-MAS) with spin counting. The elemental and functional group analyses provided input for a series of analytical constraints calculations that yield an absolute upper limit for the amount of aromatic carbon, and most probable estimates for both aromatic and non-carboxyl aliphatic carbon in each sample. Spin counting experiments demonstrate that less than 50% of the carbon in three of the fractions is observed in the NMR experiment, and even after correction for differential relaxation, the amounts of aromatic and non-carboxyl aliphatic carbon determined by ^{13}C CP-MAS NMR are dissimilar to those obtained by calculation. Unambiguous evidence is presented for the predominance of aliphatic carboxyl groups in one of the fulvic acid fractions.

Weber and Wilson [266] used anion and cation exchange resins to isolate fulvic and humic acids from soil and water.

4.17.3
Mestranol

Okuno and Higgins [267] have described a procedure for determining residual levels of mestranol (an animal damage control chemosterilant) and its 3-hydroxy homologue ethynyloestradiol in soil samples. The lower limit of detection was 0.1 ppm. After extraction in acidic medium, the samples are cleaned up by Florisil column chromatography. Soil samples are further cleaned up on a gel permeation chromatographic column so that the ethynyloestradiol fraction can be analysed by gas chromatography. The mestranol fraction is again cleaned up by gel separation and a Florisil column. Thin-layer chromatography was used to confirm the results obtained by gas chromatography.

Recoveries in soil samples averaged less than 50%, even after corrections. This may have been due to degradation of the compounds by soil microorganisms or to chemical and physical interactions with the soil. Mestranol recoveries averaged 26–30% from soils at 0.1 ppm. Recoveries of ethynyloestradiol were even lower, presumably because of its greater chemical reactivity due to the slightly acidic hydrogen in the 3-hydroxy position.

4.17.4
Enteroviruses and Antibacterial Agents

Enteroviruses [268] and antibacterial agents [269] have been determined in soil.

4.17.5
Miscellaneous

Wells and Hess [270] have reviewed the separation, clean-up and recovery of persistant organic contaminants from soils. Industrial hygiene gas detector tubes have been employed to detect severe contamination by organic volatiles in soil [271].

Di Domenico et al. [272] have described an analytical procedure for the multianalyte/multilaboratory assessment of pollutants in complex soils.

Tognotti et al. [273] studied the adsorption–desorption of contaminants on single soil particles using an electrodynamic thermogravimetric analyser.

Schulton and Sorge [274] used laser Raman spectroscopy to provide detailed information on the location, elemental composition and chemical speciation of organic compounds in soil.

4.18
Insecticides and Pesticides

4.18.1
Chlorinated Insecticides

The first step in the determination of insecticides in soil is to efficiently extract the insecticide from the soil. The insecticide content of the extract can then be determined by methods based, usually, on gas chromatography, preferably linked to a mass spectrometric detector.

Methods for the extraction of insecticides by supercritical fluid extraction are reviewed in Table 4.2. Other methods of extraction have, of course, been employed.

Novikova [282] has reviewed the literature (209 references) covering the extraction, clean-up and analysis of organochlorine (and organophosphorus) insecticides in soil. Johnson and Starr [283] and Chiba and Morley [284] have studied factors affecting the extraction of dieldrin and aldrin from different soil types; ultrasonic extraction was recommended by these workers. Lopez-Avila et al. [170] used microwave-assisted extraction to extract chlorinated insecticides from soils.

Mangani et al. [285] used Carbopack B columns to recover chlorinated insecticides in soil samples. These workers noted that, although the principles governing the adsorption and extraction process in the extraction in soil analysis are the same as those that govern liquid–solid chromatography, the

Table 4.2. Application of supercritical fluid extraction to the determination of chlorinated insecticides in soil (from author's own files)

Organochlorine insecticide	Solvent used	Comments	% Recovery	Reference
DDT, Toxaphene	CO_2	Temperature/pressure phase diagrams for CO_2–DDT and CO_2–polychlorobiphenyls	DDT: 70% Toxaphene: 75%	[277]
Dichlorvos, endrin, endrin aldehyde, pp'-DDT, mirex, decachloro-biphenyl	Subcritical CO_2 with 3% methanol	Comparison of supercritical fluid extraction, sonic extraction and Soxhlet extraction	85% of stated insecticide	[276, 278, 280]
Chlordane	CO_2	Comparison of supercritical fluid extraction, accelerated solvent extraction and Soxhlet extraction	–	[275, 277–280]
Miscellaneous	CO_2 with 3% methanol		–	[281]

Table 4.3. Gas chromatographic methods for determination of chlorinated insecticides in soil extracts (from author's own files)

Organochlorine insecticide	Extraction solvent	Comments	References
Miscellaneous	Comparison of different extraction solvents	Need efficient clean-up procedures	[288–292, 384
DDT, dieldrin	Hexane–isopropanol, hexane–acetone, hexane–isopropanol, acetone	Comparison of different extraction solvents	[293]
DDT, dieldrin, endrin, methoxychlor	Florisil column extraction	–	[294]
Dieldrin, endrin	–	Insecticides reacted with BCl_3 prior to gas chromatography. Limit of detection: 0.01 ppm	[295]
BHC isomers	Light petroleum	Comparison of GLC and thin-layer chromatography	[296]
DDT	–	Gas liquid chromatography with ECD	[297]
Miscellaneous	–	Gas liquid chromatography. Limit of detection: 0.0005 to 0.008 ppm	[298, 299]
Miscellaneous	Acetone–toluene (1:1)	Gas liquid chromatography with ^{63}Ni detector	[285]
Kepone, DDT, permethrin	–	Study of DDT breakdown	[300]
Miscellaneous	Acetone	Contribution of acetone polymers to coextracted material. Capillary gas chromatography	
Miscellaneous	Miscellaneous	Gas chromatography–mass spectrometry	[301]
Gas chromatography–mass spectrometry			
Dieldrin, heptachlor	Hexane	Resolution of chlorinated insecticides and PCB	[187]
Lindane	–	Application of gas chromatography–mass spectrometry. Limit of detection 0.1 µg/kg	[170]
Miscellaneous	Miscellaneous	Gas chromatography–ion trap tandem mass spectrometry	[302]

main feature of a chromatographic column, i.e., separation efficiency, is almost completely absent. The best results were obtained with a mixture of petroleum in ether–toluene (2:1).

Kimbrough [171] recommended the use of polar solvents with Florisil clean-up for the extraction of organochlorine insecticides from soil.

Hartonen et al. [286] have reported various extraction techniques for organochlorine insecticides in soils. Trapping efficiencies were reported for organochlorine insecticides, selected polychlorinated benzenes, polybrominated biphenyls, and polybrominated diphenyl ethers.

Steinwandter [287] studied several methods for the extraction of HCH insecticides from soil.

Schwab et al. [434] and Chirnside and Ritter [435] used extraction with methanol followed by solid-phase extraction on a C18 cartridge to extract alachlor from soil.

ELISA methods have been used to determine metochlor [456,457], alachlor and alachlor residues in soil [438].

4.18.2
Carbamate Insecticides

Again, as in the case of chlorinated insecticides, gas chromatography is the method of choice for the determination of carbamate insecticides (Table 4.4). High-performance liquid chromatography has been used more recently.

Thus Mori et al. [303] determined carbofuran by extraction of the soil with acetonitrile containing silver nitrate, partitioning with methylene chloride,

Table 4.4. Gas chromatographic determination of carbamate insecticides in soil extracts (from author's own files)

Carbamate insecticide	Extraction solvent	Comments	Reference
Carbofuran	Methanol:water (80:20) then chloroform	Derivatised with 1-fluoro-2,4, dinitrobenzene then gas liquid chromatography with N-specific detection. Limit of detection: sub μg	[308]
Carbosulfan, carbofuran	Hexane-2-propanol or methanol buffer	Gas liquid chromatography with N-specific detection	[309]
Methomyl	Chloroform	Gas liquid chromatography–mass spectrometry. Limit of detection: 1 $\mu g/kg$	[310]
Carbamyl, carbofuran	–	Converted to pentafluorobenzyl derivatisation, then gas liquid chromatography.	[311]
Oxamyl	Dichloromethane or acetone dichloromethane	Gas liquid chromatography	[312]
Methomyl, (m-s-butylphenylmethyl carbamate)	Dichloromethane	Gas liquid chromatography	[313,314]

silica gel column clean-up and silica gel high-performance liquid chromatography.

McGarvey [304] has reviewed high-performance liquid chromatographic methods of determining 31 N-methylcarbamate pesticides in soil, and Okamoto et al. [305] used automated solid-phase extraction followed by on-line high-performance liquid chromatography to determine ten carbamate pesticides in soil.

Singhal et al. [306] have determined oxamyl residues (methyl-N',N'-dimethyl-N-[methylcarbamoyl] oxy-thioaminimidate) by a method based on reaction with carbon disulfide and copper in which excess copper is added to an extract of the soil and the excess copper is determined by titration with 0.001 M EDTA to the 1-(2-pyridylazo)-2-naphthol endpoint.

N-methylcarbamate and N,N'-dimethylcarbamates have been determined in soil samples by hydrolyses with sodium bicarbonate and the resulting amines reacted with 4-chloro-7-nitrobenzo-2,1,3-oxadiazole in isobutylmethylketone solution to produce fluorescent derivatives [307]. These derivatives were separated by thin-layer chromatography on silica gel G or alumina with tetrahydrofuran–chloroform (1:49) as solvent. The fluorescence is then measured in situ (excitation at 436 nm, emission at 528 and 537 nm for the derivatives of methylamine and dimethylamine, respectively). The method was applied to natural water and to soil samples containing parts per 10^9 levels of carbamate. The disadvantage of the method is its inability to differentiate between carbamates of any one class.

4.18.3
Organophosphorus Insecticides

Organophosphorus insecticides including diazinon, ronnel, parathion ethyl, methiadathion and trichlorovinphos have been extracted from soil by subcritical carbon dioxide containing 3% methyl alcohol. At a pressure of 35.5 MPa and 50 °C, recoveries of 85% were obtained [280, 315].

Generally, a nitrogen phosphorus-specific detector is used in the determination of organophosphorus insecticides in soil [316–318]. Use of an acetone-n-hexane extraction solvent led to recoveries of 54–83%.

Abbot combined gas chromatography with mass spectrometry [317].

Skladal et al. [319] used amperometric biosensors based on acetyl or butyrylcholinesterase for the kinetic determination of organophosphorus insecticides in soil extracts.

4.18.4
Miscellaneous Insecticides

Solid-phase microextraction [320–322], microwave-assisted extraction [321, 323], accelerated solvent extraction and Soxhlet extraction have been discussed [324, 325].

Gas chromatography and high-performance liquid chromatography with NP detection [326–328], photoionisation detection and mass spectrometric detection [329–332] have all been used in general surveys of pesticides in insecticides in soils.

Puig and Barcelo [333] compared various types of gel permeation chromatographic columns for the clean-up of selected pesticides isolated from soils.

4.19
Herbicides

4.19.1
Carbamate Type

Cohen and Wheals [334] determined ten hydrolysable carbamate and substituted urea herbicides in soil in amounts down to 0.001–0.05 ppm. In this method, a solution of the herbicide-containing extract of the soil is spotted onto a silica gel G plate and developed with hexane:acetone (5:1). The plate is sprayed with 1-fluoro-1,4-dinitrobenzene in acetone and heated to 190 °C to produce the 2,4-dinitrophenyl derivative of the herbicide amine moiety; acetone extracts of the areas of interest are subjected to gas chromatography.

Acetonitrile–silver nitrate extracts of soil, partitioned with methylene chloride and cleaned up on silica gel, have been used in the high-performance liquid chromatographic determination of benfuracarb and carbofuran herbicides [204]. McGarevy et al. [336] and Lin and Cooper [336] used optimised isocratic high-performance liquid chromatography with UV detection to determine Aldicarb and its metabolites in soil.

Immunoassay- and chemiluminescence-based methods have also been used [337] to determine methyl-2-bendimadazole carbamate [338] and aldicarb and paraoxon in soil.

4.19.2
Substituted Urea Types

Gas chromatography and, more recently and to a lesser extent, high-performance liquid chromatography are the most commonly used methods for determining this type of herbicide in soil extracts. Gas chromatographic methods are reviewed in Table 4.5.

Gas chromatography of phenyl urea herbicides is difficult because of their ease of decomposition. Procedures have been reported in which careful control of conditions allows these compounds to be chromatographed intact [339, 340, 344, 345]. Alternatively, the urns can be hydrolyzed to the corresponding substituted anilines; these compounds are then determined by either gas chromatography directly [341] or as derivatives [342], or colorimetrically [343] after coupling with suitable chromospheres. Methods based on hydrolysis lack

Table 4.5. Gas chromatographic methods for the determination of substituted urea-type herbicides in soil extracts (from author's own files)

Herbicide	Extraction solvent	Comments	Reference
Miscellaneous	–	Gas chromatography/thin layer chromatography. Limit of detection: 1–50 µg/kg	[334]
Chlorbromuron and metabolites	Ethyl acetate	Gas chromatography/thin layer chromatography	[353]
Miscellaneous	Acetone	Alkaline hydrolysis, steam distillation and toluene extraction of distillate. Anilines produced are brominated and determined by gas chromatography. Limit of detection: 0.1 mg/kg	[354]
Miscellaneous	–	Pyrolysis of urea herbicide to phenylisocyanate in the injection heater of the gas liquid chromatograph, then detection by ECD	[339, 355]
Miscellaneous	–	Gas liquid chromatography with thermionic detection	
Miscellaneous	–	Alkylation then gas chromatography	[356, 357]
Miscellaneous	–	Avoidance of thermal decomposition during gas chromatography of urea herbicides	[340, 354]

specificity and involve lengthy procedures. These disadvantages can be overcome by using liquid chromatography.

Farrington et al. [346] developed a high-performance liquid chromatographic method to perform positive monitoring down to 200 µg/kg of chlorbromuron, chlorotoluron, diuron, linuron, monolinuron, chloroxuron, monouron and metobromuron in methanolic extracts of soils. Recoveries were in the range 97.5 to 102%.

Smith and Lloyd [347] have used liquid chromatography for the determination of chlorotoluron residues in soil but report that diuron and monuron interfere in the chromatographic system used.

Cotterill [348] compared two methods, high-performance liquid chromatography and gas chromatography, for the determination of diuron in soil. Cotterill [348] used the soil extraction method devised by McKone [339] in which a 25 g sample of soil was extracted with 50 ml of methanol by shaking on a wrist-action shaker for one hour. The resulting soil slurry was filtered through a Whatman No. 42 filter-paper. For gas chromatography, a 2 ml aliquot was evaporated to dryness by gently blowing air and the residue was redissolved in 2 ml of hexane. For high-performance liquid chromatography, a 25 ml aliquot was concentrated to about 1 ml under reduced pressure while warming in a water-bath at 40 °C. The remaining solvent was removed with a gentle stream of dried air, and the residue was then redissolved in 1 ml of the high-performance liquid chromatography eluent.

Although high-performance liquid chromatography is generally the most reproducible method, gas chromatography has the advantage of being more sensitive. When measuring very low residues in soils with a low organic matter content, gas chromatography could prove to be the better method but the results should be interpreted with caution due to the possible presence of unresolved metabolites.

A limit of detection of 0.04 µg/g was achieved by both methods. High-performance liquid chromatography has been used by other workers [349–352,439,440]. In particular, linuron and its metabolites [439] and isoproturon, dichlorprop [441] and, hexaflumuron [440] have been determined by this technique.

McNally et al. have applied supercritical fluid extraction chromatography to the determination of diuron and linuron in soil [442]. Schlaeppi et al. [443] have described an automated magnetic particle-based chemiluminescent immunoassay for the determination of trisulfuron in soil [462].

The presence of free anilines or other metabolites in soils and plants has been reported [358–363]. Some work has suggested that they are very strongly bound to soil components, and the findings of Caverly and Denney [354] are in agreement with these conclusions. The presence in soils of metabolites of linuron that possess the urea structure have been reported [342, 362]; these are produced mainly by microbiological degradation. The dimethyl derivative is considered to be inactive whereas the monomethyl metabolite has a phytotoxicity approaching that of the parent herbicide [362].

4.19.3
Sulfonylurea Type

McNally and Wheeler [364] used supercritical fluid extraction coupled to supercritical fluid chromatography to determine sulfonylurea herbicides in soil. Klatterback et al. [365, 366] used supercritical fluid extraction with methanol-modified carbon dioxide followed by high-performance liquid chromatography with UV detection to determine sulfonylurea herbicides obtained on a C_{18} solid-phase extraction disc. Alternatively the determination was carried out by gas chromatography of the dimethyl derivatives of the sulfonylurea herbicides, employing an electron capture or a NP detector on the gas chromatograph.

4.19.4
Triazine Type

Mills et al. [367] have described a method for the isolation of triazine metabolites from soil using automated solid-phase extraction with methanol:water

(4:1 v/v), followed by evaporation of the methanol phase and collection of the metabolites on a C_{18} octadecyl resin. Analytes are then eluted from the resin with ethyl acetate, leaving impurities in the resin.

Di Corcia [368] used subcritical water (phosphate-buffered) extraction with a graphitised carbon black cartridge to recover terbuthylazine herbicide and its metabolites from soil.

The principle methods used for the determination of triazine-type herbicides are gas chromatography (Table 4.6) and high-performance liquid chromatography (Table 4.7). Other methods that have been used include isotachoelectrophoresis [369], ELISA [370–375], spectrophotometry [376,377] and thin-layer chromatography [378] (Table 4.8).

Studies have been made of the fate of 3-amino-1,2,4-triazole herbicide in soils [379]; while adsorption of aminotriazole by clay minerals has been postulated, little is known of the interaction with pure clay minerals, particularly of the montmorillonite group. The importance of such reactions cannot be overemphasised in view of their bearing on the persistence of the herbicide in the soil.

While the high solubility of aminotriazole in water (28 g per 100 ml at 23 °C) suggests ready leaching from whole soil, Russell et al. [379] showed that, if the soil contains a montmorillonite-type mineral, the aminotriazole might be resistant to leaching as a result of adsorption by the montmorillonite.

Table 4.6. Gas chromatographic methods for the determination of triazine-type herbicides in soil extracts (from author's own files)

Triazine herbicide	Extraction method	Comments	Reference
Trifluralin, linuron, fluorochloridone, triazine types: atrazine, alachlor, metolachlor and pendimethalin	Methylene chloride or ethylacetate	Gas chromatography with N–P detection or mass spectrometry	[380, 381]
Atrazine, simazine, terbuthylazine, molinate	Solid-phase microwave-assisted extraction using methanol	Gas chromatography–mass spectrometry. Limit of detection: 1 – 10 ng/g	[382]
Atrazine, cyanazine, diethylatrazine, metochlor	Supercritical fluid chromatography with CO_2	Gas chromatography and high-performance liquid chromatography	[383, 384]
Atrazine, simazine, linuron, metribuzin, triallate, phorate	Methanol	Gas chromatography with ECD	[385, 386]
Atrazine	Hexane–acetone	Isotope dilution mass spectrometry. Limit of detection: 0.1 – 1.0 ppm	[387]

Table 4.7. High-performance liquid chromatographic methods for the determination of atrazine herbicides in soil extracts (from author's own files)

Atrazine herbicide	Extraction method	Comments	Reference
Cyanazine	–	Microbore column with diode array detection and multichannel integrator. Limit of detection: 0.25 ng absolute	[388]
Atrazine and metabolites	Cyclohexyl solid-phase extraction cartridge used to separate atrazine from soil extract	High-performance liquid chromatography with photodiode array detection	[389–391]
Triazine herbicides	Methanol, C_{18} solid-phase extraction	Gradient C_{18} high-performance liquid chromatography with UV detection at 220 nm. Limit of detection: ppb	[392]
Terbutylazine and its degradation products	Hot acetone then adsorption on cation exchange solid-phase cartridge	High-performance liquid chromatography with photodiode array detection	[393, 394]

Table 4.8. Miscellaneous methods for the determination of triazine herbicides in soil extracts (from author's own files)

Triazine herbicide	Extraction method	Comments	Reference
Atrazine, simazine, atratone, prometryn, Desmetryn, Methoprotryne	–	Capillary, isotachoelectrophoresis. Limit of detection: 10 µg/kg soil	[369]
Atrazine	–	Enzyme immunoassay (tube system)	[370]
Cyanazine, atrazine	–	Enzyme immunoassay. Limit of detection: 0.2 µg/kg	[371, 372]
Atrazine	–	Enzyme immunoassay. Limit of detection: 3.5 µg/kg	[373]
Alachlor, atrazine, capton, carbofuran, metolachlor, 2,4-D	Solid-phase microextraction	Enzyme immunoassay	[375]
Atrazine, chlorpyrifos, diuron	–	Derivative spectrophotometry	[376, 377]
Atrazine	–	Thin-layer chromatography, followed by UV absorption by scanning at 222 nm	[378]

The 3-aminotriazole molecule is protonated when adsorbed on montmorillonite surfaces to produce the 3-aminotriazolium cation. In the case of montmorillonite saturated with polyvalent cations (Ca^{2+}, Cu^{2+}, Ni^{2+}, Al^{3+}), protonation is believed to be due to the highly polarised water molecules in direct coordination to these cations. The decreasing order of extent of protonation (Ca < Mg < Al) reflects the order of decreasing polarising power of the cations. Infrared spectra indicate coordination of 3-aminotriazole to Ni^{2+} and Cu^{2+} cations. The infrared absorption band at $1696 \, cm^{-1}$ is assigned to the C=N stretching vibration of the exocyclic C=N $^+$HH group. Shifts of the $1696 \, cm^{-1}$ band to 1683 and $1666 \, cm^{-1}$ upon dehydration and deuteration, respectively, suggest that the positive charge on the protonated molecule lies on the exocyclic nitrogen. The protonated molecule undergoes normal exchange reactions with other cations.

4.19.5
Phenoxyacetic Acid Type

Supercritical fluid extraction with methanol-modified carbon dioxide has been applied to the determination of acidic herbicides such as chlorophenoxyacetic acids in soil [395].

Gas chromatography has been used extensively for the determination of phenoxyacetic acid-type herbicides in soil extracts (Table 4.9).

Phenoxyalkanoic acid herbicides are not amenable to direct gas chromatographic determination because of the high polarity or low volatility of the compounds and must be converted to their more volatile derivatives. The sensitivity of the electron capture detector towards alkyl esters of 4-chloro-2-methylphenoxy acetic acid, 4-chloro-2-methylphenoxy butyric acid, etc., is very poor. The methyl ester of 4-chloro-2-methylphenoxy acetic acid was 100 times less sensitive to electron affinity detection than the 2,4-D-methyl ester [396].

Chau and Terry [397] reported the formation of pentafluorobenzyl derivatives of ten herbicidal acids, including 4-chloro-2-methyl-phenoxyacetic acid [396]. They found that five hours was an optimum reaction time at ambient temperature with pentafluorobenzyl bromide in the presence of potassium carbonate solution. Bromination [398], nitrification [399] and esterification with halogenated alcohol [396] have also been used to study the residue analysis of 4-chloro-2-methylphenoxy acetic acid and 4-chloro-2-methylphenoxybutyric acid. Pentafluorobenzyl derivatives of phenols and carboxylic acids were prepared for detection by electron capture at very low levels [400, 401].

Yip [405] reported that the binding of the soil particles and organic matter with the herbicide residues prevented the complete extraction of them with an organic solvent. Upchurch and Mason [406] found that the extent of adsorption of herbicides is highly dependent on the type of organic matter and clay as well as on the amounts of their constituents.

Table 4.9. Gas chromatographic methods for the determination of phenoxyacetic acid herbicides in soil extracts (from author's own files)

Phenoxyacetic acid herbicide	Extraction method	Comments	Reference
2,4-Dichloro-phenoxy acetic acid, 2,4,5-trichlorophenoxy acetic acid	–	Herbicide derivatised to its methyl or 2-chlororethyl ester, then gas chromatography. 70–74% recovery	[169]
4-Chloro-2-methyl-phenoxyacetic acid and metabolites	Ether:acetone: heptane:hexane (2:1:1:1)	Derivatisation to pentafluorobenzyl derivatised after clean-up by thin layer chromatography then gas chromatography. Limit of detection 20–25 ng absolute	[402]
4-Chloro-2-methyl-phenoxyacetic acid	Dichloromethane extraction	Extracts esterified with 2,3,4,5,6-pentafluorobenzyl bromide, petroleum ether extract analysed by gas chromatography, limit of detection 0.5–2 µg/kg	[403]
2,4, Dichlorophenoxy acetic acid	–	Isotope dilution gas chromatography–mass spectrometry.	[404]

The results of fresh soil treatments show that the pH has no significant effect in the range 4.6–7.8. Also, one could conclude that the soil texture (especially the differences in clay content) has no significant influence on the amounts of residues recovered.

The results support a general conclusion that organic matter is the main factor that influences the fate of herbicides and their analyses in the soil. Sandy and clay loam soils have high but sandy loam and clay soils have very low organic matter contents. Consequently, significantly higher recoveries of the residues were generally obtained from the latter soil materials than from the former. In addition, the sandy and clay loam soils gave selectively lower 5-chloro-3-methylcatechol recoveries, related to the recoveries of 4-chloro-2-methylphenoxyacetic acid and 4-chloro-o-cresol.

Chlorinated [407, 408] and 2,4-dinitrophenoxy acid herbicides [409] have been determined. Liquid chromatography particle beam mass spectrometry has been used as an analytical finish [408]. Crescenzi et al. [410] evaluated the feasibility of selectively and rapidly extracting herbicide residues in soils by hot water and collecting analytes with a Carbograph 4 solid-phase extraction cartridge set on-line with the extraction cell. Phenoxy acid herbicides and those non-acidic and acidic herbicides that are often used in combination with phenoxy acids were selected for this study. Five soil samples were fortified with target compounds at levels of 100 and 10 ng/g (30 ng/g of clopy-ralid and picloram) by following a procedure able to mimic weathered soils. Herbicides were extracted with water at 20 °C and collected on-line by the solid-phase extraction cartridge. After the cartridge was disconnected from the extraction apparatus, analytes were recovered by stepwise elution to separate non-acidic herbicides from acidic ones. The two extracts were analysed by liquid chromatography–mass spectrometry with an electrospray ion source. At the lowest spike level considered, analyte recoveries ranged between 81 and 93%, except those for 2,4-DB and MCPB, which were 63%. The method detection limit was in the 1.7–10 ng/g range. For the analytes considered, this extraction method was more efficient overall than Soxhlet and sonication extraction techniques.

4.19.6
Imidazolinones

The imidazolinones are a relatively new class of herbicides used to control a wide spectrum of broad-leafed weeds and grasses in a variety of agricultural commodities [411]. These herbicides are very potent weed killers and are used in doses that are substantially lower than those of conventional herbicides. The members of this class of herbicides have similar structural features centred around the imidazolinone ring and an attached aromatic system bearing a carboxylic acid moiety. Imidazolinones show excellent activity against annual and perennial grasses and broad-leafed weeds when applied either pre- or post-emergence. They function by inhibiting acetohydroxy acid synthesis, the feedback enzyme in the biosynthesis of branched-chain essential acids [412–414]. This enzyme is not present in animals. The imidazolinone ring of herbicides is amphoteric and can behave as a weak base or a weak acid. The movement of the acid imidazolinones in the soil can be strongly influenced by many soil properties, the most important of which are pH, organic matter and clay content. Binding of the acid imidazolinones increases as pH decreases. Basic herbicides protonate and are adsorbed on negatively charged soil colloids. Acidic herbicide anions also become protonated as pH decreases, reducing the repulsive forces present when the molecule is dissociated, thus increasing molecular adsorption [415–419]. The typically low application rates used for imidazolinone herbicides make their chemical analysis difficult.

Reddy and Locke isolated [420] the herbicide imazaquin from soil by carbon dioxide supercritical fluid chromatography [421]; corn root bioassay and electrospray mass spectrometry have also been used to determine this herbicide.

Heber et al. [422] determined the herbicides imazethapyr and imazapyr on a 0.1 M sodium carbonate extract of soil. Clean-up was by partitioning with methylene chloride, and final analysis was by high-performance liquid chromatography with UV detection at 260 nm. Lagona et al. [423] and Krynitsky et al. [424] used the latter technique. In this method the soil utilises a combined soil column extraction and off-line solid-phase extraction for sample preparation. Analysis was by liquid chromatography–electrospray mass spectrometry in selected ion monitoring mode. Several different extractants were evaluated for the purpose of soil column extraction optimisation. The system that best optimises the extractability of imidazolines from the soil was found to be the mixture of methanol–ammonium carbonate (0.1 M, 50:50 v/v). The total recovery of each imidazoline from soil at each of the two levels investigated ranged from 87% to 95%. Under three selected ion monitoring conditions, the limit of detection (S/N = 3) was found to be 0.1–0.05 ng/g in soil samples.

Examples of this type of herbicide are imazapyr, m-imazamethabenz, p-imazamethabenz, m,p-imazamethabenzmethyl, imazethapyr and imazaquin. Imazapyr has been determined at the μg/kg level in 0.1 M ammonium acetate extracts of soil by microwave-assisted extraction using electron capture negative chemical ionisation mass spectrometry [432]. High-performance liquid chromatography with UV detection at 250 nm has been used to determine imazapyr in methanol extracts of soil [433].

4.19.7
Pre-emergent Pesticides

Dacthal is a widely used pre-emergent herbicide that is applied to many crops for the control of annual weeds. Dacthal is typically applied to agricultural soils at 6–14 kg/ha [425]. In the soil environment, Dacthal transforms to mono- and diacid metabolites that are more water-soluble than the parent herbicide [426–428]. In eastern Oregon, where Dacthal is applied to onions, the diacid metabolite is the principal form of Dacthal detected in groundwater obtained from domestic wells [430].

To assess the fate of Dacthal applied to soil, both parent and metabolite forms in water and soil should be considered. While rapid methods exist for the determination of Dacthal and its metabolites in water [429, 430], quantitative and rapid methods are needed to determine Dacthal and its metabolites in soils, since conventional methods require large volumes of solvent and time to process the extract. For example, the conventional method for extracting Dacthal and its metabolites from soil requires 200 ml of 0.4 M HCl/acetone to extract a 20 g sample and the use of hazardous diazopropane to derivatise the acids to their ester forms [431].

Field and Monohan [430] sequentially extracted Dacthal and its mono- and diacid metabolites from soils by first performing a supercritical carbon dioxide extraction to recover Dacthal, followed by a subcritical (hot) water extraction step to recover metabolites. Dacthal was recovered from soil in 15 minutes by supercritical carbon dioxide at 150 °C and 400 bar. The mono- and diacid metabolites were extracted from soil in 10 minutes under the sub-

Table 4.10. Methods for the determination of miscellaneous herbicides (from author's own files)

Herbicide	Extraction method	Comments	Reference
Dichloronitrile paraquat	Toluene	Gas chromatography, limit of detection: μg/kg	[444]
Paraquat	–	Automatic continuous flow spectrophotometry, limit of detection: μg/m	[445–447]
Paraquat, trifluralin, diphenamid	–	Gas chromatography	[448]
Paraquat, diquat	Dichloromethane	Gas chromatography	[449]
Paraquat, diquat	–	Catalytic dehydrogenation then gas chromatography	[450–453]
Paraquat	–	Enzyme-linked immunoassay, limit of detection: 0.2 mg/kg	[454]
Acarol (isopropanol-4,4'-dibromobenzylate)	–	Gas liquid radio chromatography of ^{14}C herbicide	[455, 456]
Dicloram (4-amino-3,5,6-trichloropicolinic acid)	Ethyl ether	Pyrolysis then electron capture gas chromatography	[404]
Dicamba (2-methoxy-3,6-dichlorobenzoic acid)	–	High-resolution gas chromatography–mass spectrometry, limit of detection: low ppb	
Dicamba	Aminopropyl weak ion exchanger and C_{18} solid-phase extraction	High-performance liquid chromatography	[457]
Bromacil, lenticil, terbacil	Water extraction then chloroform extraction	Gas chromatography with NP detection, limit of detection: 20 μg/kg	[458]
Bromacil, lenticil, terbacil	Miscellaneous	Miscellaneous methods	[459–469]
Fluazifop-butyl, fluazifop	Methanol–hydrochloric acid, dichloromethane	Liquid chromatography, detection at 225 and 270 nm	[470, 471]
Fluazifop-butyl, fluazifop	–	Gas chromatography	[472]
Fluazifop-butyl, fluazifop	Phenyl- and cyano-bound silica gel solid-phase extraction column	Ion-pair high-performance liquid chromatography using phenyl columns	[473]
Diclofop-methyl, diclofop	–	Gas chromatography	[474, 475]

Table 4.10. (continued)

Herbicide	Extraction method	Comments	Reference
Diclofop-methyl, diclofop	Methanol:water: ethylacetate:acetic acid (40:40:19:1)	Conversion to pentafluo-robenzyl bromide derivative then gas chromatography	[476]
Frenock, sodium-2,2,3,3-tetrafluoroproprionate	Steam distillation, solvent extraction	Mass fragmentography of 1-benzyl-3-p-polytriazine derivative	[477]
Glyphosphate	0.1 M potassium hydroxide	High-performance liquid chromatography with post-column oxidation then derivatisation with o-phthaldehyde and 2-mercaptoethanol with fluorimetric detection	[478]
Cyperquat	–	Gas chromatography–mass spectrometry	[479]
Norflurazon	Methanol	C_{18} high-performance liquid chromatography with fluorescence detection (294, 398 nm)	[480]
Propanil (3,4-dichloro-propionaniline and 3,4-dichloroaniline)	–	Infrared spectroscopy and gas chromatography–mass spectrometry	[481]
Sencor (6-t-butyl, 1,2,4-triazine-3-methylthio-2-one)	–	Gas chromatography	[482]
Trifluralin, benefin	–	Electron capture gas chromatography, limit of detection: 50 pg absolute	[483]
Miscellaneous herbicides	–	Liquid chromatography–mass spectrometry	[484–486]
Miscellaneous herbicides	–	High-performance liquid chromatography	[487]
Miscellaneous herbicides	–	Gas chromatography–mass spectrometry	[488]
Miscellaneous herbicides	–	Thin-layer chromatography	[489]
Miscellaneous herbicides	–	Enzyme-linked immunoas-say	[490]

critical water conditions of 50 °C and 200 bar. The metabolites were trapped in situ on a strong anion-exchange disk placed over the exit frit of the extraction cell. Metabolites are combined with Dacthal by placing the disk into the gas chromatograph autosampler vial containing the supercritical fluid extract.

Table 4.11. Determination of various agrochemicals in soil (from author's own files)

Agrichemical	Extraction method	Comments	Reference
Isomethiozin	–	Differential pulse polarography, limit of detection: 40 ng/g	[523]
Trichlorphon	Solvent	Gas chromatography, limit of detection: 0.002 ppm	[524]
Bromoxynil, foxylnil	Solvent	Perfluroacetylation then gas chromatography with ion trap mass spectrometry	[525]
Toxaphene	–	Electron capture negative ion mass spectrometry then high-performance liquid chromatography and capillary gas chromatography	[526]
Dimethoate	–	Spectrophotometric flotation–dissolution reaction with molybdate and methylene blue and spectrophotometric finish	[527]
Bentazone	–	High-performance liquid chromatography with photodiode array detection	[528]
Dichlorobenil	Steam distillation	High-performance liquid chromatography	[529]
Chloropyritos metabolite	Supercritical fluid extraction and subcritical water extraction	–	[530]
Flumeton	CO_2 supercritical fluid extraction	–	[531]
Danjiami acaride and metabolites	Acid-base reflux petroleum ether extraction	Derivatisation with heptafluorobutyranilide, then gas chromatography with mass spectrophotometric detection	[532]
Hexazinone and metabolites	–	Clean-up then microcolumn capillary gas chromatography	[533]
Chlorpyrifos	–	Comparison of enzyme-linked immunoassay (ELISA) methods	[534]
Metalaxyl	–	Study of chiral separations to study microbial enantioselective degradation of metalaxyl in soil	[535]

The metabolites are then simultaneously eluted from the disk and derivatised to their ethyl esters by adding 100 µl of ethyl iodide and heating the vial at 100 °C for one hour. Using this approach, only a single sample is analysed, and because the disk-catalysed alkylation reaction does not transesterify Dacthal, the speciation of Dacthal is maintained. In addition, no sample clean-up steps are required, and only a total of 5 ml of nonchlorinated organic solvent is used.

4.19.8
Miscellaneous Insecticides/Herbicides

Further information on the determination of herbicides is shown in Table 4.10. The analysis of insecticide/herbicide mixtures in soil has been reviewed by various workers [279, 491–522]. The determination of other miscellaneous agrichemicals is shown in Table 4.11.

4.20
Fungicides

Caverly and Unwin [536] have described a rapid and sensitive technique for the determination of the residues of the fungicides furalaxyl and metalaxyl in soils. The soil sample is extracted with acetone in a Soxhlet apparatus and then the extract is analysed by gas chromatography using NP-selective detection. Recoveries are generally in excess of 80% with detection limits of 0.1 mg/kg.

Dieckmann et al. [537] assayed fenpropimorph fungicide and its main metabolite fenpropimorphic acid in soil using acetone–water extraction, partitioning with methylene chloride, gel permeation clean-up, methylation of the metabolite and gas chromatography with NP detection, and gas chromatography–mass spectrometry.

Celi et al. [538] determined fenoxaprop and fenoxapropethyl in soil by solvent extraction, clean-up on Florisil or alumina cartridge and high-performance liquid chromatography with UV detection at 280 nm.

Fentin, cyhexatin and fenbutatin oxide fungicides have been determined in soil by high-performance liquid chromatography with a CN column combined with UV photoconversion and post-column morin complexation followed by fluorescence detection [539].

Singh and Chiba [540] have reviewed methods for the determination of benomyl fungicide and its degradation products in soil by chromatography.

A spectrophotometric method has been described for determining down to 2 µg of dichloro-1,4-napthaquinone fungicide in soil based on the formation of a coloured reaction product with aniline [541]. The ^{13}C-labelled fungicide cyprodinil has been investigated using NMR spectroscopy [542].

4.21
Soil Fumigants

Kerwin et al. [543] determined methyl bromide soil fumigant by cryotrapping and electron capture gas chromatography. Down to 0.23 pM of methyl bromide could be detected using this procedure. Kerwin et al. [543] found levels of methyl bromide in the stratosphere and claimed that this contributed to ozone destruction.

Gan et al. [330] have reviewed the application of static headspace analysis to the determination of fumigants such as methyl bromide in soil.

References

1. US EPA (1979) *Methods for Chemical Analysis of Water and Wastes*, EPA 600/14-79/020, US Environmental Protection Agency, Washington, DC, USA.
2. WRCB (1988) *Leaking Underground Fuel Tank (LUFT) Field Manual*, State Water Resources Control Board, Sacramento, CA, USA.
3. ASTM (2005) *Annual Book of ASTM Standards, Volume 11.02*, ASTM, West Conshohocken, PA, USA.
4. Camel V, Tambuté A, Caude M (1993) *J Chromatogr* **642**:263.
5. Liang S, Tilotta DC (1998) *Anal Chem* **70**:616.
6. Janda V, Bartle KD, Clifford AA (1993) *J Chromatogr* **642**:283.
7. Hawthorn SB (1990) *Anal Chem* **62**:633A.
8. Eckert-Tilotta SE, Hawthorne SB, Miller DJ (1993) *Fuel* **72**:1015.
9. Hawthorne SB, Miller DJ, Hegvik KM (1993) *J Chromatogr Sci* **31**:26.
10. Hawthorne SB, Hegvik KM, Yang Y, Miller DJ (1994) *Fuel* **73**:1876.
11. Yang Y, Hawthorne SB, Miller DJ (1995) *J Chromatogr* **699**:265.
12. Schafer KH, Griffiths PR (1983) *Anal Chem* **55**:1939.
13. Jinno K, Saito M (1991) *Anal Sci* **7**:361.
14. Jenkins TJ, Kaplan M, Simmonds MR, Davidson G, Healy MA, Poliakoff M (1991) *Analyst* **116**:1305.
15. Heglund DL, Tilotta DC, Miller DJ, Hawthorne SB (1994) *Anal Chem* **66**:3543.
16. Burford MD, Hawthorne SB, Miller DJ (1994) *J Chromatogr* **685**:95.
17. Lou X, Janssen H, Cramers CA (1996) *J Chromatogr* **750**:215.
18. Yang Y, Hawthorne SB, Miller DJ (1995) *J Chromatogr* **699**:265.
19. Pastor A, Vazquez E, Ciscar R, dela Guardia M (1997) *Anal Chim Acta* **344**:241.
20. Current RW, Tilotta DC (1997) *J Chromatogr* **785**:269.
21. Hawari J, Halasz A, Beiruty A, Sas I Tra HV (1997) *Environ Anal Chem* **66**:299.
22. Morel G, Samhan O, Literathy P, Al-Hashash M, Moulin L, Saeed T, Al-Matrouk K, Martin-Bowyer M, Saber A (1991) *Fresen J Anal Chem* **339**:699.
23. Hazel G, Buchholz F, Aggarwal ID, Nau G, Ewing KJ (1997) *Appl Spectrosc* **51**:984.
24. Roe VD, Lacy MJ, Stuart JD, Robbins GA (1989) *Anal Chem* **61**:2584.
25. Parr JL, Walters G, Hoffman M (1991) Sampling and Analysis of Soils for Gasoline Range Organics. In: Kostecki PT, Calabrese EJ (eds) *Hydrocarbon Contaminated Soils and Groundwater: Analysis, Fate, Environmental and Public Health Effects, Remediation, Vol. 1*, Lewis, Ann Arbor, MI, USA, pp. 105–132.
26. Xie G, Barceelona MJ, Fang J (1999) *Anal Chem* **71**:1899.
27. Remmler M, Kopinke F-D, Stottmeister U (1995) *Thermochim Acta* **263**:101.
28. White DM, Luong H, Irvine RL (1998) *J Cold Reg Eng* **12**:1.
29. Peuron P, Daugherty S (1997) *J Contam Soils* **2**:449.
30. Ostendorf DW, Leach LE, Hinlein ES, Xie Y (1991) *Ground Water Monit Rem* **11**:107.
31. Robbins GA, Bristol RD, Roe VD (1989) *Ground Water Monit Rem* **9**:87.
32. Karasek FW, Charbonneau GH, Revel GJ, Tong HY (1987) *Anal Chem* **59**:1027.
33. Geerdink MJ, Erkelens C, Van Dam JC, Frank J, Luyben KCAM (1995) *Anal Chim Acta* **315**:159.
34. Kester PE (1987) *Analysis of Volatile Organic Compounds in Soils by Purge and Trap Gas Chromatography*, Tekmar Co., Cincinnati, OH, USA.

35. de Leeuw JW, de Leer EWB, Sinninghe Damste JS, Schuyl PJW (1986) *Anal Chem* 58:1852
36. Jones S (1993) Sampling and Analysis of Soils for Gasoline Range Organics. In: Kostecki PT, Calabrese EJ (eds) *Hydrocarbon Contaminated Soils and Groundwater: Analysis, Fate, Environmental and Public Health Effects, Remediation, Vol. 1*, Lewis, Ann Arbor, MJ, USA, I, II.
37. Ilias AM, Jagner C (1993) Sampling and Analysis of Soils for Gasoline Range Organics. In: Kostecki PT, Calabrese EJ (eds) *Hydrocarbon Contaminated Soils and Groundwater: Analysis, Fate, Environmental and Public Health Effects, Remediation, Vol. 1*, Lewis, Ann Arbor, USA, 3:147.
38. Karp KE (1993) *Ground Water Monitor Rem* 13:101.
39. Picer M, Hocenski V (1994) *Water Res* 28:619.
40. Greco TG, Grob RL (1990) *J Environ Sci Heal A* 25:185.
41. Eschenboch A, Kaesner M, Bierl R, Schater G, Makro B (1994) *Chemosphere* 28:683.
42. Schwartz J, Slater D, Larson TV, Pierson WE, Koening JQ (1993) *Am Rev Respir Dis* 147:826.
43. Zemanek MG, Pollard SJ, Kenefick SL, Hrudey SE (1997) *J Air Waste Manage Assoc* 47:1250.
44. Schuetzle D, Jensen TE, Ball JC (1985) *Environ Int* 11:169.
45. Schuetzle D, Siegle WO, Jensen TE, Dearth MA, Kaiser EW, Gorse R, Kreucher W, Kulik E (1994) *Environ Health Perspect* 102:3.
46. Westerholm RN, Alsberg TE, Frommelin AB, Strandell ME, Rannug U, Winquist L, Grigoriadis V, Egeback KE (1988) *Environ Sci Technol* 22:925.
47. Alsberg T, Stenberg U, Westerholm R, Strandell M, Rannug U, Sundcvall A, Romert L, Bernson V, Pettersson B, Toftgard R, Franzen B, Jansson M, Gustafsson JA, Egeback KE, Tejle G (1985) *Environ Sci Technol* 19:43.
48. Wild SR, Waterhouse KS, McGrath SP, Jones KC (1990) *Environ Sci Technol* 24:1706.
49. Trapido M (1999) *Environ Pollut* 105:67.
50. Hayden NJ, Voice TC, Wallace RB (1997) *J Contam Hydrol* 25:271.
51. Fine P, Graber ER, Yaron B (1997) *Soil Technol* 10:133.
52. Zemanek MG, Pollard SJT, Kenefick SL, Hrudey SE (1997) *Environ Pollut* 98:239.
53. Chuang JC, Pollard MA, Chou YL, Menton RG, Wilson NK (1998) *Sci Total Environ* 224:189.
54. Marr LC, Kirchstetter TW, Harley RA, Miguel AH, Hering SV, Hammond SK (1999) *Environ Sci Technol* 33:3091.
55. Szolar OHJ, Rost H, Brauan R Loibner AP (2002) *Anal Chem* 74:2379.
56. Codina G, Vaquero MT, Comellas L, Broto-Puig F (1994) *J Chromatogr* 673:21.
57. Lundstedt S, Van Bavel B, Haglund P, Tysklind M, Öberg L (2000) *J Chromatogr* 883:151.
58. Morlaes-Munoz S, Luque-Garcia JC, Luque De Castro MD (2002) *Anal Chem* 74:4213.
59. Ramos L, Vreuls JJ, Brinkman UAT (2000) *J Chromatogr* 891:275.
60. Saim M, Dean JR, Abdullah MP, Zakaria Z (1998) *Anal Chem* 70:420.
61. Hawthorne SB, Trembley S, Moniot CL, Grobanski CR, Miller DT (2000) *J Chromatogr A* 886:237.
62. Langenfeld JJ, Hawthorne SB, Miller DJ, Rawliszyn J (1993) *Anal Chem* 65:338.
63. Reindt S, Hcffler F (1994) *Anal Chem* 66:1808.
64. Barnabas JJ, Dean JR, Tomlinson WR, Owen SP (1996) *Anal Chem* 68:2064.
65. Tena MT, Luque de Castro MD, Valcarcel M (1996) *Anal Chem* 68:2386.
66. US EPA (1995) *Test Methods for Evaluating Solid Wastes, Method 3545*, Third Edition, Update 111, SW-846, US Environmental Protection Agency, Washington, DC, USA.
67. Ezzell JL, Richter BE, Felix WD, Black SR, Meikle JE (1995) *LC–GC Int* 13:390.

68. Richter BE, Jones BA, Ezzelll JL, Porter NL, Avdalovic N, Pohl C (1996) *Anal Chem* **68**:1033.
69. Dean JR (1996) *Anal Commun* **33**:191.
70. Meyer A, Kleiböhmer W (1993) *J Chromatogr* **657**:327.
71. Dankers J, Groenenboom M, Scholtis LHA, Van der Heiden C (1993) *J Chromatogr* **641**:357.
72. Burford MD, Hawthorne SB, Miller DT (1993) *Anal Chem* **65**:1497.
73. Lee HB, Peart TE, Hang-You RL, Gere RD (1993) *J Chromatogr* **653**:83.
74. Guo F, Li QX, Alcantara-Licudine JP (1999) *Anal Chem* **71**:1309.
75. Reindl S, Hoefler F (1994) *Anal Chem* **66**:1808.
76. Lopez GA, Blanco GE, Garcia AJI, Sanz-Medel A (1992) *Chromatographia* **33**:225
77. Nondek L, Kuzilek M, Krupicka S (1993) *Chromatographia* **37**:381.
78. Huelton Lain SH, Windruch J (1996) *Int J Environ Anal Chem* **63**:245.
79. Van de Nesse RJ, Hoogland GJM, De Moel JJM, Gooljee C, Brinkma UAT, Velhorst NH (1991) *J Chromatogr* **552**:613.
80. Fernandez-Perez V, Luque de Castro MD (2000) *J Chromatogr A* **902**:357.
81. Krahn MM, Ylitalo GM, Buzitis J, Chan SL, Varanasi U, Wade TL, Jackson TJ, Brooks JM, Wolfe DA, Manen CA (1993) *Environ Sci Technol* **27**:699.
82. Robbat A, Liu TL, Abraham BM (1992) *Anal Chem* **64**:1477.
83. Hartman R (1996) *Int J Environ Anal Chem* **62**:161.
84. Dale MJ, Jones Ac, Pollard SJT, Longridge-Smith PRR, Rowley AG (1993) *Environ Sci Technol* **27**:1693.
85. Rodgers RP, Lazar Ac, Reilly PTA, Whitten WB, Ramsey JM (2000) *Anal Chem* **72**:5040.
86. McDonald PP, Almond RE, Mapes JP, Friedman SB (1994) *J AOAC Int* **77**:466.
87. Brüggemann D, Freitag R (1995) *J Chromatogr* **717**:309.
88. Lai JK, Filseth SV, Sadowski CM, Morgan F (1990) *Int J Environ Anal Chem* **40**:99.
89. Bublitz J, Christopherson A, Schade W (1996) *Fresen J Anal Chem* **355**:684.
90. Shimizu Y, Lillsetroud HM (1991) *Water Sci Technol* **23**:427.
91. Lopez-Avila V, Young R, Beckert WF (1994) *Anal Chem* **66**:1097.
92. Medina-Vera (1996) *J Appl Pyrol* **36**:27.
93. Sparrevik M, Jonassen H (1995) *Soil Environ* **5**:537.
94. Hudak RT, Melby JM, Stave JW (1994) In: *Proc 87th Annual Meeting of the Air and Waste Management Association*, Paper No. 94, 19–24 June, Cincinnati, OH, USA.
95. Fowlie PJA, Bulman TL (1986) *Anal Chem* **58**:721.
96. Brown RS, Luang JHT, Szolar DHJ (1996) *Anal Chem* **68**:287.
97. Luterman C, Dott W, Hollender J (1998) *J Chromatogr* **811**:151.
98. Lichtfouse E, Budzinski H, Garrigues P, Eglington TI (1997) *Org Geochem* **26**:353.
99. Hankin SM, John P, Simpson AW, Smith GP (1996) *Anal Chem* **68**:3235.
100. Niederer M (1998) *Environ Sci Pollut Res Int* **5**:209.
101. Meyer S, Cartellieri S, Steinhart H (1999) *Anal Chem* **71**:4023.
102. Martin JH, Siebert AM, Loehr RC (1991) *J Environ Chem* **117**:291.
103. US EPA (1979) *Toxic Substances Control Act*, US Environmental Protection Agency, Washington, DC, USA.
104. Hawthorne SB, Miller DJ (1994) *Anal Chem* **66**:4005.
105. Reighard TS, Olesik SV (1996) *Anal Chem* **68**:3612.
106. Futter JE, Wall P (1993) *J Planar Chromatogr* **6**:372.
107. Hawthorne SB, Yang Y, Miller DJ (1994) *Anal Chem* **66**:2912.
108. Dean JR, Santamaria-Rekondo A, Ludkin E (1996) *Anal Commun* **33**:413.
109. Fisher JA, Scarlett MJ, Stott AD (1997) *Environ Sci Technol* **31**:1120.
110. Llompart MP, Lorenzo RA, Cela R, Pare JR (1997) *Analyst* **122**:133.

111. Hottenstein CS, Jourdan SW, Hayes MC, Rubio FM, Herzog DP, Lawruk TS (1995) *Environ Sci Technol* **29**:2754.
112. Voznakova Z, Podehradska J, Kohlickova M (1996) *Chemosphere* **33**:285.
113. Crespin MA, Gallego M, Valcarcel M (1999) *Anal Chem* **71**:2687.
114. Llopart-Vizoso MP, Lorenzo-Ferreira RA, Cela-Torrijos R (1996) *J High Res Chromatogr* **19**:207.
115. Danis TG, Albanis TA (1996) *Toxicol Environ Chem* **53**:9.
116. Talsky G (1983) *Int J Environ Anal Chem* **14**:81.
117. Langbehn A, Steinhart H (1994) *J High Res Chromatogr* **17**:293.
118. Mendoza CA, Frind EO (1990) *Water Resour Res* **26**:379.
119. Little JC, Daisey JM, Nazaroff WW (1992) *Environ Sci Technol* **26**:2058.
120. Kotiainen R (1995) *Atmos Environ* **29**:693.
121. Janku J, Kulies V, Machakova Z, Kuras M (1994) *Fresen Environ Bull* **3**:345.
122. Hawthorne SB, Miller DJ, Lagenfeld JJ (1990) *J Chromatogr Sci* **26**:2.
123. Hewitt AD, Miyares PH, Leggett DC, Jenkins TF (1992) *Environ Sci Technol* **26**:1932.
124. Milana MR, Maggio A, Denaro M, Feliciani R, Gramiccioni L (1991) *J Chromatogr* **552**:205.
125. Maggio A, Milana MR, Denaro M, Feliciani R, Gramiccioni L (1991) *J High Res Chromatogr* **14**:618.
126. Pavlostathis SG, Mathavan GN (1992) *Environ Technol* **13**:23.
127. Stuart JD, Miller ME, Williams-Burnett ML (1997) *J Soil Contam* **6**:439.
128. Kawata K, Tanabe A, Saitos S, Sakai M, Yasuhara A (1997) *Bull Environ Contam Toxicol* **58**:893.
129. Hewitt AD (1998) *Environ Sci Technol* **32**:143.
130. Papaefsthathicn E, Luque de Castro MD (1997) *J Chromatogr* **779**:352.
131. James KJ, Stack MA (1996) *J High Res Chromatogr* **19**:515.
132. Askari MDF, Maskarinec MP, Smith SM, Beam DM, Travis CC (1991) *Anal Chem* **68**:3431.
133. Hiatt MH (1995) *Anal Chem* **67**:4044.
134. Bianchi AP, Varney MS, Phillips J (1991) *J Chromatogr* **542**:413.
135. Yokouchi Y, Sano M (1991) *J Chromatogr* **555**:297.
136. Krock KA, Wilkins CL (1996) *J Chromatogr* **726**:167.
137. Barrio ME, Lliberia JL, Comellas L, Broto-Puig F (1996) *J Chromatogr* **719**:131
138. Drodz J, Novák J (1979) *J Chromatogr* **165**:141.
139. Voice TC, Kolb B (1993) *Environ Sci Technol* **27**:70.
140. Kurán P, Soják L (1996) *J Chromatogr A* **733**:119.
141. Kostiainen R, Kotiaho T, Mattila I, Mansikka T, Ojala M, Ketola RA (1996) *Anal Chem* **70**:3028.
142. Budzinski H, Letellier M, Thompson S, LeMenach R, Garriques P (2000) *Fresen J Anal Chem* **367**:165.
143. Yang J, Her J-W (1999) *Anal Chem* **71**:4690.
144. Hewitt AD (1996) *Volatile Organic Compounds in the Environment*, ASTM STP 1261, ASTM, Washington, DC, USA, p. 170–180.
145. Hewitt AD (1996) *Volatile Organic Compounds in the Environment*, ASTM STP 1261, ASTM, Washington, DC, USA, p. 181–191.
146. Liikala TL, Olsen KB, Teel SS, Lanigan DC (1996) *Environ Sci Technol* **30**:3441.
147. Ball WP, Xia G, Durfee DP, Wilson RD, Brown MJ, Mackey DM (1997) *Ground Water Monit Rem* **17**:104.
148. Deetman AA, Demeulemeester P, Garcia M (1976) *Anal Chim Acta* **82**:1.

149. Neumayr V (1986) In: Milde G, Leschber R (eds) *Soil and Groundwater Protection*, Gustav Fischer Verlag, Stuttgart, Germany, p. 65–84.
150. De Leon IR, Maberry MA, Overton EB, Raschke CK, Remele PC, Steele CF, Laseter JL (1980) *J Chromatogr Sci* **18**:85.
151. Kerfoot HB (1987) *Environ Sci Technol* **21**:1022.
152. Mehran M, Olsen R, Rector BM (1987) *Groundwater* **25**:275.
153. Bowadt S, Hawthorne SB (1995) *J Chromatogr A* **703**:549
154. Deuster R, Lubahn N, Friedrich C, Kleiböhmer W (1997) *J Chromatogr A* **785**:227.
155. Hawthorne Sb, Grabanski CB, Hageman KJ, Miller DJ (1998) *J Chromatogr A* **814**:151.
156. Arthur CL, Pawliszyn J (1990) *Anal Chem* **62**:2145.
157. Louch D, Moltagh S, Pawliszyn J (1992) *Anal Chem* **64**:1187.
158. Zhang Z, Yang MJ, Pawliszyn J (1994) *Anal Chem* **66**:844A.
159. Yang J, Her J-W (2000) *Anal Chem* **72**:878.
160. Tavendale MH, Wilkins Al, Langdon AG, Mackie KL, Stuthridge T, McFarlane PN (1995) *Environ Sci Technol* **29**:1407.
161. Passivirta J, hakaal H, Knuutinen J, Otollinene T, Särkkä J, Welling L, Paukku R, Lammi R (1990) *Chemosphere* **21**:1355.
162. Alonso MC, Puig D, Silgoner I, Grasserbauer M, Barcelo D (1998) *J Chromatogr A* **823**:231.
163. Höfler F, Ezzell J, Richter B (1995) *LaborPraxis* **4**:58.
164. Santos FJ, Jáuregui O, Pinto FJ, Galceran MT (1998) *J Chromatogr A* **823**:249.
165. Yang Y, Belghazi M, Lagadec A, Miller DJ, Hawthorne SB (1998) *J Chromatogr A* **810**:149.
166. Hageman KJ, Mazeas L, Grabansky CB, Miller DJ, Hawthorne SB (1996) *J Chromatogr A* **68**:3892.
167. Wennrich L, Popp D, Möder M (2000) *Anal Chem* **72**:546.
168. Stark A (1969) *J Agric Food Chem* **17**:871.
169. Renberg L (1974) *Anal Chem* **46**:459.
170. Lopez-Avila V, Hirata P, Kroska S, Flanagan M, Taylor JH, Hern SC (1985) *Anal Chem* **57**:2797.
171. Kimbrough DE, Chin R, Wakakuwa J (1994) *Analyst* **119**:1283.
172. Kimbrough DE, Chin R, Wakakuwa J (1994) *Analyst* **119**:1277.
173. Lopez-Avila V, Benedicto J, Charon C, Young R, Beckert VF (1995) *Environ Sci Technol* **29**:2709.
174. Llompart M, Li K, Fingas MF (1999) *J Microcolumn Sep* **11**:397.
175. Yang Y, Bowadt S, Hawthorne SB, Miller PJ (1995) *Anal Chem* **67**:4571.
176. Hawthorne SB, Lagenfeld JJ, Miller JJ Burford MD (1992) *Anal Chem* **64**:1614.
177. Fuoco R, Griffiths PR (1992) *Ann Chim (Rome)* **82**:235.
178. Brady BO, Kao CC, Doolan KM (1987) *Ind Eng Chem* **26**:261.
179. Von Bavel B, Järemo M, Karlsson L, Lindström G (1996) *Anal Chem* **68**:1279.
180. Abraham BM, Liu TY, Robbat A (1993) *Hazard Waste Hazard* **10**:461.
181. Robbat A, Liu TY, Abraham BM (1992) *Anal Chem* **64**:358.
182. Alford-Stevens AL, Eichelbergr JW, Budde WL (1988) *Environ Sci Technol* **22**:304.
183. Benicka E, Novakovsky R, Hronzek J, Krupsik J, Soudra D, De Zeeuw J (1996) *J High Res Chromatogr* **19**:95.
184. Jensen G, Renberg L Reutergard L (1977) *Anal Chem* **49**:316.
185. Southwest Water Laboratory (1971) *Sediment Extraction Procedures Method No .SP 8/71*, Southwest Water Laboratory, Athens, GA, USA.
186. Chiarenzelli J, Scrudato R, Arnold G, Wunderlich M, Rafferty D (1996) *Chemosphere* **33**:899.

187. Teichman J, Bevenue A, Hylin JW (1978) *J Chromatogr* **151**:155.
188. Lopshire RF, Watson JT, Enke CG (1996) *Toxicol Ind Health* **12**:375.
189. Glausch A, Blanch GP, Schurig V (1996) *J Chromatogr A* **723**:399.
190. Johnson JC, Van Emon JM (1996) *Anal Chem* **68**:162.
191. Schuetz AJ, Weller MG, Niessner R (1999) *Fresen J Anal Chem* **363**:777.
192. Pullen S, Haiber G, Schöeler HH, Hock B (1996) *Int J Environ Anal Chem* **65**:127.
193. Vo-Dihn T, Pal A, Pal T (1994) *Anal Chem* **66**:1264.
194. Watts W, Pal A, Ford L, Miller GH, Vo-Dinh T, Eastwood D, Liberg R (1992) *Appl Spectrosc* **46**:1235.
195. Klingston PW, Henry MA, Aldrin KJ, Pryde SD (1994) *J Chromatogr B* **632**:383.
196. Alcock RE, Halsall CJ, Harris CA, Johnston AE, Lead WA, Sanders G, Jones KC (1994) *Environ Sci Technol* **28**:1838.
197. Hellman H (1985) *Deutsch Gewässer Kundlich Mitteilungen* **29**:111.
198. Yu Ma C, Bayne CK (1993) *Anal Chem* **65**:772.
199. Tucker RE, Young AL, Gray AP (eds) (1983) *Human and Environmental Risks of Chlorinated Dioxins and Related Compounds*, Plenum, New York, NY, USA.
200. Kimbrough RD (ed) (1980) *Halogenated Biphenyls, Terphenyls, Naphthalenes, Dibenzodioxins and Related Products*, Elsevier, Amsterdam, The Netherlands.
201. Hutzinyer O, Frei RW, Merian E, Pocchiara F (eds) (1982) *Chlorinated Dioxins and Related Compounds. Impact on the Environment*, Pergamon, New York, USA.
202. Nicolson WJ, Moore JA (eds) (1979) *Health Effects of Halogenated Aromatic Hydrocarbons*, New York Academy of Sciences, New York, USA.
203. Lee DHK, Falk HL (eds) (1973) *Environmental Health Perspectives Experimental Issue No 5*, US Department of Health, Education and Welfare, Publication No. (NIH) 74-218, September.
204. Huff JR, Moore JA, Saracci DR, Tomatis L (1980) *Environ Health Perspect* **36**:221–240.
205. Rappe C, Buser HR, Bosshardt HP (1979) In: Nicolson WJ, Moore JA (eds) *Health Effects of Halogenated Aromatic Hydrocarbons*, New York Academy of Sciences, New York, USA, p. 1–18.
206. Esposito MP, Tiernan TO, Dryden FE (1980) *Dioxins*, US EPA Report No. EPA-600/2-80-197, US EPA Washington, DC, USA.
207. Olie K, Vermeullen PL, Hutzinger O (1977) *Chemosphere* **6**:455.
208. Ahling B, Lindskog A, Jansson B, Sundstrom G (1977) *Chemosphere* **8**:461.
209. Buser HR, Bosshardt HR, Rappe C, Lindahl R (1978) *Chemosphere* **7**:419.
210. Crosby DG, Wong AS (1976) *Chemosphere* **5**:327.
211. Lamparski LL, Stehl RH, Johnson RL (1980) *Environ Sci Technol* **14**:196.
212. McConnell FE (1980) In: Kimbrough RD (ed) *Halogenated Biphenyls, Terphenyls, Naphthalenes, Dibenzodioxins and Related Products*, Elsevier, Amsterdam, The Netherlands, pp. 109–150.
213. Goldstein JA (1980) In: Kimbrough RD (ed) *Halogenated Biphenols, Terpyhenyls, Napthalenes. Dibenzodioxins and Related Products*, Elsevier, Amsterdam, The Netherlands, pp. 151–190.
214. Di Domenico A, Vivcano G, Zapponi G (1982) In: Hutzinger O, Frei RW, Merian E, Pocchiari F (eds) *Chlorinated Dioxins and Related Compounds – Impact on the Environment*, Pergamon, New York, USA, pp. 105–114.
215. Ward CT, Matsumura F (1978) *Arch Environ Contam Toxicol* **7**:349.
216. Young AL (1983) In: Tucker RE, Young AL, Gray AP (eds) *Human and Environmental Risks of Chlorinated Dioxins and Related Compounds*, Plenum, New York, USA, pp. 173–190.

217. Bickel MH, Muhlback S (1982) In: Hutzinger O, Frei RW, Merian E, Pocchiari F (eds) *Chlorinated Dioxins and Related Compounds – Impact on the Environment*, Pergamon, New York, USA, pp. 303–306.
218. Decad GM, Birnbaum LS, Matthews S (1982) In: Hutzinger O, Frei RW, Merian E, Pocchiari F (eds) *Chlorinated Dioxins and Related Compounds – Impact on the Environment*, Pergamon, New York, USA, pp. 307–315.
219. Isensee AR (1978) *Ecol Bull* **27**:255.
220. Masuda Y, Kuroki H (1980) In: Kimbrough RD (ed) *Halogenated Biphenyls, Terphenyls, Naphthalenes, Dibenzodioxins and Related Compounds*, Elsevier, Amsterdam, The Netherlands, pp. 561–569.
221. Walters RW, Guiseppi-Elie A (1988) *Environ Sci Technol* **22**:819.
222. Onuska FI, Terry KA (1989) *J High Res Chromatogr* **12**:357.
223. Richter BE, Ezzell JL, Knowles DE, Hoefler F, Mattulat AKR, Scheutwinkel M, Waddell DS, Gobran T, Khurana V (1997) *Chemosphere* **34**:975.
224. Kjeller LO, Kulp SE, Jonsseon B, Rappe C (1993) *Toxicol Environ Chem* **39**:1.
225. Hengstmann R, Haman R, Weber H, Keltrup A (1989) *Fresen J Anal Chem* **335**:982.
226. Garner FE, Homsher MT, Pearson JG (1986) *ASTM STP* **925**:132.
227. Tong HY, Monson SJ, Gross ML, Huang LQ (1991) *Anal Chem* **63**:2697.
228. Smith LM, Stalling DL, Johnson JL (1984) *Anal Chem* **56**:1830.
229. Di Domenico A, Merli F, Bonforti L, Camoni I, Di Muccio A, Taggi F, Vergori L, Colli G, Elli G, Gorni A, Grassi P, Invernizzi G, Jemma A, Luciani L, Cattabeni F, De Angelis L, Galli G, Chiabrando C, Fanelli R (1979) *Anal Chem* **51**:735.
230. Wellington TJ, Schneider WF, Worsnop DR, Nielsen OJ, Schested J, Debruyn WJ, Shorter JA (1994) *Environ Sci Technol* **28**:320A.
231. Bowden DJ, Clegg SL, Brimblecombe P (1996) *Chemosphere* **32**:405.
232. Kotamarthi VR, Rodriguez JM, Ko MKW, Tromp TK, Sze ND, Prather M (1998) *J Geophys Res* **103**:5747.
233. Tromp TK, Ko MKW, Rodriguez JM, Sze ND (1995) *Nature* **376**:327.
234. Schwazbach SE (1995) *Nature* **376**:297.
235. Emptage M, Tabinowski J, Odom JM (1997) *Environ Sci Technol* **31**:732.
236. Boutonnet JC, Bingham P, Calmari D, de Rooij C, Franklin J, Kawano T, Libre J, McCulloch A, Malinverno G, Odom JM, Rusch GM, Smythe K, Sobolev I, Thompson R, Tiedje JM (1999) *Hum Ecol Risk Assess* **5**:59.
237. Jordan A, Frank H (1999) *Environ Sci Technol* **33**:522.
238. Wujcik CE, Cahill TM, Seiber JN (1999) *Environ Sci Technol* **33**:1747.
239. Frank H, Klein A, Renschen D (1996) *Nature* **382**:34.
240. Forrest C (ed) (1994) *Alternative Fluorocarbons Environmental Acceptability Study (AFEAS) Workshop on the Environmental Fate of Trifluoroacetic Acid*, AFEAS, Washington, DC, USA.
241. Cahill TM, Benesch JA, Gustin MS, Zimmerman ZJ, Seiber JN (1999) *Anal Chem* **71**:4465.
242. Oostdyk TS, Grob RL, Snyder JL, McNally ME (1993) *Anal Chem* **65**:596.
243. Peters RJ, van Renesse von Duivenbode JAD (1994) *Fresen J Anal Chem* **348**:249.
244. Brumley WC, Brownrigg CR, Brilis GM (1991) *J Chromatogr* **558**:223.
245. Kido A, Shinohara R, Eto S, Koga M, Hori T (1979) *Jpn J Water Pollut Res* **2**:245.
246. Krone CA, Burrows DW, Brown DW, Robisch PA, Friedman AJ, Malins DC (1986) *Environ Sci Technol* **20**:1144.
247. Tsukiokai T (1988) *Analyst* **113**:193.
248. De Geus H, Zegers BN, Lingeman H, Brinckman UAT (1994) *Int J Environ Anal Chem* **56**:119.

249. Hewitt AD, Jenkins TF, Ranney TA (2001) *Field Anal Chem Technol* **5**:228.
250. Walsh ME, Jenkins TF, Thorne PG (1995) *J Energy Mater* **13**:357.
251. Wormhoudt J, Shorter JH, McManus JB, Kebabian PL, Zahniser MS, Kolb CE, Davis WM, Cespedes ER (1996) *Appl Opt* **35**:3992.
252. Sylvia JM, Janni JIA, Klein JD, Spencer KM (2000) *Anal Chem* **72**:5834.
253. Albert KJ, Myrick ML, Brown SB, James DL, Milanovich FD, Walt DR (2002) *Environ Sci Technol* **35**:3193.
254. Emery AP, Chesler SN, MacCrehan WA (1992) *J Chromatogr* **606**:221.
255. Surugiu I, Svitel J, Ye L, Haupt K, Danielsson B (2001) *Anal Chem* **73**:4388.
256. Gauger PR, Holt D, Patterson CH, Charles PT, Shriver-Lake L, Kusterbeck AW (2001) *J Hazard Mater* **83**:51.
257. Grant CL, Jenkins TF, Myers KF, McCormick EF (1995) *Environ Toxicol Chem* **14**:1865.
258. Garlucci G, Airoldi L, Fanelli R (1984) *J Chromatogr* **287**:425.
259. David MJ, Seiber JN (1996) *Anal Chem* **68**:3038.
260. Ingram JC, Groenewald GS, Appelhans AD, Dahl DA, Delmore JE (1996) *Anal Chem* **68**:1309.
261. Jacobs LW, Chou S-F, Tiedje JM (1976) *J Agric Food Chem* **24**:1198.
262. Norstrom A, Andersson K, Rappe C (1976) *Chemosphere* **4**:255.
263. Nowack B, Kari FG, Hilger SU, Sigg L (1996) *Anal Chem* **68**:561.
264. Saar RA, Weber JH (1980) *Anal Chem* **52**:2095.
265. Wilson MA, Vassallo AM, Perduc EM, Reuter JH (1987) *Anal Chem* **59**:551.
266. Weber JH, Wilson SA (1975) *Water Res* **9**:1079.
267. Okbunu I, Higgins A (1977) *Bull Environ Contam Toxicol* **18**:428.
268. Golet EM, Strehler A, Alder AC, Giger W (2002) *Anal Chem* **74**:5455.
269. Straub TM, Pepper IL, Abbaszadegon M, Gerba CP (1994) *Appl Environ Microbiol* **60**:1014.
270. Wells DE, Hess F (2000) *Tech Instrum Anal Chem* **21**:73.
271. Crouch MS (1990) *Chem Eng Prog* **86**:41.
272. Di Domenico A, De Filip E, Ferri F, Iacovella N, Miniero R, di Tella ES, Tafani P, Baldassarri LT (1992) *Microchem J* **46**:48.
273. Tognotti L, Flytzani-Stephanopaulos M, Sarofim AF, Kopsinis H, Stankides M (1991) *Environ Sci Technol* **25**:104.
274. Schulton HR, Sorge C (1995) *Eur J Soil Sci* **46**:567.
275. Brady BO, Kao CC, Dooley KM (1987) *Chem Res* **26**:261.
276. Snyder JL, Grob RL, McNally ME, Oostdyk TS (1992) *Anal Chem* **64**:1940.
277. Wong JM, Li QX, Hammock BD, Seiber JN (1991) *J Agric Food Chem* **39**:1802.
278. Snyder JL, Grob RL, McNally ME, Oostdyke TS (1993) *J Chromatogr Sci* **31**:183.
279. Snyder JL, Grob RL, McNally ME, Oostdyke TS (1992) *Anal Chem* **64**:1940.
280. Schantz MM, Bowadt S, Benner BA Jr, Wise SA, Hawthorne SB (1998) *J Chromatogr A* **816**:213.
281. Brumley WC, Latorre E, Kelliher V, Marcus A, Knowles DE (1998) *J Liq Chromatogr R T* **21**:1199.
282. Novikova KE, Zhur USE (1972) *Khim Obstich* **18**:562.
283. Johnson RE, Starr RI (1972) *J Agric Food Chem* **20**:48.
284. Chiba M, Morley HV (1968) *J Agric Food Chem* **16**:916.
285. Mangani F, Crescentini G, Bruner F (1981) *Anal Chem* **53**:1627.
286. Hartonen K, Bowadt S, Hawthorne SB, Riekkola M-L (1997) *J Chromatogr* **774**:229.
287. Steinwandter H (1987) *Fresen Z Anal Chem* **327**:309.
288. Miller JM, Singh J (1976) *Bull Environ Contam Toxicol* **16**:483.
289. Cooke BK, Western NM (1980) *Analyst* **105**:490.

290. Goodwin ES, Goulden R, Reynolds JG (1961) *Analyst* **86**:697.
291. Stringer A, Pickard JA, Lyons CH (1974) *Pestic Sci* **5**:587.
292. Mills PA, Bong BA, LaVerna RK, Burke JA (1972) *J Assoc Off Anal Chem* **55**:39.
293. Deubert KH (1970) *Bull Environ Contam Toxicol* **5**:379.
294. Hesselberg RJ, Johnson JL (1972) *Bull Environ Contam Toxicol* **7**:115.
295. Woodham DW, Loftis CD, Collier CW (1972) *J Agric Food Chem* **20**:163.
296. Mahel'ova H, Sackmanereva M, Szokolav A, Kovac J (1974) *J Chromatogr* **89**:177.
297. Gooding PH, Philip HG, Tawnk HS (1972) *Bull Environ Contam Toxicol* **7**:288.
298. Suzuki M, Yaomoto Y, Wanatabe Y (1977) *Environ Sci Technol* **11**:1109.
299. Suzuki M, Yaomoto Y, Wanatabe Y (1973) *Nippon Nogei Kagaku Kaishi* **47**:1.
300. Gambrell RP, Reddy CN, Collard V, Green G, Patrick WH Jr (1984) *J Water Pollut Cont* **56**:174.
301. Suzuki M, Morimoto M (1986) *J High Res Chromatogr Chromatogr* **9**:692.
302. Pyle S, Marcus AB (1997) *J Mass Spectrom* **32**:897.
303. Mori H, Kobayashi M, Yagi K, Takahasi M, Gondu T (1987) *Nippon Noyaku Gakkaishi* **12**:491.
304. McGarvey BD (1993) *J Chromatogr* **642**:89.
305. Okamoto HS, Wijekoon D, Esperaza CE, Chang JC, Park SL (1992) *Water Test Quality Assurance*, Vol. 3, ASTM STP 1075, ASTM, West Conshohocken, PA, USA.
306. Singhal JP, Khan S, Bansal OP (1978) *Analyst* **103**:872.
307. Lawrence JF, Frei RW (1972) *Anal Chem* **41**:2046.
308. Bilikova A, Kuthan A (1983) *Vodni Hospodarstvi Ser B* **33**:215.
309. Leppert BC, Markle JC, Helt RC, Fujie GH (1983) *J Agric Food Chem* **31**:220.
310. Fung KKH (1976) *Pestic Sci* **7**:571.
311. Coburn JA, Ripley BD, Chau ASY (1976) *J Assoc Off Anal Chem* **59**:188.
312. Bromilow RH (1976) *Analyst* **101**:982.
313. Westlake WE, Morika I, Gunther FA (1972) *Bull Environ Contam Toxicol* **8**:109.
314. Reeves RG, Woodham DW (1974) *J Agric Food Chem* **22**:76.
315. Snyder JL, Grob RL, McNally ME, Oostdyk TS (1992) *Anal Chem* **64**:1940.
316. Singh Sb, Kulshrestha G (1993) *J Chromatogr* **637**:109.
317. Robbat A Jr, Liu C, Liu T-Y (1992) *J Chromatogr* **625**:277.
318. Kjolholt J (1985) *J Chromatogr* **325**:231.
319. Skladal P, Fiala I, Erejei J (1996) *Int J Environ Anal Chem* **65**:139.
320. Levy JM, Dolata L, Ravey RM, Storozynsky E, Holowczak KA (1993) *J High Res Chromatogr* **16**:368.
321. Stent SJ, Babbit BW, Da Lunha AR, Safarpour MM (1998) *J AOAC Int* **81**:1054.
322. Tuguka AM, Sarna LP, Webster GRB (1997) *Int J Environ Anal Chem* **68**:137.
323. Silgoner I, Krska R, Lombas E, Gans O, Rosenberg E, Grasserbauer M (1998) *Fresen J Anal Chem* **362**:120.
324. Miellet A (1986) *Ann Falsif Expert Chim* **79**:245.
325. Lopez-Avila V, Young R, Beckert WF (1993) *J AOAC Int* **76**:864.
326. Redondo MJ, Ruiz MT, Bolunda R, Font G (1993) *Chromatographia* **36**:187.
327. Driscoll JN, Atwood SE (1993) *J Chromatogr* **642**:435.
328. Sanchez-Rasero F, Matallo MB, Dios G, Romero E, Pena A (1998) *J Chromatogr A* **799**:355.
329. Marek Le E, Koskinen WC (1997) *Am Environ Lab* **9**:26
330. Gan J, Ppapiernik S, Yates SR (1998) *J Agric Food Chem* **46**:986.
331. Sanchez-Brunete C, Perez RA, Miguel E, Tadeo JL (1998) *J Chromatogr A* **823**:17.
332. Perez RA, Sanchez-Brunete C, Miguel E, Tadeo JL (1998) *J Agric Food Chem* **46**:1864.
333. Puig D, Barcelo D (1994) *J Chromatogr A* (1994) **673**:55.

334. Cohen IK, Wheals RB (1969) *J Chromatogr* **43**:233.
335. McGarvey BD, Chiba M, Broadbent AB (1986) *J AOAC Int* **62**:852.
336. Linly LY, Cooper WTJ (1987) *J Chromatogr* **390**:285.
337. Bushway RJ, Fan TS, Young BES, Paradis LR, Perkins LB (1994) *J Agric Food Chem* **42**:1138.
338. Roda A, Rauch P, Ferri P, Girotti S, Ghini S, Carrea G, Bovara R (1994) *Anal Chim Acta* **294**:35.
339. McKone CE (1969) *J Chromatogr* **44**:60.
340. Kahn SU, Greenhalgh R, Cockrane WP (1975) *Bull Environ Contam Toxicol* **13**:602.
341. Kirkland JJ (1962) *Anal Chem* **34**:428.
342. Lokke H (1974) *Pestic Sci* **5**:749.
343. Friestead HO (1974) *J AOAC Int* **57**:221.
344. Parouchais C (1973) *J AOAC Int* **56**:831.
345. Onley JH, Yip G (1971) *J AOAC Int* **54**:1366.
346. Farrington DS, Hopkins RG, Ruzicka JHA (1977) *Analyst* **102**:377.
347. Smith AE, Lord KA (1975) *J Chromatogr* **107**:407.
348. Cotterill EG (1980) *Analyst* **105**:987.
349. Sidwell JA, Ruzicka JHA (1976) *Analyst* **101**:111.
350. Lawrence JF (1976) *J AOAC Int* **59**:1066.
351. Byast TH (1977) *J Chromatogr* **134**:216.
352. Pribyl J, Herzel F (1978) *J Chromatogr* **166**:272.
353. Katz SE, Strusz RF (1968) *Bull Environ Contam Toxicol* **3**:258.
354. Caverly DJ, Denney RC (1978) *Analyst* **103**:368.
355. Spengler D, Hamroll B (1970) *J Chromatogr A* **49**:205.
356. Jarczyk HJ (1975) *Pflanzenschutz-Nachr Bayer* **28**:334.
357. Lawrence JF, Laver GJ (1975) *J Agric Food Chem* **23**:1106.
358. Bieser H, Grolimmund K (1974) *J AOAC Int* **57**:1294.
359. Grossbard E, Marsh JAP (1974) *Pestic Sci* **5**:609.
360. Sheets TJ (1964) *J Agric Food Chem* **12**:30.
361. Bartha R (1971) *J Agric Food Chem* **19**:385.
362. Geissbuhler H (1969) In: Kearney PC, Kaufman DD (eds) *Degradation of Herbicides*, Marcel Dekker, New York, USA, p. 79.
363. Burge WD (1972) *Soil Biol Biochem* **4**:379.
364. McNally MEP, Wheeler JR (1988) *J Chromatogr* **435**:63.
365. Klaferbach P, Holland PJ (1993) *J Agric Food Chem* **41**:396.
366. Klafferbach P, Holland PJ (1993) *Mass Spectrom* **22**:565.
367. Mills MS, Thurman EM (1992) *Anal Chem* **64**:1985.
368. Di Corcia A, Caracciolo AB, Crescenzi C, Guilano G, Murtas S, Smperi R (1999) *Environ Sci Technol* **33**:3278.
369. Stransky Z (1983) *J Chromatogr* **320**:219.
370. Bushway RJ, Perkins SP, Savage SA, Ferguson BS (1988) *Bull Environ Contam Toxicol* **40**:647.
371. Lawruk TS, Lachman CE, Jourdan SW, Fleeker JR, Herzog DP, Rubio FM (1993) *J Agric Food Chem* **41**:747.
372. Wittman C, Schmidt RD (1994) *J Agric Food Chem* **42**:1041.
373. Dankwardt A, Seifert J, Hock B (1993) *Acta Hydrochim Hydrobiol* **21**:110.
374. Dankwardt A, Hock B (1993) *GIT Fachz Lab* **37**:839.
375. Lopez-Avila V, Charan C, Beckert WF (1994) *Trends Anal Chem* **13**:118.
376. Martinez-Galera M, Martinez-Vidal JL, Garrido Frenich A (1994) *Anal Lett* **27**:807.
377. Martinez-Vidal JL, Martinez-Galera M (1994) *Ann Chim (Rome)* **84**:177.

378. Johnson RM, Halaweish F, Fuhrmann JJ (1992) *J Liq Chromatogr* **15**:2941.
379. Russell JD, Cruz MI, White JL (1968) *J Agric Food Chem* **16**:21.
380. Miellet A (1988) *Ann Falsif Expert Chim* **80**:467.
381. Sanchez-Brunete C, Martinez L, Tadeo JL (1994) *J Agric Food Chem* **42**:2210.
382. Hernandez F, Beltran J, Lopez FJ, Gaspar JV (2000) *Anal Chem* **72**:2313.
383. Pappiloud S, Haerdi W (1994) *Chromatographia* **38**:514.
384. Steinheimer TR, Pfeiffer RL, Scoggin KD (1994) *Anal Chem* **66**:645.
385. Cotterill EG (1979) *Analyst* **104**:878.
386. Byast TH, Cotterill EG, Hance RJ (1977) *Methods for the Analysis of Herbicide Residues*, Second Edition, Technical Report No. 15, Agricultural Research Council, Weed Research Organisation, Yarnton, UK.
387. Lopez-Avila V, Hirata P, Kroska S, Flanagan M, Taylor JH, Hern SC (1985) *Anal Chem* **57**:2797.
388. Sanchez-Rasero F, Dios GC (1988) *J Chromatogr* **447**:426.
389. Xu Y, Lorenz W, Pfister G, Bahadir M, Korte F (1986) *Fresen Z Anal Chem* **325**:377
390. Papilloud S, Haerdi W, Chiron S, Barcelo D (1986) *Environ Sci Technol* **30**:1822.
391. Steinheimer TR (1993) *J Agric Food Chem* **41**:588.
392. Qiao X, Durnog R, Hummel HE (1991) *Meded Fac Landbouww Risksuniv Gent* **56**:949.
393. Schewes R, Maidl FX, Fischbeck G, Lepschy von Gleissenthall J, Süss A (1993) *J Chromatogr* **641**:89.
394. Schewes R, Wüst S, Lepschy von Gleissenthall J, Maidl FX, Süss A, Hock B, Fischbeck G (1994) *Anal Lett* **27**:487.
395. Hawthorne SB, Miller DJ, Nivens DE, White DC (1992) *Anal Chem* **64**:405.
396. Gutenmann WH, Lisk DS (1964) *J Assoc Off Anal Chem* **47**:353.
397. Chau ASY, Terry K (1975) *J Assoc Off Anal Chem* **58**:1294.
398. Gutenmann WH, Lisk DS (1963) *J Assoc Off Anal Chem* **46**:859.
399. Bache CA, Lisk DJ, Loos MA (1964) *J Assoc Off Anal Chem* **47**:348.
400. Kawahara FK (1968) *Anal Chem* **40**:1009.
401. Kawahara FK (1971) *Environ Sci Technol* **5**:235.
402. Sattar MA, Paasivirta J (1979) *Anal Chem* **51**:598.
403. Waliszewski SH, Szymczynski G (1985) *Fresen Z Anal Chem* **322**:510.
404. Lopez-Avila V, Hirta P, Kraske S, Taylor JH (1986) *J Agric Food Chem* **34**:530.
405. Yip G (1975) *J Chromatogr Sci* **13**:225.
406. Upchurch RP, Mason DD (1962) *Weeds* **10**:9.
407. Li C, Magee RJ, James BD (1991) *Anal Chim Acta* **255**:167.
408. Kim IS, Sasinos FI, Stephens RD, Wang J, Brown MA (1991) *Anal Chem* **63**:819.
409. Rochette EA, Harsh JB, Hill HH Jr (1993) *Talanta* **40**:147.
410. Crescenzi C, D'Ascenzo G, Di Corcia A, Nazzari M, Marchese S, Samperi R (1999) *Anal Chem* **71**:2157.
411. Shaner DL, O'Connor SL (eds) (1991) *The Imidazolinone Herbicides*, CRC Press, Boca Raton, FL, USA.
412. Shaner DL, Anderson PC, Stidham MA (1984) *Plant Physiol* **76**:545.
413. Anderson PC, Hibert KA (1985) *Weed Sci* **33**:479.
414. Shaner DL, Millipudi NM (1991) In: Shaner DL, O'Connor SL (eds) *The Imidazolinone Herbicides*, CRC Press, Boca Raton, FL, USA.
415. Wehtje GR, Dickens R, Wilcut JW, Hajek BF (1987) *Weed Sci* **35**:858.
416. Stoughaard RN, Shea PJ, Martin AR (1990) *Weed Sci* **38**:67.
417. Che M, Loux MM, Traina SJ, Logan TJ (1992) *J Environ Qual* **21**:698.
418. Loux MM, Liebl RA, Slife RW (1989) *Weed Sci* **37**:712.
419. Renner KA, Meggit WF, Penner D (1988) *Weed Sci* **36**:78.

420. Reddy KN, Locke MA (1994) *Weed Sci* **42**:249.
421. O'Bryan KA, Brecke BJ, Shilling DG, Colvin DL (1994) *Weed Technol* **8**:203.
422. Heber V, Siebers J, Holting HG, Vetten HJ, Krenzig R, Bahadire M (1998) *Fresen J Anal Chem* **360**:739.
423. Lagana A, Fago G, Marino A (1998) *Anal Chem* **70**:121.
424. Krynitsky AJ, Stout SJ, Nejad H, Cavolier TC (1999) *J Assoc Off Anal Chem* **82**:956.
425. Worthing CR (ed)(1983) *The Pesticide Manual: A World Compendium*, Seventh Edition, British Crop Protection Council, London, UK.
426. Gershon H, McClure GW (1966) *Contrib Boyce Thompson Inst* **23**:291.
427. Miller JH, Keeley PE, Thullen RJ, Carter CH (1976) *Weed Sci* **26**:20.
428. Ross LJ, Nicosia S, McChesney MM, Hefner KL, Gonzales DA, Siebers JN (1990) *J Environ Qual* **19**:715.
429. Monohan K, Tinsley IJ, Shephers SF, Field JA (1995) *J Agric Food Chem* **43**:2418.
430. Field JA, Monohan K (1995) *Anal Chem* **67**:3357.
431. Wettasinghe A, Tinsley IJ (1993) *Bull Environ Contam Toxicol* **50**:226.
432. Stout SJ, da Cunha AR, Allardice DG (1996) *Anal Chem* **68**:653.
433. Liu W, Pusino A, Gessa C (1992) *Sci Total Environ* **123–124**:39.
434. Schwab AP, Splichal P, Sonon LS (1993) In: Hoddinott KB, O'Shay TA (eds) *Applications of Agricultural Analysis in Environmental Studies*, ASTM STP 1162, ASTM, West Conshohocken, PA, USA, pp. 86–91.
435. Chirnside AEM, Ritter WF (1993) In: Hoddinott KB, O'Shay TA (eds) *Applications of Agricultural Analysis in Environmental Studies*, ASTM STP 1162, ASTM, West Conshohocken, PA, USA, pp. 92–97.
436. Lawruk TS, Lachman CE, Jourdan SW, Fleeker JR, Herzo DP, Rubio FM (1993) *J Agric Food Chem* **41**:1426.
437. Hall JC, Wilson LK, Chapman RA (1992) *J Environ Health B* **27**:523.
438. Casino P, Morais S, Puchades R, Aquiera A (2001) *Environ Sci Technol* **35**:4111.
439. Henze G, Meyer A, Hauser J (1993) *Fresen J Anal Chem* **346**:761.
440. Khoshab A, Teasdale R (1994) *J Chromatogr* **660**:195.
441. Liegeois E, Dehon Y, De Brabant B, Perry P, Portetelle D, Copin A (1992) *Sci Total Environ* **123–124**:17.
442. MacNally MEP, Wheeler JR (1988) *J Chromatogr* **447**:53.
443. Schlaeppi J-MA, Kessler A, Foery W (1994) *J Agric Food Chem* **42**:1914.
444. Herzel F (1980) *J Chromatogr* **193**:320.
445. Agudo M, Rios A, Valcarcel M (1993) *Anal Chim Acta* **281**:103.
446. Calderbank A, Yuens O (1965) *Analyst* **90**:99.
447. Pope JD, Benner JE (1974) *J AOAC Int* **57**:202.
448. Payne WR, Pope JD, Benner JE (1974) *J Agric Food Chem* **22**:79.
449. Khan SL (1974) *J Agric Food Chem* **22**:863.
450. Niewola Z, Benner JP, Swaine H (1986) *Analyst* **111**:399.
451. Niewola A, Walsh ST, Davies GE (1983) *Int J Immunopharmacol* **5**:211.
452. Niewola Z, Hayward C, Symington BA, Robson RT (1985) *Clin Chim Acta* **148**:149.
453. Kohler G, Milstein C (1975) *Nature* **256**:495.
454. Cannizzara RD, Cullen TE, Murphy RT (1970) *J Agric Food Chem* **18**:728.
455. Abbot SD, Hall RC, Giam GS (1969) *J Chromatogr* **45**:317.
456. Hall RC, Giam GS, Merkle MG (1970) *Anal Chem* **42**:432.
457. Krzyszowska AJ, Vance GF (1994) *J Agric Food Chem* **42**:1693.
458. Caverly DJ, Denney RC (1977) *Analyst* **102**:576.
459. Wheeler WB, Thompson NP, Ray BR, Wilcox M (1971) *Weed Res* **19**:307.
460. Jolliffe UA, Day BE, Jordan RS, Mann JD (1967) *J Agric Food Chem* **15**:174.

461. Gutamann WH, List DJ (1971) *J AOAC Int* **54**:975.
462. Bevenue A, Ogata JH (1970) *J Chromatogr* **46**:110.
463. Pease HL (1966) *J Agric Food Chem* **17**:121.
464. Pease HL (1966) *J Agric Food Chem* **14**:94.
465. Gawronski S, Skapsi H (1974) *Zesz Nauk Akad Roln Warsz Ogrodnictwo* **8**:59.
466. Hamilton DJ (1968) *J Agric Food Chem* **16**:152.
467. Von Stryke FG, Zajacs GF (1965) *J Chromatogr* **41**:125.
468. Maier-Bode H, Riedmann M (1975) *Residue Rev* **54**:113.
469. Jarcjyk HJ (1975) *Pflanzeuschutz-Nachr Bayer* **37**:319.
470. Negre M, Gennari M, Cignetti A (1987) *J Chromatogr* **387**:541.
471. Patuni M, Marucchini C, Businelli M, Vischetti C (1987) *Pestic Sci* **21**:193.
472. Clegg BS (1987) *J Agric Food Chem* **35**:269.
473. Zanco M, Pfeister G, Kettrup A (1992) *Fresen J Anal Chem* **344**:39.
474. Blau K, King G (1978) *Handbook of Derivatives for Chromatography*, Heyden, London.
475. Gaynor JD, MacTavish DC (1981) *J Agric Food Chem* **29**:626.
476. Gaynor JD, MacTavish DC (1982) *Analyst* **107**:700.
477. Tsukioka T, Shimizu S, Murakami T (1985) *Analyst* **110**:39.
478. Spann KP, Hargreaves PA (1994) *Pest Sci* **40**:41.
479. Khan SU, Lee KS (1976) *J Agric Food Chem* **24**:684.
480. Willan WT, Mueller TC (1994) *J AOAC Int* **77**:752.
481. Oda M, Yulimoto M, Zesso O (1975) *Kenkyu* **20**:12.
482. Prestel D, Weisgerber I, Klein W, Korte F (1976) *Chemosphere* **5**:137.
483. Downer GB, Hall M, Mallen ONB (1976) *J Agric Food Chem* **24**:1223.
484. Stout SJ, DaCuhna Ar, Sararpour MM (1997) *J AOAC Int* **80**:426.
485. Smith AE (1992) *Int J Environ Anal Chem* **46**:111.
486. Turin HJ, Bowan RS (1993) *J Environ Qual* **22**:332.
487. Michels K (1993) *GIT Fachz Lab* **37**:28.
488. Sanchez-Brunete C, Garcia-Valcarcel AI, Tadeo JL (1994) *J Chromatogr A* **675**:213.
489. Kovac J, Tekel J, Kurucova M (1987) *Z Lebensm Unters Forsch* **184**:96.
490. Bekheit HKM, Lucas AD, Szurdoki F, Gee SJ, Hammock BD (1993) *J Agric Food Chem* **41**:2220.
491. Cotterill EG (1982) *Analyst* **107**:76.
492. Gambrell RP, Reddy CN, Collard V, Green G, Patrick WH Jr (1984) *J Water Pollut Control* **56**:174.
493. Kavetskii VN, Bublik LI, Fuzik GV (1987) *J Anal Chem USSR* **42**:1037.
494. Khan S (1975) *J AOAC Int* **58**:1027.
495. Bache CA, Lisk DJ (1966) *Anal Chem* **38**:783.
496. Abbott DC, Egan H, Hammond EW, Thomson J (1964) *Analyst* **89**:480.
497. Fahling TM, Paulatis ME, Johnson PM, McNally MEP (1993) *Anal Chem* **65**:1462.
498. McKone CE, Cotterill EG (1974) *Bull Environ Contam Toxicol* **11**:233.
499. Coterill EG (1978) *Bull Environ Contam Toxicol* **19**:471.
500. Smith AE, Hayden BJ (1979) *J Chromatogr* **171**:482.
501. Johnson ER, Yu T, Montgomery MI (1977) *Bull Environ Contam Toxicol* **17**:369.
502. Howard SF, Yip G (1971) *J AOAC Int* **54**:970.
503. St John LE, Lisk DJ (1967) *J Dairy Sci* **50**:582.
504. Gutenmann WH, Lisk DJ (1963) *J Agric Food Chem* **11**:301.
505. Thio AP, Kornet MJ, Tan HIS, Tomkins DH (1979) *Anal Lett* **12**:1009.
506. McKone CE, Hance RJ (1972) *J Chromatogr* **69**:204.
507. Cotterill EG (1975) *J Chromatogr* **106**:409.
508. Gutenmann WH, Lisk DJ (1977) *J AOAC Int* **60**:1070.

509. Agemian H, Chau ASY (1977) *Arch Environ Contam Toxicol* **6**:69.
510. Mierzwa S, Witak S (1977) *J Chromatogr* **136**:105.
511. Cotterill EG (1979) *J Chromatogr* **171**:478.
512. Chau ASY, Terry K (1976) *J AOAC Int* **59**:633.
513. Sattar MA, Hattula ML, Lahtipera M, Passivirta J (1977) *Chemosphere* **11**:747.
514. Garbrecht TP (1970) *J AOAC Int* **53**:70.
515. Bache CA, St John LE, Lisk DJ (1968) *Anal Chem* **40**:1241.
516. Garle M, Petters I (1977) *J Chromatogr* **140**:165.
517. Gyllenhaal O, Naslund B, Hartvig P (1978) *J Chromatogr* **156**:330.
518. Gyllenhaal O, Ehrsson H (1975) *J Chromatogr* **107**:327.
519. Hartvig P, Fagerlund C (1977) *J Chromatogr* **140**:170.
520. Abbott DC, Wagstaff PJ (1969) *J Chromatogr* **43**:361.
521. Smith AE, Fitzpatrick A (1971) *J Chromatogr* **57**:303.
522. Kearney PC, Plimmer JR, Wheller WB, Konston A (1976) *Pestic Biochem Phys* **6**:229.
523. Valentin JFA, Diez-Caballero RB, Altuna MAG (1988) *Analyst* **113**:629.
524. Devine M (1973) *J Agric Food Chem* **21**:1095.
525. Sanchez-Brunete C, Garcia-Valcarcel AI, Tadeo JL (1994) *Chromatographia* **38**:624.
526. Onuska FI, Terry KA, Seech A, Antonic M (1994) *J Chromatogr A* **665**:125.
527. Vanisha Das J, Ramachandran KN, Gupta VK (1994) *Analyst* **119**:1387.
528. Sanchez Rasero F, Baez ME, Dios CG (1992) *Sci Total Environ* **123–124**:57.
529. Schmidt M, Hamman R, Kettrup A (1988) *Int J Environ Anal Chem* **33**:1.
530. Jiminez-Carmona MM, Manclus JJ, Montoya A, Luque de Castro MD (1997) *J Chromatogr* **785**:329.
531. Locke MA (1993) *J Agric Food Chem* **41**:1081.
532. Lu Y, Xue Y, Xu J, Won W (1992) *J AOAC Int* **75**:1100.
533. Fenig JC (1992) *Can J Chem* **70**:1087.
534. Novakova O (1994) *Chromatographia* **39**:62.
535. Buser H-R, Müler MD, Poiger T, Balmer ME (2002) *Environ Sci Technol* **36**:221.
536. Caverly DJ, Unwin J (1981) *Analyst* **106**:389.
537. Dieckmann H, Stockmaier M, Kreuzig R, Bahadir M (1993) *Fresen J Anal Chem* **345**:784.
538. Celi L, Negre M, Gennari M (1993) *Pestic Sci* **38**:43.
539. Stäb JA, Rozing MJM, van Hattum V, Cofino W-P, Brinkman UAT (1992) *J Chromatogr* **609**:195.
540. Singh RP, Chiba M (1993) *J Chromatogr* **643**:249.
541. Burket SE, Medvedeva NY, Ivanov BG (1969) *Z Analit Khim* **24**:264.
542. Dec J, Haider K, Benesi A, Rangaswamy V, Schäffer A, Plüken U, Bollag J-M (1997) *Environ Sci Technol* **31**:1128.
543. Kerwin RA, Crill PM, Tabot RW, Hines ME, Shorter JH, Kolb CE, Harriss RC (1996) *Anal Chem* **68**:899.

5 Determination of Organometallic Compounds in Soils

A very limited amount of work has been carried out on the determination of organometallic compounds in soils. It is expected that work in this area will increase in the future.

5.1
Organoarsenic Compounds

Barshick evaluated glow discharge mass spectrometry and gas chromatography–mass spectrometry for total element assays in soil [1]. Glow discharge mass spectrometry is of limited value for volatile elements such as arsenic (i.e. mercury) or when the element is not an inorganic salt but a volatile organometallic compound. A solid-phase microextraction fibre was shown to be an effective sampling medium for several organometallic compounds.

Odanaka et al. [2] has reported the application of gas chromatography with multiple ion detection after hydride generation with sodium borohydride to the determination of methylarsenic and dimethylarsenic compounds, trimethylarsenic oxide and inorganic arsenic in soil and sediments. Recoveries in spiking experiments were 100–102% (methylarsenic and dimethylarsenic compounds and inorganic arsenic) and 72% (trimethylarsenic oxide).

Dithiol derivatisation with solid-phase microextraction and gas chromatography–mass spectrometry has been used to determine organoarsenic compounds in soil [3].

Arsenic specks have been determined in soil using inductively coupled plasma mass spectrometry coupled with secondary ion mass spectrometry and by ion exclusion chromatography coupled with plasma mass spectrometry [4].

Maher [6] has described a method for the determination of down to 0.01 mg/kg of organoarsenic compounds in marine sediments. In this procedure, the organoarsenic compounds are separated from an extract of the sediment by ion exchange chromatography, and the isolated organoarsenic compounds are reduced to arsines with sodium borohydride and collected in a cold trap. Controlled evaporation of the arsine fractions and detection by atomic absorption spectrometry completes the analysis.

Hydride generation atomic absorption spectrometry has been used to determine arsenic species in water [1].

Naidu et al. [8] showed that separation of arsenic species from soil solutions could be performed in less than five minutes using capillary electrophoresis. The detection limit is 0.1 to 0.5 mg/l.

Thomas et al. [9] coupled HP–LC with IC–PMS to determine volatile forms of arsenic in soil.

Soderquist et al. [10] determined hydroxydimethylarsine oxide in soil by converting it to iododimethylarsine using hydrogen iodide followed by determination at 105 °C on a column (450 cm × 2.8 mg) packed with 10% DC-200 on Gas-Chrom Q (60–80 mesh), with nitrogen as carrier gas (20–30 min^{-1}) and electron capture direction. The recovery of hydroxydimethylarsine oxide (0.15 ppm) added to soil was 91.3 ± 5.1%.

Because of the wide usage of organic arsenicals, and because little information exists on the fate of these compounds in soils, Von Endt et al. [11] studied monosodium methane arsenic acid (MSMA) as a model for examining the metabolism of this class of compounds by soil microorganisms. Experiments involving the release of radioactive carbon dioxide from MSMA-^{14}C-treated soils were conducted in a system consisting of two test tubes connected in series. Days after treatment with MSMA-^{14}C it was observed that from 1.7 to 10.0% of the MSMA-^{14}C was degraded in nonsterile soil, as compared with 0.7% in steam-sterilised controls. Four soil microorganisms isolated in pure culture degraded from 3–20% of the MSMA-^{14}C to $^{14}CO_2$ when grown in liquid culture containing 10 ppm of monosodium methane arsenic acid and 1 g per litre of yeast extract. Thin-layer chromatography on silica gel G-coated plates effected the separation of monosodium methane arsenic acid, arsenate and arsenite. Only arsenate and monosodium methane arsenic acid were detected after thin-layer chromatography of extracts from the soil and microbial growth experiments. These data indicate that soil microorganisms are at least partly responsible for monosodium methane arsenic acid degradation in soil. Thin-layer chromatography was carried out on 20 × 20 cm glass plates coated 0.25 mm-thick with a suitable support and dried overnight. Silica gel G, silica gel H and cellulose were examined as the solid phases for chromatography of methane arsonate, arsenite and arsenate. Several sprays for the visualisation of the arsenicals on plates were tested.

5.2
Organolead Compounds

Blais et al. [12] has described a method using HPLC coupled with AAS for the determination of ionic alkyl lead compounds in soils. They demonstrated that previously published methods gave poor recoveries of lead and the formation of artifacts during the isolation and derivatisation procedures. An alternative procedure is described involving a series of selective extractions of tetraalkyl leads, ionic alkyl leads, and inorganic ionic lead salts from soils and street dusts. Alkyl lead salts were selectively extracted complexometrically from samples containing up to 1000 mg ionic lead per kg. The extracts were then butylated and analysed by gas chromatography–atomic absorption spectroscopy. Re-extraction of the sample with methylisobutylketone–dithizone permitted the recovery of ionic lead. In the samples tested, ethyl lead salts were detected, but not methyl lead salts. Concentrations of these analytes were significantly correlated with levels of extractable ionic lead, but not with total lead.

5.3
Organomercury Compunds

Water extraction followed by derivatisation with sodium tetraethyl boron then solid-phase microextraction/gas chromatography/mass spectrometry has been used to determine down to 200 ng/l of methyl mercury in soil [13].

Kimura and Miller and coworkers [5–18] determined low levels of inorganic plus volatile mercury. The soils were digested with sulfuric acid, hydrogen peroxide and potassium permanganate and then mercury swept of this mixture with air into an absorbing solution of potassium permanganate and sulfuric acid. Mercury is then determined in this solution by a dithizone-based spectrophotometric procedure at 605 nm.

A disadvantage of these above procedures is that the lowest concentration of mercury that can be determined in the soil or sediment samples is of the order of 0.05 – 1.0 mg/kg. These high detection limits are in part due to high blanks caused by the multiplicity of digestion reagents used in the procedures. Several investigators have liberated mercury from soil and sediment samples by application of heat to the samples and collection of the released mercury on gold surfaces. The mercury was then released from the gold by application of heat or by absorption in a solution containing oxidising agents [19].

Bretthauer et al. [20] and Anderson et al. [21] described a method in which samples were ignited in a high-pressure oxygen-filled bomb. After ignition, the mercury was absorbed in a nitric acid solution. Pillay et al. [22] used a wet-ashing procedure with sulfuric acid and perchloric acid to digest samples. The released mercury was precipitated as the sulfide. The precipitate was then redigested using aqua regia.

Feldman [23, 31] digested solid samples with potassium dichromate, nitric acid, perchloric and sulfuric acid. Bishop et al. [25, 32] used aqua regia and potassium permanganate for digestion. Jacobs and Keeney [33] oxidised sediments using aqua regia, potassium permanganate and potassium persulfate.

The approved US Environmental Protection Agency [25] digestion procedure requires aqua regia and potassium permanganate as oxidants. These digestion procedures are slow and often hazardous because of the combination of strong oxidising agents and high temperatures. In some of the methods, mercuric sulfide is not adequately recovered. The oxidising reagents, especially the potassium permanganate, are commonly contaminated with mercury, which prevents accurate results at low concentrations.

Longbottom et al. [26] has described gas chromatographic methods for the determination of alkyl mercury in soils and sediments.

Earlier work on the determination of total mercury in river sediments also include that of Iskander et al. [27]. Iskander applied flameless atomic absorption to a sulfuric acid–nitric acid digest of the sample following reduction with potassium permanganate, potassium persulfate and stannous chloride. A detection limit of one part in 10^9 is claimed for this somewhat laborious method.

As Umezaki and Iwamoto [28] have reported that organic mercury can be reduced directly with stannous chloride in the presence of sodium hydroxide and copper(II), the determination of organic mercury can be simplified, particularly if the reagent used for back-extraction does not interfere with the reduction of organic mercury. Matsunaga and Takahasi [29] found that extraction with an ammoniacal glutathione solution was satisfactory.

Langmyhr et al. [30] have applied cold vapour atomic absorption spectrometry (AAS) to the determination of organomercury compounds in soils and sediments.

5.4
Organotin Compounds

Procedures have been described for the analysis of methyltins [34] butyltins [35], mixed methylbutyltins [36], various alkyltins [37] cyclohexyltins [38] and phenyltins [39]. Alkylation also offers the possibility of selecting the volatility range of the derivatives, which are in most cases analysed by gas chromatography. However, there are few methods for the sensitive determination of a broad range of organotin compounds in environmental samples. Recently, the sensitive determination of butyltin residues in sediment and surface water was described on the basis of extraction/methylation and high-resolution gas chromatography with flame photometric detection [40].

To determine methyltin, butyltin and inorganic tin in Great Bay soils and sediments, Randall et al. [41] extracted the freeze-dried sediment with 2.5 mol/l calcium chloride and 2.5 mol/l hydrochloric acid and analysed it by hydride generation AAS. Detection limits for inorganic tin and tributyltin were 2.2 ng/kg and 0.6 ng/kg, respectively. Recoveries of methyltin and butyltin species from spiking experiments were greater than $70 \pm 10\%$. Tributyltin was found in all sampled sites, probably originating from tributyltin-based antifouling paints.

Lietal [42] demonstrated that, whereas AAS is usually insensitive to organotin compounds, the addition of tributyl phosphate enhances sensitivity considerably. Tributyltin at $1000\,^\circ$C converts organotin to SnP_2O_7 and $Sn_2P_2O_7$.

In situ derivatisation and supercritical fluid extraction have been applied to the determination of butyl and phenyltin compounds in soils [43].

Sinex et al. [44] have described a method for the determination of methyltin compounds based on reaction with sodium borohydride to form tin hydrides and then purge and trap analysis followed by gas chromatography with mass spectrometric detection. Down to 3–5 pg absolute (as tin) of methyltin compounds (equivalent to the sub μg/kg range) can be determined by this procedure.

Lobinski et al. [45] optimsed conditions for the comprehensive speciation of organotin compounds in soils and sediments. They used capillary gas chromatography coupled to helium microwave-induced plasma emission spectrometry to determine mono-, di-, tri- and some tetraalkylated tin compounds. Ionic organotin compounds were extracted with pentane from the sample as the organotin–diethyldithiocarbamate complexes and then converted to their pentabromo derivatives prior to gas chromatography. The absolute detection limit was 0.5 pg as tin, equivalent to 10–30 μg/kg.

Adinarayana et al. [46] determined triphenyltin compounds in plants and soil by thin-layer chromatography with biological detection.

Lucero et al. [47] has previewed methods for the determination of triphenyltin compounds in soils.

References

1. Barshick CM, Barshick S-A, Mohill ML, Britt PF, Smith DH (1996) *Rapid Commun Mass Sp* **10**:341.
2. Odanaka Y, Tsuchiya W, Matano O, Goto S (1983) *Anal Chem* **55**:929.
3. Szostek B, Aldstadt JH (1998) *J Chromatogr A* **807**:253.
4. Koellensperger G, Nurmi J, Hann S, Stingeder G, Fitz WJ, Wenzel WW (2002) *J Anal Atom Spectrom* **17**:1047.
5. Nakazato T, Tao H, Tamiguchi T, Isshiki K (2002) *Talanta* **58**:121.
6. Maher WA (1981) *Anal Chim Acta* **126**:157.
7. Capelo JL, Lavilla I, Bendicho C (2002) *Anal Chem* **73**:3732.
8. Naidu R, Smith J, McLaren RG, Stevens DP, Sumner ME, Jackson PE (2000) *J Soil Sci Soc Am* **64**:122.
9. Thomas P, Finnie JK, Williams JG (1997) *J Anal Atom Spectrom* **12**:1367.
10. Soderquist CJ, Crosby DG, Bowers JB (1974) *Anal Chem* **46**:155.
11. Von Endt DW, Kearney PC, Kaufmann DD (1968) *J Agric Food Chem* **16**:17.
12. Blais JS, Marshall WD (1989) *J Anal Atom Spectrom* **4**:271.
13. Beichert A, Pacberg S, Wenclawiak BW (2000) *Appl Organomet Chem* **14**:493.
14. Kimura Y, Miller VL (1962) *Anal Chim Acta* **37**:325.
15. Kimura Y, Miller VL (1960) *Anal Chem* **32**:420.
16. Kimura Y, Miller VL (1962) *Anal Chem* **34**:325.
17. Kimura Y, Miller VL (1964) *J Agric Food Chem* **15**:253.
18. Polley D, Miller VL (1955) *Anal Chem* **27**:1162.
19. Leong PC, Ong HO (1971) *Anal Chem* **43**:940.
20. Bretthauer EW, Moghissi AA, Snyder SS, Mathews NW (1974) *Anal Chem* **46**:445.
21. Anderson DH, Evans JH, Murphy JJ, White WW (1971) *Anal Chem* **43**:1151.

22. Pillay KKS, Thomas Jr. CC, Sondel JA, Hyche CM (1971) *Anal Chem* **43**:1419.
23. Feldman C (1974) *Anal Chem* **46**:99.
24. Bishop JN, Taylor LA, Nearby BO (1975) *The Determination of Mercury in the Environment*, US Environmental Protection Agency, Cincinnati, OH, USA, p. 120.
25. US EPA (1974) *Methods for Chemical Analysis of Water and Wastes*, US Environmental Protection Agency, Cincinnati, OH, p. 134–138.
26. Longbottom JE, Dressman RC, Lichtenberg JJ (1973) *J Assoc Off Anal Chem* **56**:1297.
27. Iskander IK, Syers JK, Jakobs LW, Keeney DR, Gilmour JT (1972) *Analyst* **97**:388.
28. Umezaki Y, Iwamoto K (1971) *Japan Analyst* **20**:173.
29. Matsunaga K, Takahashi S (1976) *Anal Chim Acta* **87**:487.
30. Langmyhr FJ, Aamodt J (1976) *Anal Chim Acta* **87**:483.
31. Feldman C (1974) *Anal Chem* **46**:1606.
32. Bishop JN, Taylor LA, Neary BP (1973) In: *The Determination of Mercury in Environmental Samples*, Ministry of the Environment, Ontario, Canada.
33. Jacobs LW, Keeney DR (1976) *Environ Sci Tech* **8**:267.
34. Lu KL, Pulford ID, Duncan HJ (1979) *Anal Chim Acta* **106**:319.
35. Ure AM, Hernandez-Artiga MI, Mitchell MC (1978) *Anal Chim Acta* **96**:37.
36. Pederson B, Willems M, Jorgensen JS (1980) *Analyst* **105**:119.
37. Henn EL (1977) In: *Flameless Atomic Absorption Analysis*, ASTM STP 618, American Society of Testing Materials, Philadelphia, PA, USA, pp.54–64.
38. Jones JM (1977) *Comm Soil Sci Plant Anal* **8**:340.
39. Dahlquist RL, Knoll JW (1978) *Appl Spectrosc* **32**:1.
40. MAFF (Ministry of Agriculture, Fisheries and Food) (1979) Method 54: Nickel, Nitric, Perchloric Acid Soluble in Soil. In: *The Analysis of Agricultural Material, Second Edition*, RB 427, HMSO, London, UK.
41. Randall L Hons JS, Ucher JH (1988) *Environ Tech Lett* **74**:71.
42. Li H, Gong B, Matsumoto K (1996) *Anal Chem* **68**:2277.
43. Cai Y, Azaga R, Bayona JM (1994) *Anal Chem* **66**:1161.
44. Sinex SA, Cantillo AY, Helz GR (1980) *Anal Chem* **52**:2342.
45. Lobinski R, Dirkx WMR, Ceulemans M, Adams FC (1992) *Anal Chem* **64**:159.
46. Adinarayana M, Singh US, Dwivedi TS (1988) *J Chromatogr* **435**:210.
47. Lucero RA, Otieno MA, May L, Eng G (1992) *Appl Org Chem* **6**:273.

6 Determination of Anions in Soil

6.1
Arsenate/Arsenite

See Sect. 2.5.

6.2
Borate

A method has been described [1] for the determination of borate in soils based on the conversion of borate in a hot water extract to fluoroborate by the action of orthophosphoric acid and sodium fluoride. The concentration of fluoroborate is measured spectrophotometrically as the blue complex formed with methylene blue which is extracted into 1,2-dichloroethane. Nitrates and nitrites interfere, but these can be removed by reduction with zinc powder and orthophosphoric acid.

Aznarez et al. [2] used curcumin as a chromogenic reagent to estimate borate in soils. The borate is extracted from the sample with 2-methylpentane-2,4-diol into methyl isobutyl ketone. The selectivity of the extraction of boric acid with 2-methylpentane-2,4-diol into methyl isobutyl ketone provides a preconcentration method and the simultaneous elimination of numerous interferents. In this procedure, 0.2 – 1 g of finely ground soil is digested with 5 ml of concentrated nitric acid–perchloric acid (3 + 1) in a PTFE liquid pressure bomb at 150 °C for two hours. The solution is cooled and diluted and any residue is filtered off through Albet 242 filter paper. Acidity is neutralised with 6 M sodium hydroxide and the solution diluted to 100 ml with hydrochloric acid (1 + 1) in a calibrated flask. A portion of this solution containing 10 to 100 µg boron is extracted three times with 10 ml portions of methyl isobutyl ketone in order to eliminate iron interference, then extracted with 10 ml of 20% v/v 2-methylpentane-2,4 diol in methyl isobutyl ketone. The extract is shaken for five minutes and, finally, dried by the addition of 1 g of anhydrous sodium sulfate. To carry out spectrophotometry, 3 ml of the organic phase is transferred into a polyethylene test tube and 2 ml of a 0.1% m/v solution of curcumin in glacial acetic acid and 2 ml concentrated phosphoric acid are added. The mixture is shaken and heated to 70 ± 3 °C for one hour. After cooling, the absorbance of the solution is measured at 510 nm against a reagent blank (Fig. 6.1). A calibration graph is prepared by adding various volumes of standard borate solution containing 10–100 µg boron, and then adding an equal volume of hydrochloric acid (1 + 1) to this. The solution is then extracted with 100 ml of 20% v/v 2-methylpentane-2,4-diol and spectrophotometric measurements carried out as described above.

The calibration graph at 510 nm is a straight line and Beer's law is obeyed from 0.5 to 5 µg/ml of boron in the final measured solution (corresponding to 10–110 µg of boron in the aqueous phase). The molar absorptivity, calculated from the slope of the statistical working calibration graph at 510 nm, was 2905 l/mol/cm. The Sandell sensitivity was 0.011 µg cm^2 of boron. The precision of the method for ten replicate determinations was 0.6%. The absorbance of the reagent blank solution at 510 nm was 0.010 ± 0.003 for ten replicate determinations. Therefore, the detection limit was 0.04 µg/ml of boron in the final measured solution.

Aznarez et al. [2] also applied molecular fluorescence spectroscopy to the determination of borate in soils. The boron is extracted from the soil with 2-methylpentane-2,4 diol into methyl isobutyl ketone and 0.1% m/v dibenzoylmethane in methyl isobutyl ketone and concentrated phosphoric acid is

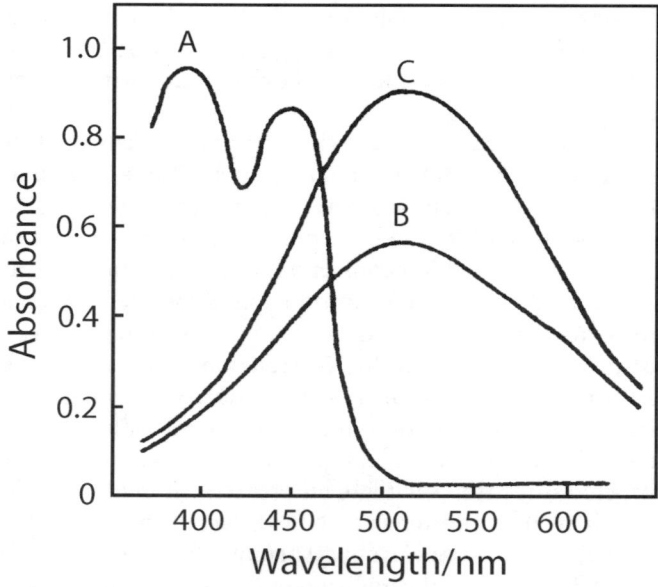

Figure 6.1. Adsorption spectra of A: reagent blank solution measured against IBMK as reference, B: Boron curcumin compound against reagent blank, 3 µg/l boron, and C: as B, 5 µg/l of boron. From [2]

added. After heating the solution to 80 °C for three minutes the selective fluorescence intensity is measured at 400 nm (excitation 390 nm). The procedure is calibrated against standard borate solutions containing 0.5 – 5.0 µg of boron.

6.3
Bromide

The widespread use of methyl bromide as a soil fumigant for crop protection has necessitated the development of a specific method for the determination of bromide and total bromine in soils subsequent to fumigation.

Roughan et al. [3] has described a gas chromatographic method for carrying out this analysis in which the soil is mixed with sodium hydroxide solution and then treated with ethanol prior to evaporation to dryness. After muffling, the residue is digested with sulfuric acid and to this solution is added to acetonitrile and ethylene oxide:

$$H^+ + Br^- + C_2H_4O \rightarrow HO-CH_2-CH_2-Be$$

The 2-bromoethanol produced in this reaction is analysed by gas chromatography using an election capture detector. At the 10 mg/kg level in soil a standard deviation of ±0.34 mg/kg was obtained, i.e., a coefficient of variation of ±3%. Recoveries from soil were 81 – 94%.

Van Staden [4, 5] employed flow injection analysis coupled with a coated tubular solid-state bromide-selective electrode for the determination of bromide in soils. Soil-extracted samples are injected into 10 mol/l potassium nitrate carrier solution containing 1000 mg/l chloride as an ionic strength adjustment buffer. The sample buffer zone formed is transported through the bromide selective electrode onto the reference electrode. The method is applicable in the range 10 – 50 000 mg/l bromide. The coefficient of variation of this method is better than 1.6%.

Care had to be taken in the preparation of suitable homogeneous coated tubular bromide-selective electrodes from silver bromide. Maximum contact area is obtained by using a tubular electrode that is well-coated. The maximum sensitivity was obtained when the electrode was coated, tested, left in ca. 500 mg/dm^3 bromide solution, recoated, etc., until maximum response was obtained, and then conditioned overnight in 20 mg/dm^3 bromide solution. It was also necessary to carry out an actual test run of about ten minutes.

Gladney and Perrin [6] used epithermal neutron activation analysis to determine down to 50 ppb bromine in the US Geological Survey Reference Soils GXR-2, GXR-5 and GXR-6, and the Canadian Certified Reference Soils SO-1, SO-2, SG-3 and SO-4. The values reported in Table 6.1 indicate that good agreement was obtained between neutron activation analysis results and recommended values. The relative standard deviation was on the order of ±10% over the concentration range 1 – 15 ppm bromine:

Table 6.1. Bromine concentrations in Canadian Certified Reference Soils (from [6])

Soil Ref. No.	Soil type	Bottle No.	Br, ppm X $\pm \alpha$	Recommended value
SO-1	Regosolic	133	1.3 ± 0.1	1.4 ± 0.2
	clay	711	1.4 ± 0.2	
SO-2	Podzolic	97	14.8 ± 1.0	15 ± 2
	B horizon	903	15.2 ± 1.5	
SO-3	Calcareous	495	5.5 ± 0.7	5.2 ± 0.8
	C horizon	1023	4.8 ± 0.2	
SO-4	Chernozemic	103	5.5 ± 0.3	5.6 ± 0.5
	A horizon	441	5.8 ± 0.5	

6.4
Carbonate

Collins [7] has described a gasometric method for the determination of carbonate in soil based on reaction with hydrochloric acid and subsequent measurement of the volume of carbon dioxide produced. This is also the basis
of a standard HMSO method for the determination of carbonate in soils [8]
(Fig. 6.2).

6.5
Chlorate

A method for the determination of chlorate in water extracts of soil is based on
its conversion to free chlorine upon reaction with hydrochloric acid, followed
by spectrophotometric evaluation of chlorine at 448 nm by the spectrophotometric o-toluidine method [9]. A correction is made for interference by iron
(III), nitrite, free chloride derived from hypochlorites and strong oxidising
agents by subtracting the absorbance of a modified blank, containing a lower
concentration of hydrochloric acid, from that obtained in the test.

6.6
Chloride

An official procedure [10, 11] describes a method for the determination of
chloride in a saturated calcium sulfate extract of soil. The extract is acidified
and the concentration of chloride is determined by titration with mercuric
nitrate using diphenylcarbazone as indicator. Mercuric ion in the presence of
chloride forms mercuric chloride, which, although soluble, provides insuffi-
cient mercuric ion to form the mercuric–diphenylcarbazone complex. When
all of the chloride has been removed in this way, addition of further mercuric
ion produces the violet complex.

Davy and Bembrick [12] have described a method for the determination of chloride in water extracts in soils based on measurement of the EMF developed between two silver–silver chloride electrodes in a cell with a liquid function and suitable electrolyte.

McLeod et al. [13] has developed an apparatus by which measurements of chloride, pH and electrolytic conductivity are obtained simultaneously for sus-

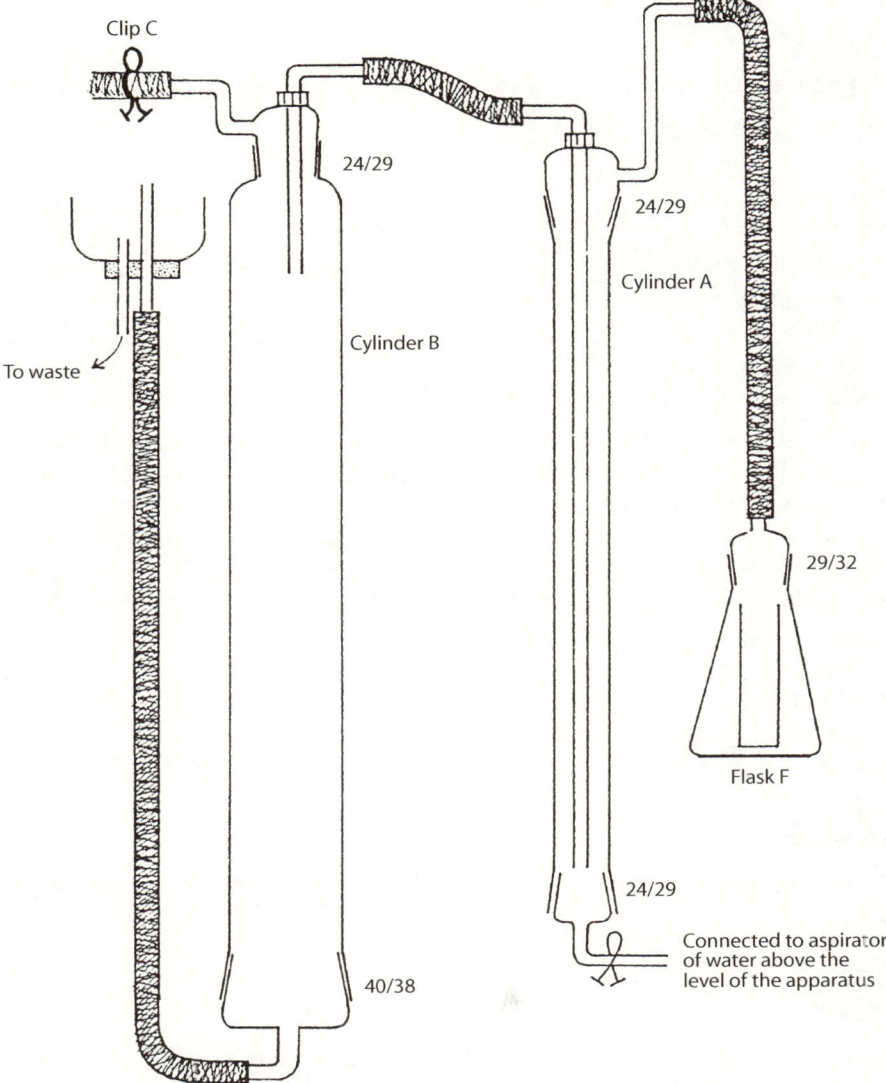

Figure 6.2. Gasometric method for the determination of carbonate in soil. From [7]

pensions of soil in water through the use of a triple electrode system mounted in a single unit. A glass electrode and a silver–silver chloride electrode with a common reference electrode and two pH meters are used for the determination of pH and chloride, respectively. Electrolytic conductivity is measured by an ohmmeter principle using silver electrodes. The outputs of the three meters are recorded on a three-pen recorder with electrically independent channels. The pH (range 0–10) is read from the chart while the values for chloride and electrolytic conductivity are obtained from graphs or tables.

6.7
Chromate

See Sect. 2.13.

6.8
Cyanide

Tecator [14] produced apparatus based on distillation and titration or spectrophotometry for the determination of cyanide in soil.

6.9
Iodide

Van Vliet et al. [15] have described a semi-automated procedure for the determination of iodine in soils. The soil sample is digested with 2 N sodium hydroxide, and then the soil is centrifuged off. The resulting solution is digested with perchloric and nitric acid (2:1 v/v) at 265 °C until clear. Iodine is determined in this solution by a method based on the oxidation of arsenic III by cerium IV: 3–5 ppm mg/kg added to soil was recovered at the 98% level.

6.10
Manganate

See Sect. 2.16.

6.11
Molybdate

Mehra and Frankenberger [16] used ion chromatography to determine molybdate in soils.

6.12
Nitrate

Techniques used to determine nitrates in soils include titration [17], spectrophotometry [18–26, 29–31], flow injection analysis [20, 21], ion selective electrodes [27, 28], and ion chromatography [28, 31–44].

A method [17] has been described for the determination of nitrate and nitrite nitrogen and ammonium ions in the 2 M potassium chloride extracts of moist soils. Firstly, an aliquot of the extract is made alkaline and the released ammonia determined via an ammonia-selective probe or titrimetrically. The nitrate in the ammonia-free extract is then reduced to ammonia with Devarda's alloy and the ammonia removed by distillation and determined titrimetrically. The concentration of nitrite in the extract is then determined spectrophotometrically as the red dye formed by coupling diazotised sulfanilic acid with N naphthylethylenediamine hydrochloride.

Henrickson and Selmer-Olson [18] applied an autoanalyser to the determination of nitrate and nitrite in soil extracts. In an autoanalyser, the water sample, buffered to pH 8.6 with aqueous ammonia–ammonium chloride, is passed through a copperised cadmium reductor column. The nitrite formed is reacted with sulfuric acid and N-1-naphthylethylenediamine, and the extinction of the azo dye is measured at 520 nm. For soil extracts, the range and standard deviation are 0.5 – 1.0 and 0.007 mg/l, respectively.

Garcia Gutierrez [19] has described an azo coupling spectrophotometric method for the determination of nitrite and nitrate in soils. Nitrite is determined spectrophotometrically at 550 nm after treatment with sulfuric acid and N-1-naphthylethylenediamine to form an azo dye. In another portion of the sample, nitrate is reduced to nitrite by passing a pH 9.6 buffered solution through a cadmium reductor and proceeding as above. Soils were boiled with water and calcium carbonate, treated with freshly precipitated aluminium hydroxide and active carbon, and filtered prior to analysis by the above procedure.

Tecator [20] has described a spectrophotometric method for the determination of nitrate and ammonium employing sulfanilamide and N-1-naphthylethylenediamine in 2 M potassium chloride extracts of soil samples.

Lindau and Spalding [21] have studied the effects of 2 M potassium chloride extractant ratios of between 1:1 to 1:10 on nitrate recovery in nitrate and nitrite extractions from soil. Preliminary data indicated that concentrations of extractable nitrate and nitrogen isotopic values were influenced by the volume of extractant. The 1:1 extractions showed decreasing nitrogen isotope values with increasing nitrate levels, whereas in the 1:10 extractions these values were independent of each other. Incomplete extraction occurred for the 1:1 ratios. The ratio required for maximal recovery was not determined.

Elton-Bott [22] and Osibanjo and Ajaya [23] determined nitrate in soil by a spectrophotometric method based on 3,4-xylenol. In one of these procedures [28], nitration of the 3,4-xylenol is carried out instantaneously at about 0 °C in 80% sulfuric acid and the nitration product is extracted into toluene, the excess of the reagent remaining in the aqueous layer. The toluene layer is then treated with sodium hydroxide solution to form a coloured product (the sodium salt of the nitrophenol in the aqueous layer), the absorbance of which is measured at 432 nm. Interferences from common anions, including chloride and nitrate, were investigated.

In this method the soil extract and the blank and standards are evaporated almost to dryness by gentle heating and then cooled in an ice pack. Five millilitres of 80% sulfuric acid and then 1 ml of a 2% ethanolic solution of 3,4-xylenol are added. This solution is transferred to a separatory funnel with 80 ml ice-cooled distilled water. Toluene (10 ml) is added to the isolated toluene extract, and 5 ml of 1% sodium hydroxide is added to convert the phenol to the phenoxide. The lower aqueous phase is separated and evaluated spectrophotometrically at 432 nm using matched silica cells with distilled water in the reference cell.

Keay and Menage [24] carried out an automated determination of nitrate and ammonia in 2 M potassium chloride extracts of soils. The sample is reacted in an autoanalyser with a 0.25% suspension of magnesium oxide; the ammonia liberated from ammonium ion is absorbed in 0.1 M hydrochloric acid and determined spectrophotometrically at 625 nm by the indophenol blue method. The sum of ammonium and nitrate is determined similarly but with the addition of 4.5% titanous sulfate solution before distillation, thereby reducing nitrate but not nitrite to ammonia. The nitrate content of the soil can then be obtained by difference.

Hadjidemetriou [25] has carried out a comparative study of the determination of nitrates in calciferous soils by the phenoldisulfonic acid and the chromotropic acid spectrophotometric methods. He used 0.02 N cupric sulfate as soil extractant. Silver sulfate was added to remove chlorides. Nitrites, if present, were eliminated by acidifying the extract with N in sulfuric acid. The phenol disulfonic acid method is subject to interference by other ions. Details of the chromotropic acid method are given below.

In this method 50 ml of 0.02 N copper(II) sulfate is added to 10 g of the air-dried soil sample. A 3 ml volume of the nitrate is moved to a 25 ml flask and cooled in ice. To this solution is added 1 ml 0.1% chromotropic acid sodium salt dissolved in concentrated sulfuric acid.

Concentrated sulfuric acid (6 ml) is then added and the solution left for 45 min for colour to fully develop prior to spectrophotometric evaluation at 430 nm. Standard solutions (0–35 mg/l NO_3) and blanks were subject to the same treatment.

Table 6.2. Regression equations and correlation coefficients between the three methods for nitrate-nitrogen determination (from [25])

Regression equation	Correlation coefficient
Chromotropic method = 1.92 + 0.99 (phenoldisulfonic acid method)	0.9998
Ion-selective electrode method = 1.58 + 0.96 (phenoldisulfonic acid method)	0.9998
Chromotropic method = 0.31 + 1.03 (ion-selective electrode method)	0.9996

Good agreement was obtained between results obtained on soil extracts by this method, the more lengthy phenolic disulfonic method and an ion-selective electrode method in the nitrate nitrogen range 3 – 200 mg/kg. The relationships between the three methods are shown in Table 6.2. There is a very close relationship between the methods; the correlation coefficients are almost unity, indicating that the phenoldisulfonic acid method could be replaced with the ion-selective electrode or the chromotropic acid method.

Flow Injection Analysis

Tecator [20] has described a flow injection system for the determination of nitrate and nitrite in 2 mol/l potassium chloride extracts of soil samples. Nitrate is reduced to nitrite with a copperised cadmium reductor and this nitrite is determined by a standard spectrophotometric procedure in which the soil sample extract containing nitrate is injected into a carrier stream. Upon the addition of acidic sulfanilamide a diazo compound is formed which then reacts with N-(l-naphthyl)ethylenediamine dihydrochloride provided from a second merging stream. A purple azo dye is formed, the intensity of which is proportional to the sum of the nitrate and the nitrite concentration. Nitrite in the original sample is determined by direct spectrophotometry of the soil extract without cadmium reduction.

Microdiffusion Method

Waughman [26] has described a simple microdiffusion method for estimating nitrate (and ammonia) in soils. In this method, nitrate in the soil extract is reduced to ammonia by titanous sulfate and the ammonia is then released from the solution and diffused and absorbed onto a nylon square impregnated with dilute sulfuric acid. The nylon is then dipped into a solution of a chromogenic reagent for ammonia and the colour evaluated spectrophotometrically.

Nitrate-Selective Electrode

The nitrate ion-selective electrode has been used extensively, even though there are interferences from other ions [32–34,38,41,42]. The rapidity and the good accuracy achieved using this electrode [28,31–41] have made it suitable for use in routine analysis and in soil agrochemical research [34,38,43].

Various workers [28,34,41] have used different extraction solutions in the ion-selective electrode method, depending on the soil being analysed. The most important are [23,32,35–37]: potassium sulfate [39], aluminium sulfate [30], copper (II) sulfate [32], calcium hydroxide [33], and copper sulfate(II) with aluminium and silver resins [41].

Hadjidemietriou [25] used an Orion Model 93-07 nitrate ion-selective electrode with a 1×2 sensing module construction, and an Orion Model 90-02 double junction reference electrode fitted on a pH meter. The outer chamber was filled with 0.04 M ammonium sulfate solution and the inner chamber with Orion 90-00-02 solution.

Goodman [28] described an automated procedure for the determination of nitrate in soils. The apparatus automatically extracts and analyses batches of up to 60 soil samples. Analysis is performed electrochemically by means of an ion-selective electrode and reference electrode. Corning ion-selective electrodes were found to be superior to those produced by Orion in this application. Recoveries of nitrate in this method were between 94 and 95%. The calibration curve was linear down to 2.5 mg/l nitrate. A plan of the general arrangement is shown in Fig. 6.3.

It consists of a rail-mounted carriage, which carries rows of sample beakers past three "stations" where each sample receives an aliquot of extractant (usually water), is thoroughly stirred, and has an electrode or electrodes lowered into it. The electrical output from the electrode(s) is passed through an amplifier to a flatbed recorder. Control of the sequence of operations is completely automatic, involving a system of three interlocking motor-driven cam timers, thereby ensuring that each sample receives identical treatment.

The apparatus used in this method consisted of a Corning liquid junction nitrate ion-selective electrode operating through a Pye Model 291 pH meter. This electrode has a flat end incorporating the sensing membrane. Also used

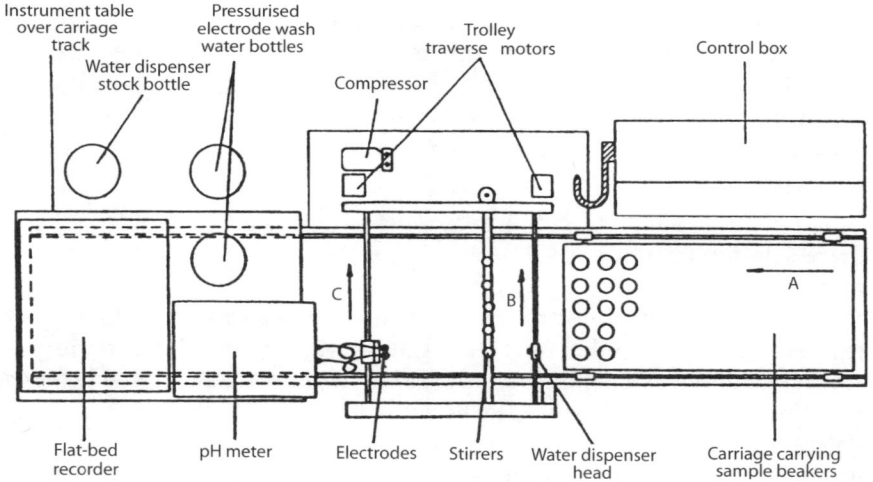

Figure 6.3. Layout of apparatus for the determination of nitrate. *Arrows* show direction of movement of carriage (A) and water and electrode trolleys (B) and (C), respectively (not to scale). From [28]

were a Philips R44/2/-SD/1, double junction reference electrode, containing 0.02 M potassium chloride solution in the outer chamber.

In a series of experiments to test the Corning electrode with the apparatus, Goodman [28] added a range of standard nitrate solutions to weighed samples of a sandy loam soil. The nitrate contents of these modified samples were then determined by three different methods: (a) extracting 20 g of soil with 50 ml of water, filtering and analysing the filtrate by Kjeldahl distillation with alkaline titanium(III) sulfate; (b) preparing the filtrate as in (a) and determining the nitrate concentration in the extract by manually inserting the Corning electrode; and (c) extracting 10 g of soil with 25 ml of water on the apparatus and determining the nitrate content of the extract by automatic insertion of the Corning electrode into the soil suspension.

The slopes of the regression lines show that the electrode recorded 94–95% of the added nitrate, the response of the electrode in soil suspension being substantially linear down to 2.5 µg/l of nitrate in the extract.

Ion Chromatography

Bradfield and Cooke [45] (Bradfield EG, Private Communication) give details of a procedure for the determination of nitrate (and chloride, phosphate and sulfate) in aqueous extracts of soil by an ion chromatographic technique with ultraviolet light. Recoveries ranged from 84 to 108%.

6.13
Nitrite and Nitrate

Bhuchar and Amar [46] determined nitrite in soil by acidifying the sample to pH 4 and adding mercaptoacetic acid to produce a red-coloured complex, which is extracted into tributylphosphate from a solution of 2 N acid. The red colour is evaluated spectrophotometrically at 322 nm. The method is applicable in the range 2–40 mg nitrite per litre of extract.

Chaube et al. [47] investigated the determination of ultratrace concentrations of nitrite in sulfuric acid extracts in soil. In their method, the nitrite is used to diazotise o-nitroaniline and the o-nitrophenyldiazonium chloride produced is coupled with N-naphthylethylenediamine hydrochloride. The red-violet dye produced is extracted into isoamyl alcohol and evaluated spectrophotometrically at 545 nm. Beer's law is obeyed in the range 0.1 to 0.6 mg nitrate per litre of solution.

Wu and Lin [43] have described a spectrophotometric method for the determination of micro amounts of nitrite in soil. The chromogenic reagents were p-aminoacetophenone and resorcinol in sodium carbonate–sodium acetate medium at pH 9, which form a golden-coloured complex with nitrite at 435 nm.

$$CH_3CO-C_6H_4-NH_2^+ + NO_2^- + 2H^+ \rightarrow CH_3CO-C_6H_4-N^+\equiv N + 2H_2O$$

$$CH_3-COC_6H_4-N^+\equiv N + \underset{\underset{}{\bigcirc}}{\overset{HO}{}}-OH \overset{pH\,4}{\rightarrow}$$

$$CH_3CO-C_6H_4-N{=}N-\bigcirc-OH + H^+$$

Foreign ions are masked with a composite EDTA–sodium hexametaphosphate reagent and interference by sulfide is overcome by the addition of mercuric chloride, which also mitigates interference by thiosulfate, sulfate, tetrathionate and iodide, and the precipitated mercuric sulfide is filtered off prior to the addition of chromogenic reagents and spectrophotometry. Beer's law is obeyed up to levels of 20 µg nitrite in a 60 ml test solution.

Garcia Gutierrez [19] has described an azo-coupled spectrophotometric procedure for the determination of nitrite and nitrate in soil extracts, prepared by boiling the soil with water and calcium carbonate, treating it with freshly precipitated aluminium hydroxide and active carbon and then filtering. Nitrite is determined spectrophotometrically at 550 nm after treatment with sulfuric acid and N-(1-naphthyl)ethylenediamine to form an azo dye. In another portion of soil extract, nitrate is reduced to nitrite by passing a pH 9.6 buffered solution though a cadmium reductor and proceeding as above.

6.14
Nitric Oxide and Nitrous Oxide

Both indirect [5, 50] and direct [51, 55–57] evidence indicate that gaseous forms of nitrogen can be lost from soil during the nitrification of ammonium or ammonium-forming fertilisers by soil microorganisms. It appears that evolution of nitrogen, dinitrogen oxide and nitrogen oxide or its oxidative derivative, nitrogen dioxide, can occur, resulting in poor fertiliser efficiency.

Smith and Chalk [49] have described a simple method for determining nitrogen oxide and nitrogen dioxide evolved from soils in closed systems. These gases are absorbed by an acidic solution of potassium permanganate, and the resulting nitrate is determined by a steam distillation method. Excess permanganate is reduced with iron(II) sulfate and neutralised with sodium hydroxide solution. Ammonium in solution is removed by distillation with magnesium oxide, and nitrate is determined by distillation after reduction to ammonium by Devarda's alloy. Nitrogen and dinitogen oxide evolved from soils are measured using gas chromatography on a single 0.61 mm column with a molecular sieve of 5 A, temperature programmed to 250 °C at 39 °C/min after an initial period of one minute at 35 °C. A complete analysis requires 20 min and 2 μg of nitrogen can be determined quantitatively for each gas.

Smith and Chalk [49] achieved a complete separation of oxygen, nitric oxide, nitrous oxide and carbon dioxide on a column packed with a molecular sieve of 5 A (100–120 mesh), programmed between 35 °C and 250 °C at 39 °C/m.

6.15
Phosphate

Spectrophotometric evaluation at 880 nm of the phosphomolybdate complex has been used to determine phosphates at pH 8.5 in soil [58].

Bickford and Willett [59] have pointed out that the filtration of aqueous calcium chloride extracts of soils containing phosphate through Gelman 9A6 cellulose acetate membranes which contain a wetting agent caused low results in methods for determining phosphate, due to the presence of some contaminant in the membrane. Gelman TCM-450 or Whatman No 42 membrane, on the other hand, does not interfere in the determination of phosphate.

6.16
Selenite

Karlson and Frankenberger [60] have developed a simple column ion-chromatographic column method for the determination of selenite in soil extracts with the simultaneous determination of chloride, nitrite, nitrate and phosphate. Separation of the anions was conducted on a low-capacity ion-exchange column, and anions were quantified by conductiometric detection. The eluent stream consisted of 1.5 mmol/l phthalic acid and adjusted to pH 2.7 with formic acid.

The method requires minimal sample pretreatment, allowing for precise measurements of trace levels of selenite in the presence of high background levels of chloride, nitrites and phosphate. Interfering chloride ions were removed by reduction with a silver-saturated cation exchange resin. The detection limit for selenite was 3 μg/l of soil extract and the standard deviation was 6.7% with soil extracts of (0.5 mg/l). Between 0.5 and 99.6 μg/l of selenite were found in soil extracts.

6.17
Sulfate

Ogner and Haugen [61] have described a technique for the automated determination of sulfate in water samples and soil extracts containing large amounts of humic compounds. This technique can be applied to the determination of sulfate in concentration ranges of $0-60$ and $0-3000$ mg/l (as sulfate) in the aqueous extract.

A turbidimetric method has been described for the determination of water-soluble and acid-soluble sulfate in hydrochloric acid extracts of soils [62]. Spectrophotometry has also been employed to carry out the determination of total water-soluble sulfate in soil [63,64].

Landers et al. [65] have described a digestion procedure for the determination of phosphate buffer extractable sulfate in soils. The amount of sulfate extracted from a given substrate depends on the extraction procedure as well as the adsorptivity and solubility of the sulfate constituents. A phosphate buffer solution (2.23 g/l Na_2HPO_4 H_2O in double-distilled water) will remove sulfate due to the higher affinity of phosphorus for anion exchange sites. Soil samples of up to 5 grams are placed in 0.25 ml flasks and extracted in 100 ml of phosphate buffer solution by shaking vigorously for an hour. The suspension is centrifuged to remove suspended particulates. The supernatant is then placed into the barrel of a 10 ml disposable syringe fitted with a filter adapter and the sample filtered through a GF/C filter (Whatman, 98% retention of 1.2 μm). Filtrate (up to 2 ml) is added to the digestion flasks and the hydriodic acid reduction procedure is followed to convert sulfate to sulfide. Sulfide is then estimated by the digestion–distillation procedure followed by spectrophotometry of the p-aminodimethylaniline–ferric ammonium sulfate complex.

Flow Injection Analysis

Krug et al. [66, 67] used flow injection turbidimetry to determine sulfate in natural waters and plant digests. They described an improved flow injection system with alternative streams of reagents. Samples were injected into an inert carrier comprising 0.3% EDTA disodium salt and 0.2 mol/l sodium hydroxide. The inert carrier is mixed with 5% barium chloride containing 0.05% polyvinyl alcohol to form a barium sulfate. The range of the method can be extended to low concentrations by continuously adding sulfate to the sample carrier stream. System performance is improved by automatic pumping of the reagent stream and an alkaline EDTA solution at high flow rate. All operations were controlled by an electronically operated proportional injector commutator. No baseline drift was observed even after analysing 3000 samples. The method is capable of analysing 120 samples per hour with a relative standard deviation of less than 1% for sulfate concentrations in the range $1-30$ mg/l. Analytical recovery was $97-102$%.

Atomic Absorption Spectrometry

In an indirect method for determining sulfate in soil extracts, Little et al. [68] precipitate sulfate as the lead salt in 40% ethanol medium. Unconsumed soluble lead is determined by atomic absorption spectrometry. The method is applicable to soil samples containing as little as 4 mg/kg sulfate.

Molecular Emission Cavity Analysis

Molecular emission cavity analysis has been used to determine soluble sulfate in soil [69].

Ion Chromatography

The Bradfield and Cooke [45] procedure (Sect. 6.12) determines sulfate in soils.

Miscellaneous

Bolan et al. [70] used packed columns of soil with high and low sulfate adsorption capacity in laboratory studies of the movement of sulfate through the soil. Adsorption isotherms were obtained by batch experiments, using ^{35}sulfide as a tracer for the movement of the applied ^{32}sulfide. The differences in movement could be explained by reference to the adsorption isotherms. Breakthrough curves were obtained for varying concentrations of the applied sulfate solution. These were in good agreement with curves obtained by numerical solution of the dispersion–convection equation assuming a Freundlich absorption isotherm and instantaneous reversible adsorption. However, calculations based on these assumptions failed to account for sulfate distributions in the soil columns after leaching of a pulse of sulfate added to the surface soil. The reasons for this are discussed.

6.18
Sulfide and Other Sulfur Functions

Sulfur is an important component of both natural and anthropogenic processes. Sulfur's role in atmospheric, aquatic and terrestrial systems has been investigated due to its importance both in the formation of acidic precipitation and as a macronutrient required by all organisms [71, 72]. Sulfur has a vast array of both inorganic and organic chemical species. Our understanding of sulfur dynamics has been restricted due to a lack of information on the role of specific sulfur constituents in affecting sulfur fluxes and transformations. For example, a knowledge of organic sulfur is very important when evaluating forest soils. Previous work on such substrates has generally ignored the organic sulfate constituents, with most work focusing only on inorganic sulfate and sulfide.

Landers et al. [65] have combined and modified various analytical methods used to determine the major sulfur constituents in soils. Independently, these methods are useful. However, in combination with the same digestion–distillation apparatus, they provide a reliable and convenient group of analytical methods that can be used in investigations of sulfur dynamics. Landers et al. [65] have also described a set of analytical methods for the determination of sulfide and sulfate in soils. Landers et al. [65] used a digestion flask to determine hydrochloric acid-digestible sulfur (method 1), zinc hydrochloric acid-reducible sulfur (method 2) and hydroidic acid-reducible sulfur (method 3) in soils.

Method 1: Hydrochloric acid digestion sulfur apparatus (for acid-digestible inorganic sulfur) with digestion–distillation flask [73].

The following reagents are added to the trapping flask and the gas washing column, respectively:

An acetate *trapping solution.*

Stock solution: dissolve 50 g of zinc acetate in 12.5 g of sodium acetate in double-distilled water, dilute to one litre and filter.

Diluted solution: before each analysis mix 100 ml of stock solution with 200 ml of double-distilled water, and add 80 ml of this solution to each gas trapping flask.

Pyrogallol–sodium phosphate solution: Dissolve 10 g sodium dihydrogen phosphate (NaH_2PO_4 H_2O) and 10 g of pyrogallol in 100 ml of double-distilled water, bubble with nitrogen to dissolve. Prepare daily and discard when brown coloration develops. Add 10 ml of this solution to the gas washing column.

The soil sample (0.05 – 2 g) is placed in the digestion flask and 10 ml of 1:1 hydrochloric acid is added. All connections are closed quickly, the nitrogen flow is commenced, and the samples are refluxed. The treatment reduces various sulfur constituents to hydrogen sulfide, which is moved by the stream of nitrogen into the trapping flask where it forms zinc sulfide. Colorimetric reagents, *p*-amino dimethylaniline sulfate [76] and 12.5% ferric ammonium sulfate in 97.5% sulfuric acid, are added to the gas trapping flask and then acid-digestible inorganic sulfur is estimated via the 670 nm absorption maximum.

Method 2: Zinc hydrochloric and reducible sulfur (non-sulfate inorganic sulfur) [74].

The trapping solution is prepared as for Method 1 and the same reagents are used in the trapping, the flask and the gas washing column. A wet sample (0.05 – 0.2 g) is placed into a digestion flask containing about 2 g of granulated zinc metal. The system is flushed with nitrogen, 10 ml of 1.1 M hydrochloric acid is added, and the solution is boiled for an hour. The gas flow is continuous when adding the reagent in order to prevent liberated hydrogen gas from causing the sample to enter the gas import tubes. Extreme foaming has been a problem with some soil samples but the addition of 1.5 ml of an antifoam spray (AH Thomas Co., Philadelphia, PA) has solved the problem and no interference has been found. Sodium thiosulfate is used as a standard.

The hydrogen sulfide produced is estimated by the *p*-amino dimethylaniline–ferric ammonium sulfate spectrophotometric method [76], utilising the

670 nm absorption maximum. Reagent blanks are run. Reagent blanks are low (< 0.03 μM s).

Method 3: Hydriodic acid-reducible sulfur (non-carbon-bonded sulfur) [76].

The trapping system is prepared as described in Method 1, and the same reagents are used in the trapping flask and the gas washing column. Prepare the hydriodic acid reducing agent as follows: combine 300 ml of hydriodic acid, 75 ml hypophosphorus acid (50%) and 150 ml of 88% formic acid. Boil gently with a nitrogen gas stream for ten minutes after reaching 115 °C. During the ten-minute period the temperature is kept between 115 °C and 117 °C. Upon completion, the reagent may appear bright yellow or brown, apparently depending on the quality of the reagents used. The mixed reagent has a shelf-life of about two weeks.

A wet sample (0.01 – 0.1 g) is added to the digestion flask and 4 ml of mixed reagent are added. The gas flow is started and the sample is refluxed for an hour. Potassium sulfate is used for a standard. The hydrogen sulfide produced is estimated by the *p*-amino dimethylaniline–ferric ammonium sulfate spectrophotometric method [76] utilising the 670 nm absorption maximum. Reagent blanks are run. Reagent blanks are low (< 0.03 μM s).

Examples of soil analyses carried out by Landers et al. [65] for acid-digestible inorganic sulfur (HCI-S) non-sulfate inorganic sulfur (Zn–HCI-S) non-carbon-bonded sulfur (HI–S), as well as total sulfur, sulfate and carbon-bonded sulfate ($C–O–SO_3$), are shown in Table 6.3.

Ester sulfate and carbon-bonded sulfur are the main sulfur constituents of these soils.

Clark and Lesage [76] have described a method for the determination of elemental sulfur in soils using gas chromatography with flame photometric detection after the sulfur is reacted to form Ph_3PS.

Ray et al. [77] used an indirect method based on AAS for the determination of sulfide in flooded acid sulfate soils. Hydrogen sulfide, evolved during the anaerobic metabolism of sulfate, is readily converted into insoluble metal sulfides, chiefly iron sulfide, in flooded acid sulfate soils. A method for determining sulfide is based on the precipitation of the sulfide as zinc sulfide and subsequent determination by methylene blue formation or iodine titrimetry.

Table 6.3. Sulfur constituents in soils in mol/gb (from [65])

	Total S	HCI–S	Zn–HCI–S	HI–S	SO_4^{-2}–S	C–S	C–O–SO$_3$
Forest Soil 1	50 ± 9.3	nd	0.62 ± 1.1	8.25 ± 0.93	0.48 ± 0.08	41.9	7.16
Forest Soil 2	16.5 ± 1.6	nd	0.69 ± 0.03	5.00 ± 0.62	0.72 ± 0.13	11.5	3.59

nd: Not detectable

Ray et al. [77] have also described a method for determining sulfide in soil extracts involving the precipitation of zinc sulfide by the action of zinc on the hydrogen sulfide-flooded acid sulfate soil, and then indirect determination of sulfide by determining the zinc in the precipitate and also the zinc remaining in solution, after the precipitation by AAS. Over 85% of the sulfide was recovered in this procedure.

6.19
Thallate

See Sect. 2.25.

6.20
Tungstate

Mehra and Frankenberger [78] used ion chromatography to determine tungstate in soil.

6.21
Uranate

See Sect. 2.28.

6.22
Vanadate

Abbasi [75] determined metavanadate in solution by a method based on the formation of a violet colour with vanadium(V) on addition of a chloroform solution of N-(p-NN dimethylanilino-3-methoxy-2-naphtho)hydroxamic acid to the acidified (4–6 mol/l hydrochloric acid) sample. This solution was evaluated spectrophotometrically at 570 nm. The detection limit was 0.05 µg vanadium at a dilution ratio of 1:10^7. Very few interferences occur in this procedure. The method was also applied to extracts of soils, plants and geological samples. See also Sect. 2.29.

6.23
Multiple Anions

Bradfield and Cooke [45] have described an ion-chromatographic method employing a UV detector for the determination of nitrate sulfate and phosphate in water extracts of soils (see Fig. 6.4). Soils are leached with water and Dowex 50-X4 ion exchange resin added to the aqueous extract, which is then passed

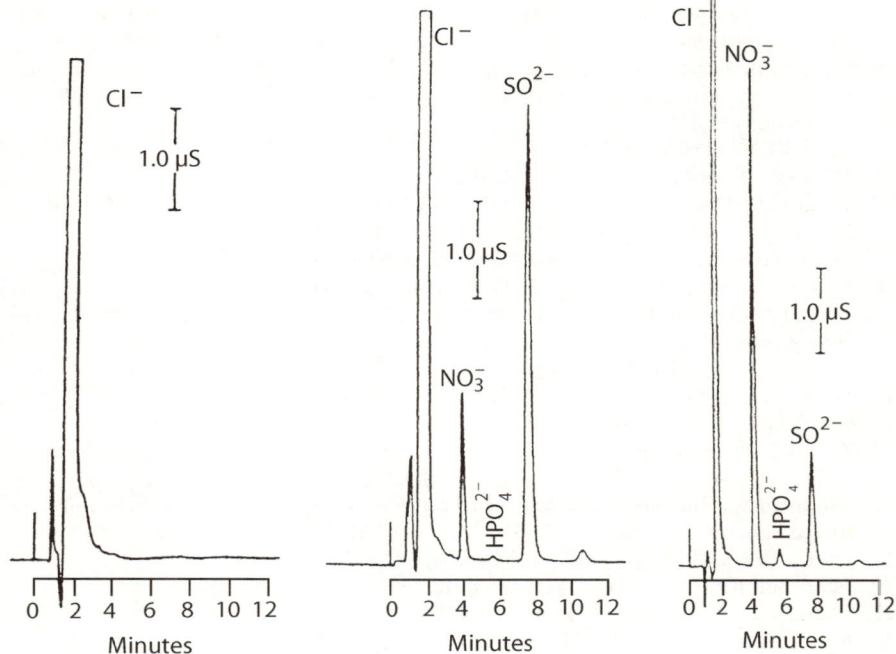

Figure 6.4. Determination of anions in soil extracts. *left*: Blank, 10 mol/l KCl; *middle*: Soil sample A, 10 mmol/l KCl extract; *right*: soil sample B, 10 mmol/l, KCl extract (1:500 dilution), AMPIC–NGI should also be used in series to remove humic acids. From [45]

through a Sep-Pak C_{18} cartridge and the eluate passed through the ion chromatographic column. The best separation of these anions was obtained using a 5×10^{-4} mol/l potassium hydrogen phthalate solution in 20% methanol at pH 4.9. A reverse-phase system was employed. Detection times were 5.5, 7.9, 12.6 and 18 minutes for chloride, nitrate, phosphate and sulfate, respectively. Recoveries ranged from 84 to 108% with a mean of 97%.

References

1. Ducret L (1957) *Anal Chim Acta* **17**:213.
2. Aznarez J, Bonilla A, Vidal JC (1983) *Analyst* **108**:368.
3. Roughan JA, Rothan PA, Wilkins JPG (1983) *Analyst* **108**:742.
4. Van Staden JF (1987) *Analyst* **112**:595.
5. Van Staden JF (1986) *Anal Chim Acta* **179**:407.
6. Gladney ES, Perrin DR (1979) *Anal Chem* **51**:2015.
7. Collins SHJ (1906) *Soc Chem Ind Lond* **25**:518.
8. Ministry of Agriculture, Fisheries and Food (1973) *The Analysis of Agricultural Materials – Carbonate in Soil, Method 15*, Technical Bulletin 27, HMSO, London, UK.

9. Ministry of Agriculture, Fisheries and Food (1979) *The Analysis of Agricultural Materials – Chlorate in Soil, Method 16*, Technical Bulletin RB 427, HMSO, London, UK.
10. Ministry of Agriculture, Fisheries and Food (1973) *The Analysis of Agricultural Materials – Chloride in Soil, Method 19*, Technical Bulletin RB 427, HMSO, London, UK.
11. Clarke FE (1950) *Anal Chem* **22**:553.
12. Davy BG, Bembrick MA (1969) *Proc Soil Sci Soc America* **33**:385.
13. McLeod S, Stace HCT, Tucker BM, Bakker P (1974) *Analyst* **99**:193.
14. Tecator Ltd. (1987) *Cyanides in Waste Waters, Soils and Sludges using the 1026 Distilling Unit*. Application Note AN 89/87 and AN 86/87, Tecator Ltd, Hoganes, Sweden.
15. Van Vliet H, Basson WD, Böhmer RG (1975) *Analyst* **100**:405.
16. Mehra HC, Frankenberger Jr. WT (1989) *Anal Chim Acta* **217**:383.
17. Ministry of Agriculture, Fisheries and Food (1979) *The Analysis of Agricultural Materials – Ammonium, Nitrate and Nitrite Nitrogen, Potassium Chloride Extractable in Soil Method 60*, Technical Bulletin RB427, Second Edition, HMSO, London, UK.
18. Henrickson A, Selmer-Olson AR (1970) *Analyst* **95**:514.
19. Garcia Gutierrez G (1973) Infeion *Quin Analet Pura Apl Ind* **27**:171.
20. Tecator Ltd. (1983) *Determination of Nitrate and Ammonia in Soil Samples, Extractable with 2 M Potassium Chloride* Application Note AN65/83 (1983) and *Determination of Nitrate in Soil Samples, Extractable with 2 M Potassium Chloride using, Flow Injection Analysis*, Application Note AN65-31/83, Tecator Ltd., Hoganes, Sweden.
21. Lindau CW, Spalding RF (1984) *Groundwater* **22**:273.
22. Elton-Bott RR (1977) *Process Technology* **8**:215.
23. Osibanjo D, Ajayi SO (1980) *Analyst* **105**:908.
24. Keay J, Menage PMA (1970) *Analyst* **95**:379.
25. Hadjidemetriou DG (1982) *Analyst* **107**:25.
26. Waughman A (1981) *Environ Res* **26**:529.
27. Bremner JM, Bundy LG, Agarwal AS (1968) *Anal Lett* **1**:837.
28. Goodman D (1976) *Analyst* **101**:943.
29. Myers RJK, Paul EA (1962) In: Jackson ML (ed) *Soil Chemical Analysis*, Constable, London, UK, p. 197.
30. Bremner JM (1965) In: Black CA (ed) *Methods of Soil Analysis, Pt. 2*, American Society of Agronomy, Madison, WI, USA, p. 1191.
31. Sims JR, Jackson GD (1971) *Soil Sci America Soc Proc* **35**:603.
32. Mahendrappa MK (1969) *Soil Sci* **108**:132.
33. Qien A, Selmer-Olsen AR (1969) *Analyst* **94**:888.
34. Fiskell JGA, Breland HA (1969) *Soil Crop Sci Soc Fla Proc* **29**:63.
35. Milham PJ, Awad AS, Paull AS, Bull JH (1970) *Analyst* **95**:751
36. Smith GR (1975) *Anal Lett* **8**:503.
37. Krupsky NK, Alexandrova AM, Gubareva DN, Varenik VA (1978) *Agrokhimya* **10**:133.
38. Revek A (1973) *Soil Science* **116**:388.
39. Tchagina EG, Dubinina RI, Golovin VA, Materova EA, Grekovitch AA (1980) *Agrokhimiya* **5**:134.
40. Houba VJG, van Schowenburg JC, Walinga I, Novozamsky J (eds) (1979)*Soil Analysis II: Methods of Analysis For Soils* ,Agricultural University, Wageningen, The Netherlands, p. 43, 83.
41. Bound GF (1977) *J Sci Food Agric* **28**:501.
42. Orion Research (1978) *Instruction Manual for Nitrate Ion Electrode, Model 93-07*, Orion Research, Cambridge, MA, USA.
43. Nasko BS, Alexandrova AM, Gubareva AM, Razday VC (1980) *Agrokhimy,* **4**:131.

44. Ministry of Agriculture, Fisheries and Food (1979) *The Analysis of Agricultural Materials – Nitrate Nitrogen, Calcium Sulfate, Extractable in Soil Method 59,* Technical Bulletin RB427, Second Edition, HMSO, London, UK.
45. Bradfield EG, Cooke DT (1985) *Analyst* **110**:1409.
46. Bhuchar VM, Amar UK (1972) *Indian J Tech* **10**:433.
47. Chaube A, Baveja AK, Gupta VK (1984) *Talanta* **31**:391.
48. Wu Q-F, Liu P-F (1983) *Talanta* **30**:374.
49. Smith CJ, Chalk PM (1979) *Analyst* **104**:538.
50. Gerretsen FC, de Hoop H (1957) *Can J Microbiol* **3**:359.
51. Wagner GH, Smith GE (1958) *Soil Sci* **85**:125.
52. Soulides DA, Clark FE (1958) *Proc Soil Sci Soc Am* **22**:308.
53. Clark FE, Beard WE, Smith DH (1960) *Proc Soil Sci Soc Am* **24**:50.
54. Khan MFA, Moore AW (1968) *Soil Sci* **106**:232.
55. Schwartzbeck RA, MacGregor JM, Schmidt EL (1961) *Proc Soil Sci Soc Am* **25**:186.
56. Meek BD, MacKenzie AJ (1965) *Proc Soil Sci Soc Am* **29**:176.
57. Steen WC, Stojanovic BJ (1971) *Proc Soil Sci Soc Am* **35**:277.
58. Murphy J, Riley JP (1962) *Anal Chim Acta* **27**:31.
59. Bickford GP, Willet IR (1981) *Water Res* **15**:511.
60. Karlson U, Frankenberger WT (1986) *Anal Chem* **58**:2704.
61. Ogner G, Haugen A (1977) *Analyst* **102**:453.
62. Ministry of Agriculture, Fisheries and Food (1979) *The Analysis of Agricultural Materials – Acid Soluble and Water Soluble Sulfate Sulfur in Soil Being Considered for Tile Draining Method 87,* Technical Bulletin RB427, Second Edition, HMSO, London, UK.
63. Ministry of Agriculture, Fisheries and Food (1979) *The Analysis of Agricultural Materials – Sulfate Sulfur Total Water Soluble in Soil Method 76,* Technical Bulletin RB427, Second Edition, HMSO, London, UK.
64. Chauhan PPS, Chauhan CPS (1979) *Soil Sci* **128**:193.
65. Landers DR, David MB, Mitchell MJ (1981) *Int J Environ Anal Chem* **14**:245.
66. Krug FJ, Zagatto EAG, Reis BF, Bahia FOO, Jacintho AO, Jorgensen SS (1983) *Anal Chim Acta* **145**:179.
67. Krug FJ, Filho HB, Zagatto EAG, Jorgensen SS (1977) *Analyst* **102**:503.
68. Little LP, Reeve R, Proug GM, Luchan AJ (1969) *Sci Food Agric* **20**:673.
69. Al-Ghabsha TS, Bogdanski SL, Townshend A (1980) *Anal Chim Acta* **120**:383.
70. Bolan NS, Scotter DR, Syers JK, Tillman RW (1986) *Soil Sci Soc Am J* **50**:1419.
71. Shriner DS, Henderson GSJ (1978) *Environ Qual* **7**:392.
72. Smittenberg J, Harinsen GW, Quispel A, Otzen D (1951) *Plant Soil* **3**:535.
73. Aspiras RB, Keeney DR, Chesters G (1972) *Anal Lett* **5**:425.
74. Johnson CM, Ulrich A (1959) *Calif Agric Exp Station Bull No.766.*
75. Abbasi SA (1981) *Int J Environ Stud* **18**:51.
76. Clark PD, Lesage KL (1989) *J Chromatogr Sci* **27**:259.
77. Ray RC, Nayar PK, Misra AK, Sethunathan N (1980) *Analyst* **105**:984.
78. Mehra HC, Frankenberger WT (1989) *Analyst* **144**:707.

7 Determination of Cations in Plant Materials, Vegetables and Fruit

7.1
Aluminium

Inductively coupled plasma (ICP) methods have been used to determine aluminium in fruit [1]. See Sect. 7.34.2.

7.2
Antimony

See Sects. 7.34.2 and 7.34.4.

7.3
Arsenic

Lisk [2] has described a molybdenum blue spectrophotometric method for the determination down to 10 μg of arsenic in potatoes. See also Sects. 7.34.1, 7.34.2 and 7.34.7.

7.4
Barium

See Sect. 7.34.2.

7.5
Beryllium

See Sect. 7.34.2.

7.6
Bismuth

See Sects. 7.34.2, 7.34.4 and 7.34.7.

7.7
Cadmium

Atomic absorption spectrometry [AAS] has been used to determine cadmium in fruit [3] and vegetables [3–5]. Detection limits of 0.06 mg/kg[5] and 0.1 mg/kg were achieved. Culver determined down to 2×10^{-12} g of cadmium in potatoes and spinach.

A standard official method [6] has been used for the determination of cadmium in plant material. In this method the sample is digested with 1:4 v/v perchloric acid:nitric acid at 200 °C and the residue dissolved in hydrochloric acid. Cadmium is then converted to the diethyldithiocarbamate. A chloroform extract of this solution is used for the determination of cadmium at the 228.8 nm emission line. See also Sects. 7.34.1 and 7.34.4.

7.8
Calcium

A standard official method has been published for the determination of calcium in plant material [7]. The plant material is ashed and the residue dissolved in hydrochloric acid. Following the addition of strontium chloride releasing solution, calcium is determined by AAS at 422.7 nm. See Sect. 7.34.1.

7.9
Chromium

See Sect. 7.34.1.

7.10
Cobalt

A standard official method has been published for the determination of cobalt in plant material [8]. The samples are digested with 1:4 v/v perchloric acid:nitric acid and the residue dissolved in nitric acid. Cobalt is then extracted into chloroform as the diethyldithiocarbonate. The latter complex is decomposed by bromine and cobalt extracted into dilute hydrochloric acid. Following the addition of a borate buffer, cobalt is then extracted as the o-nitrocresol complex [9]. Excess coupling agent is removed by repeated extraction with copper acetate solution and cobalt determined spectrophotometrically at 360 nm. See Sects. 7.34.1, 7.34.3 and 7.34.4.

7.11
Copper

AAS is the method of choice for the determination of copper in plant material.

In an official method [10], the residue obtained by combustion of the plant material is dissolved in hydrochloric acid and the concentration of copper in this solution is determined either by AA spectroscopy [11] at 324.8 nm or spectrophotometrically as the yellow-brown dibenzyldithiocarbonate complex in carbon tetrachloride at 440 nm.

Simmons and Loneragan [12] point out that the concentration of copper present in plant material is of considerable interest, since in many parts of the world it is deficient for the growth of either the plant itself or for animals grazing on the plant. Levels of 3–5 μg copper/g of plant material are considered to be inadequate for the copper requirements of animals. These workers describe a L'vov platform atomic absorption spectrometric method using a heated graphic atomiser for the determination of copper in plant material such as grass, wheat and lupin. The sample is digested with 1:9 perchloric acid:nitric acid at temperatures up to 200 °C. The residue is dissolved in 3% perchloric acid and, following the addition of ammonium pyrrolidine carbodithionate, copper is extracted into methyl isobutyl ketone prior to atomic adsorption spectrometric determination at 324.7 or 327.4 nm. Simmons [13] showed that background adsorption is a potentially serious source of error when analysing perchloric acid digests of plant materials such as kale for copper by AAS. This effect was mainly due to the presence of calcium. The magnitude of the background adsorption was modified by flame conditions, calcium concentration and the levels of other major elements present. Use of a lean flame, a low total gas flow rate and an observation height of 11 mm minimised the effect of high calcium levels. Background correction or chemical separation of the copper from the matrix is recommended when very low concentrations of copper accompany high calcium levels. See also Sects. 7.34.1, 7.34.4 and 7.34.7.

7.12
Germanium

See Sect. 7.34.2.

7.13
Gold

Schiller et al. [14] used neutron activation analysis to examine the gold contents of soil and plants in an attempt to correlate the two.

7.14
Iron

A standard official method has been published for the determination of iron in plant material [15]. The iron remaining after the destruction of the organic material is dissolved in 0.4 m hydrochloric acid. The concentration of iron in this solution is determined spectrophotometrically at 570 nm following reduction to the ferrous state with quinol and reaction with 1:10 phenanthroline.

Mortatti et al. [16] has described a method for the determination of total iron in extracts of perchloric acid–nitric acid digests of plant materials. Their method involves flow injection analysis of the 1:10 phenanthroline complex.

Effects of mixing coil lengths, sample volume, flow rates, reagent concentrations and interfering species were investigated. The proposed procedure allows the determination of iron concentrations in the range 0.1–30 ppm at a rate of up to 180 samples per hour with relative standard deviations of lower than 1%. The results agree with those obtained by AAS, and an iron concentration identical to the certified value was found by analysing NBS standard reference orchard leaves.

Jaymen et al. [17] and other workers [18–20] found that aluminium in plant digests enhances the iron–1,10-phenanthroline colour, leading to high results in the determination of iron. Both the iron and aluminium complexes of phenanthroline exhibit identical absorption characteristics. Attempts to mask the aluminium in solution with sodium fluoride have been unsuccessful, as the fluoride ions suppress the colour formed with iron and reagent. The determination of iron after the separation of aluminium and phosphates is simple and rapid. This method is reliable and recoveries are quantitative.

The separate aluminium and phosphate from iron the finely ground samples left overnight in a furnace at 450 °C. The residue is dissolved in nitric acid:hydrochloric acid:water (20:25:50 v/v) and then evaporated to dryness. The residue is dissolved in 0.05 N hydrochloric acid. Portions of this extract are used for the determination of iron.

See also Sects. 7.34.1, 7.34.4 and 7.34.7.

7.15
Lead

A standard official method is available for the determination of lead in plant material [21]. Lead is determined in an acid extract of the plant material by reaction with ammonium pyrrolidine diethyldithiocarbomate, extraction with chloroform, and AA spectrometric evaluation at 217 nm.

Thomas et al. [22] used chelation followed by AAS with electrothermal atomisation to determine down to 2.5 mg/kg lead in fruit and vegetables and in apples [23]. Stafilov and Rizova [24] used AAS to determine lead in cereals.

Stephens et al. [25] used a slurry technique to determine lead in spinach by AAS. In this method, the powdered spinach is suspended in a thixotropic thickening agent, Viscalex HV30, and the slurry is injected directly into the electrothermal atomiser. Oxygen is introduced during the ashing stage to allow the use of higher ash temperatures and to avoid the build-up of carbonaceous residue in the tube. Concentrations of up to 10% m/v of powdered spinach can be tolerated in the suspension. Good agreement was achieved between results obtained by a standard additions procedure and by direct calibration with aqueous standards, and also by an alternative wet digestion procedure.

Approximate detection limits were calculated from the mean absorbance and relative standard deviation values obtained for each spinach slurry concentration, and the results are presented in Table 7.1. Although the values cannot be directly compared as they have been calculated at different absorbance values, and they do not represent true detection limits as the concentrations used are too high owing to the level of lead in the spinach sample used, they clearly indicate that an increase in spinach concentration allows lower detection limits to be achieved.

To check the overall precision of the procedure, five 3% m/v spinach slurries were individually prepared and analysed. The relative standard deviation was calculated from the means of the five absorbance values obtained from each spinach sample. The precision of the method was 3%.

The slurry technique is an attractive compromise offering simplicity of sample preparation along with the convenience of a permanent or semi-permanent sample "solution" for repetitive analysis. The application of higher powder concentrations is a practical possibility, at least as far as lead is concerned, and then lower detection limits are then feasible.

Workers at PerkinElmer (Perkin Elmer Ltd., Beaconsfield, UK, Private Communication) claim that sensitivity and precision in lead determinations in spinach were both improved by using the L'vov platforms as opposed to normal "off the wall" techniques. This is because dependence on temperature is reduced. Johns et al. [26] has described a quantitative thin-layer chromatographic procedure for the determination of lead in plant tissues. The method is based on the use of ammonium pyrrolidinedithiocarbamate for the extraction and enrichment of lead. Instead of the previously reported conversion and visualisation of the lead complex in the short wavelength region using dithizone, the lead carbamate was converted to lead sulfide with the aid of a 6% solution of sodium sulfide in methanol/water (3:1) on silica plates, after development of the plates with toluene. See also Sects. 7.34.1, 7.34.4 and 7.34.5

Table 7.1. Detection limits for lead with increasing concentrations of spinach powder in the slurry (from [25])

Concentration of spinach powder in slurry, % m/V	Mean absorbance	Relative standard deviation, %	Calculated detection limit*, µg/g
1	0.055	9.8	0.360
3	0.175	2.2	0.104
6	0.338	4.5	0.105
10	0.462	2.1	0.047

*: Calculated on the basis of 2σ using the data in columns 2 and 3 (see text for explanation)

7.16
Magnesium

An official method has been published for the determination of magnesium in plant material [27]. A hydrochloric acid digest of the sample is treated with strontium chloride perchloric acid releasing agent and magnesium is determined by AAS using the 285 nm emission line. See Sect. 7.34.1.

7.17
Manganese

This element has been determined in perchloric acid digests of plant materials by a spectrophotometric procedure as permanganate ion obtained by oxidation with periodic acid or by AAS using the 279.5 nm emission line [28].

Gine et al. [29] has described a semi-automatic determination of manganese in plant digests using flow injection analysis. This technique utilises the introduction of the sample into a continuously flowing carrier stream of formaldoxime reagent. When injected, the sample is pushed by this stream and dispersed into the reagent stream, whereupon the required reaction takes place. The coloured complex is then carried into a spectrophotometric flow cell, where the absorbance is measured after an exactly defined time interval.

The system employed for this work is shown in Fig. 7.1. Polyethylene tubing of 0.86 mm i.d. was used and all connections were made from Perspex®. The pump was a Technicon AAII peristaltic pump, fitted with Tygon pumping tubes. The samples were injected by means of a proportional injector. The sample was aspirated to fill a loop, which exactly defined the injected volume; this loop was then placed as part of a carrier stream.

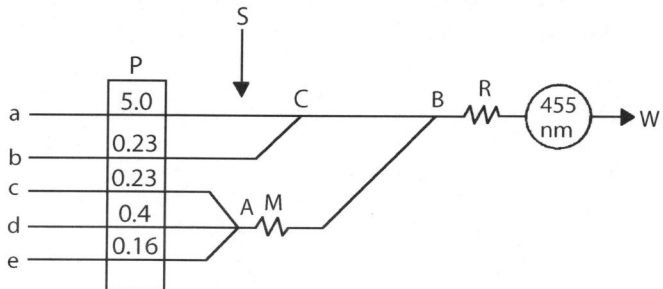

Figure 7.1. Flow diagram of the system for determination of manganese. *P*: peristaltic pump; *S*: injection port; *R*: reaction coil (length 30 cm); *M*: mixing coil (length 60 cm); and *W*: waste. The numbers in the pump are the flow rates in ml/min of the carrier, reduction, reagent, neutralisation and masking streams, which correspond to (a), (b), (c), (d) and (e), respectively. The distances S–C and C–B are approximately 2 cm. From [29]

Water samples were injected without any pretreatment other than preservation (5 ml of 2 m sulfuric acid were added per 1000 ml of sample, immediately after collection) and plant samples were digested with nitric–perchloric acid using a Technicon BD-40 block digestor [30].

The formaldoxime, masking and neutralisation streams are added to each other at point A (Fig. 7.1) and mixed by passage through coil M. At point B, the reagents meet the sample zone, which had previously received a reduction stream of ascorbic acid at point C. The formation of the coloured complex takes place in the reduction coil (R) and the absorbance is measured at 455 nm in a Beckman Model 25 spectrophotometer connected to a Beckman Model 24–25 ACC recorder and equipped with a Hellma Type 178 flow cell, light path 10 mm, volume 0.08 ml.

For the analysis of plant digests, the acidic conditions of the samples were maintained in the carrier stream. Consequently, the sodium hydroxide concentration in the neutralisation stream was changed in order to achieve the required alkaline conditions in the final stream.

In order to prevent the reduction between iron(II) and formaldoxime occurring, another iron complexing agent (potassium cyanide) was used in the presence of a reductant (ascorbic acid) that reduces iron(III) to iron(II). Aluminium, titanium, uranium, molybdenum and chromium also form light-coloured complexes that normally do not interfere in the determination of manganese in water or plant material by this method. If the aluminium or titanium concentrations are higher than 40 ppm an additional masking flow of tartrate is recommended [31].

The effect of the presence of suspended and coloured materials in the sample was evaluated by replacing the formaldoxime with water and running the sample again, so as to obtain blank values.

A series of determinations were performed using the system in Fig. 7.1, with an injected volume of 0.35 ml. Under these conditions, about 135 samples per hour could be analysed, with a standard deviation of better than 1% over the range 0.2–2 ppm of manganese.

In order to check the accuracy of the proposed method, a comparison of this method with AAS was made. In general, good agreement between the results was obtained [32]. See also Sects. 7.34.1, 7.34.4 and 7.34.7.

7.18
Mercury

Atomic Absorption Spectrometry (AAS)

In an official method [33] for the determination of mercury in plant material, the sample is digested at 150 °C with 70% m/v concentrated nitric acid in a pressure vessel. Potassium permanganate is added to prevent loss of mercury and remove nitrogen oxides, and the concentration of mercury was determined either by flameless AAS at 253 nm or by flameless atomic fluorescence.

Hon et al. [34] describe a simple piece of equipment for the determination of down to 80 µg/l of mercury by AAS using a static cold vapour procedure. In this method [35], the sample was digested with the sulfuric acid, a measured portion pipetted into the reduction vessel, and the vessel immediately capped. The reductant, comprising 1% stannous chloride, was introduced. The evolved elemental mercury in the headspace was then introduced into the absorption cell by water displacement. Maximum sensitivity is obtained when the volume of the displaced air is equal to the internal volume of the absorption cell, and the mercury solution is 9 M in sulfuric acid. The peak absorbance at 253.7 nm exhibited a marked decline for hydrochloric acid concentrations above 1.5 M and for nitric acid concentrations above 3 M. The calibration graph obtained for mercury(II) in 9 M sulfuric acid is linear from 0 to 17 ng/ml, and the sensitivity is 0.08 ng/ml. A windowless absorption cell can also be used with a narrower linear calibration range.

The instrument used was a Perkin-Elmer Model 360 atomic-absorption spectrometer with the burner removed. A Varian Techtron mercury hollow-cathode lamp was used as the light source and the wavelength and band width were set at 253.7 and 0.2 nm, respectively. The lamp current was 4.5 mA.

The absorption cell was made of rectangular-section Plexiglas of dimensions 17 cm × 3.3 mm × 5 mm (the 5 mm section being mounted vertically) (see Fig. 7.2). The cell had an internal volume of 2.8 ml and two quartz windows were attached to its ends with epoxy cement. The cell was fixed on a metal support, which in turn was fitted to the burner support hole. The inlet and outlet ports (2 mm id) were about 2 mm from each end of the absorption cell. A windowless cell with similar dimensions was also made, but the inlet port was in the middle of the cell.

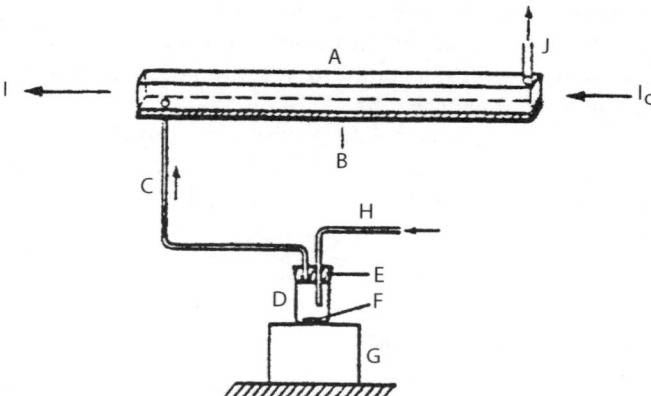

Figure 7.2. Schematic diagram of the static mercury vapour apparatus. *A*: Absorption cell; *B*: metal support; *C*: PTFE tubing; *D*: reduction vessel; *E*: silicone rubber; *F*: magnetic bar; *G*: magnetic stirrer; *H*: PTFE tubing; I_0: incident beam intensity; *I*: transmitted beam intensity and *J*: exhaust. From [34]

The reduction vessel was a glass vial (3.6 × 1.5 cm id) with a screw-cap, whose liner was removed and replaced with a layer of silicone rubber. Two pieces of Teflon tubing passed through the screw-cap; the longer one (25 cm × 2 mm od) connected the reaction vial to the inlet port of the absorption cell, and the shorter one (11 cm × 1 mm od) had about 2.5 cm of its length inside the vial and was used for the injection of reductant and distilled water (via syringes) into the vial.

Using one-gram samples of leaves with a certified 0.155 ± 0.015 µg/g of mercury, values of between 0.157 and 0.170 µg were obtained by this method.

Kimura and Miller [36] used spectrophotometry of the dithizonate to determine down to 0.1 µg of mercury in ten-gram samples of grain.

This method uses a concentrating aeration procedure at room temperature following digestion of the samples with sulfuric acid, hydrogen peroxide and potassium permanganate. Its chief advantages are the elimination of the need for filtration, its applicability to mercury solutions in sulfuric acid to 22 N, nitric and sulfuric mixtures to 4 N and 8 N, respectively, and its ability to concentrate dilute mercury solutions during the process of mercury separation in order that the entire sample may be taken for analysis. Final analyses are made from solutions of constant composition and volume regardless of the original material and volume of the digest or solution.

For turf and grain samples containing mercury(II), methyl mercury, ethyl mercury and phenyl mercury compounds, weigh a turf core of size $\frac{3}{4} \times 2$ inches or a five-gram sample into a conical flask. Add 1 ml of potassium dichromate (equivalent 10% Cr) solution in case of grains. Attach the flask to a 300 mm West-type standard taper condenser. Add 20–30 ml of 1.8 N sulfuric acid through the condenser and mix vigorously. Place the turf samples onto the stream bath with intermittent agitation for 30 minutes and cool to room temperature. Add 50% hydrogen peroxide in 0.5 ml portions with vigorous mixing after each addition. The decomposition of hydrogen peroxide being exothermic, the temperature will rise gradually to about 150 °C. When the danger of foaming subsides, increase the amount of each addition to 1 ml. Continue the peroxide addition at a rate sufficient to keep the solution bubbling gently. When the temperature drops below the proper reaction temperature and two successive additions of peroxide fail to decompose, a microburner is used to raise the temperature. The addition of peroxide is discounted after the solution turns blue-green, or in the absence of chromium, light yellow, and residual peroxide is allowed to decompose with the use of low heat after the condenser has been washed down with water. When the decomposition of peroxide has apparently ceased, add 5% permanganate slowly to give a 5 ml excess while the temperature is maintained. The permanganate is added in 5 ml portions or less until the mixed colour persists for 15 minutes. Cool the sample and add 20 ml of a solution comprising 5 N sodium chloride (250 ml) and 25% hydroxyl ammonium sulfate (75 ml). Into this solution insert an agitator–aerator, as shown in Fig. 7.3. This flask contains a reagent compris-

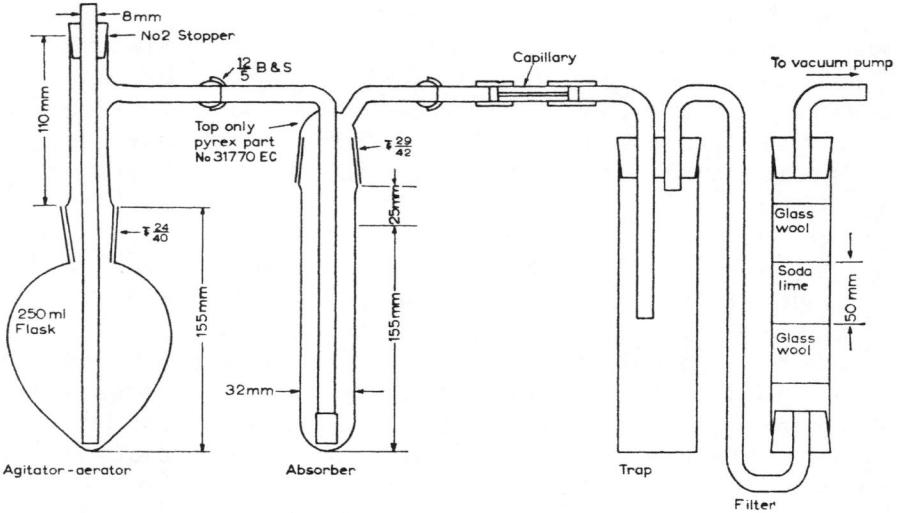

Figure 7.3. Mercury reduction–aeration apparatus. From [36]

ing 5% potassium permanganate and 18 N sulfuric acid followed later in the aeration by stannous chloride. Finally the solution is reduced with a solution comprising 25% hydroxyl ammonium sulfate and 5 N sodium chloride. In the final determination 1.8 N sulfuric acid and 1 mg/ml dithizone in chloroform are added. The chloroform phase is evaluated spectrophotometrically at 605 nm.

Recoveries of mercury ranged from 83% at the 1 µg mercury level to 96.1% at the 100 µg mercury level in turf, 94% at the 0.5 µg mercury level to 99.5% at the 4 µg mercury level in cracked whole barley, and 101% at the 0.1 µg mercury level to 94% at the 0.5 µg mercury level in wheat. For five-gram barley samples containing less than 5 µg mercury, the standard deviation of a sample determination was 0.12.

The determination of down to 0.02 – 0.25 ppm of mercury in cereals by differential pulse anodic stripping voltammetry has been discussed by Lugowska et al. [37].

Muscat et al. [38] used atomic fluorescence spectroscopy to determine down to 0.6 ng of mercury in 20 – 30 mg samples of wheat.

Tatton and Wagstaff [39] determined down to 0.01 mg/kg of methyl mercury in potatoes, apples and tomatoes by gas chromatography.

This method is automated and the autoanalyser manifold used is shown in Fig. 7.4. A Technicon sampler plate and proportioning pump were used in conjunction with a Cecil CE 212 ultraviolet monitor and a Servoscribe RE 511 recorder. Glass tubing must be used to make connections from the double

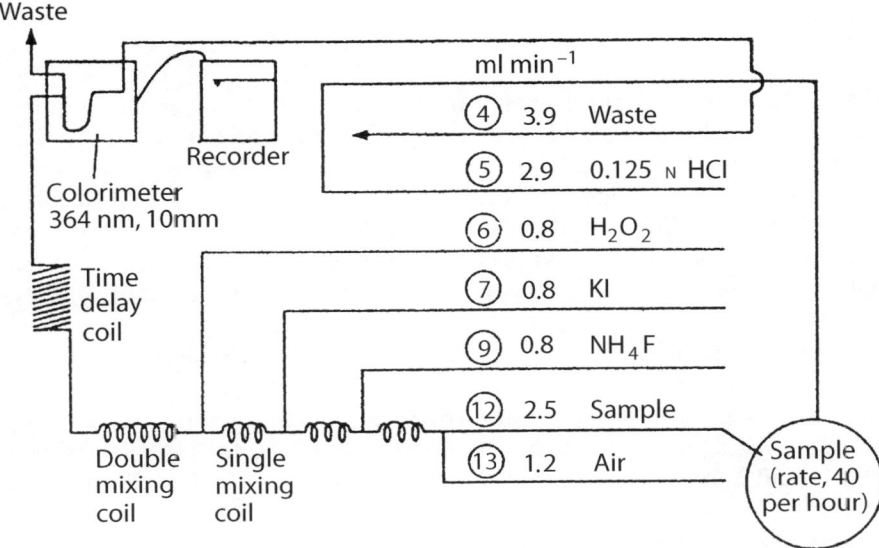

Figure 7.4. Flow diagram for autoanalyser used in the determination molybdenum. From [40]

mixing coil to the delay coil and from the delay coil to the flow cell, as iodine is absorbed onto the surface of the plastic tubing and this causes unsteady baselines.

The sample is prepared for analysis by first washing it by immersion in 0.1% Teepol solution for 15 seconds with gentle wiping of the surface. It is then rinsed rapidly in distilled water and dried at 65 to 70 °C for 24 hours. This material is ground to a powder.

For molybdenum concentrations of 0.05 – 0.5 ppm, 2 g is weighed and placed into a 50 ml conical flask. The material is ashed overnight in a muffle furnace at 450 °C. To the cooled ash 2 ml of a nitric acid–perchloric acid mixture are added and a digestion reflux funnel (as described by Bradfield [43]) is placed in the flask. Digestion occurs on a hot plate until dense white fumes of perchloric acid are evolved. The contents of the flask are evaporated to dryness.

Kosta and Byrne [42] used neutron activation analysis to determine down to 1 ng/g of mercury in 0.59 samples of flour.

Wang and Wai [43] showed that bioaccumulated mercury in plants can be recovered using a methanol-modified supercritical-fluid carbon dioxide containing a dichromate liquid.

Haller et al. [44] has reviewed methods for the determination of low levels down to 13 ppb in plant tissues.

See Sect. 7.34.1.

7.19
Molybdenum

This element is essential for plant nutrition and plant molybdenum requirements are species-dependent. A standard official method has been published
for the determination of molybdenum in plant material [45, 46].

The plant sample is digested with 60% m/m perchloric acid: 70% m/m nitric
acid, 1:4 v/v. The acids are then removed by volatilisation and any silica present
is dehydrated to render it insoluble. The concentration of molybdenum in this
solution is then determined spectrophotometrically at 470 nm on a diisopropyl
ether extract.

Alternatively, to reduce interference from any iron and tungsten present in
the sample, the powdered plant material is boiled under reflux with 5.5 N hydrochloric acid. An ethonolic solution of 2% α-monoxime is added and a chloroform extract prepared. To the chloroform phase is added nitric acid:per-
chloric acid 3:1 v/v, and the mixture is heated until perchloric acid fumes are
evolved.

The residue is dissolved in 0.25 N hydrochloric acid and molybdenum is
determined by the iodide catalytic method. Molybdenum was determined in
some plant material using both methods and the results are shown in Table 7.2.

Results obtained by direct extraction method 1 were less precise than those
obtained by the benzion α-monoxime method 2.

Hoenig et al. [47] also studied factors influencing the determination of
molybdenum in plant material by electrothermal AAS. These workers showed
that the progressive degradation of the pyrolytic graphite surface of atomisers
provides variable and misleading results for molybdenum peak height measurements. The changes in the peak shapes produce no analytical problems
during the lifetime of the atomiser ($\sim$300 firings) when integrated absorbance

Table 7.2. Molybdenum concentrations (ppm) in reference samples of plant material (from [46])

Material	Extraction (Method II) Mean	Standard deviation‡	Direct (Method I) Mean	Standard deviation‡	Reported values Mean	Ranges
Lucerne	0.18	0.023 (6)	0.20	0.043 (14)	0.20	0.19–0.23
Oat	0.18	0.016 (8)	0.20	0.043 (15)	0.15	0.12–0.18
Kale	1.10	0.176 (8)	0.90	0.126 (13)	0.79	0.76–0.81
Tomato leaf	0.45	0.035 (8)	0.28	0.074 (7)	0.42	0.36–0.46
Kale	2.48	0.205 (8)	2.31	0.165 (10)	2.33	1.86–2.80
Strawberry leaf	1.11	0.155 (8)	1.08	0.147 (14)	–	–

‡ Figures in parentheses are the number of determinations made.
Method I: Without benzoin α-monoxime
Method II: With benzoin α-monoxime

(signals) is considered and the possible base-line drifts are controlled. This was demonstrated on plant samples mineralised by simple digestion with a mixture of nitric acid and hydrogen peroxide. The value of this method was assessed by comparison with a standard dry oxidation method and by molybdenum determination on National Bureau of Standards reference plant samples. The relative standard deviations ($n = 5$) of the full analytical procedure do not exceed 7%.

Recoveries of molybdenum obtained by the nitric acid–hydrogen peroxide digestion procedure ranged from 96.4 ± 2.9% for lettuce to 100.9 ± 5.4% for beet leaves compared to 101.1 ± 5.2% for lettuce to 98.8 ± 4.95 for beet leaves obtained by dry oxidation. Recoveries obtained for reference plant materials are listed in Table 7.3.

Figure 7.5 demonstrates the dependence of peak height measurements on the state of the tube coating. In a single-element nitric acid solution, the peak height values decrease strongly after about 100 firings (curve a). Similarly, many workers have previously reported a gradual decrease in the sensitivity of the atomic absorption determination of molybdenum when a furnace coated with pyrolytic graphite was used. This suggests a progressive destruction of the tube coating by repeated exposure to high temperatures and strong acids, resulting in increased porosity of the graphite surface [48, 49]. See also Sects. 7.34.1, 7.34.3 and 7.34.4.

Bradfield and Stickland [40, 41] determined molybdenum in plant tissue by its catalytic effect on the liberation of iodine from the reaction between potassium iodide and hydrogen peroxide. The detection limit is 0.003 μg/ml of molybdenum. Interference from iron and tungsten can be overcome by addition of ammonium fluoride, but for the greatest precision and accuracy a preliminary separation of molybdenum as its benzoin α-monoxime complex is recommended.

Ten millilitres of 0.25 N hydrochloric acid is added to the residue, which is then boiled under reflux for a few minutes. Materials with high manganese content may yield pink or brown solutions at this stage. If this occurs, dilute hydrogen peroxide solution is added until the colour disappears. The molybdenum contents of the sample are then determined by the iodine catalytic method on the autoanalyser.

Table 7.3. Recovery of molybdenum from reference plant materials (from [47])

	Molybdenum content μg/g (dry matter)		
SRM (NBS)	Certified	Found after dry ashing[a]	Found after digestion[b]
1571, Orchard leaves	0.30 ± 0.10	0.340 ± 0.025	0.360 ± 0.018
1572, Citrus leaves	0.17 ± 0.09	0.240 ± 0.020	0.235 ± 0.015

SRM (NBS) Standard reference material (National Bureau of Standards)
[a] Direct oxidation method, $n = 3$
[b] HNO_3–H_2O_2, $n = 5$

Figure 7.5. Peak absorbances in molybdenum measurements during the graphite tube lifetime: (*a*) 200 pg of Mo, 6% HNO₃; (*b*) plant matrix; (*c*) plant matrix + 200 pg of Mo. The amounts of molybdenum in the same plant sample calculated by the addition method were found to be 140 and 240 pg (0.7 and 1.2 μg/g dry matter) at age 1 and 2, respectively. From [47]

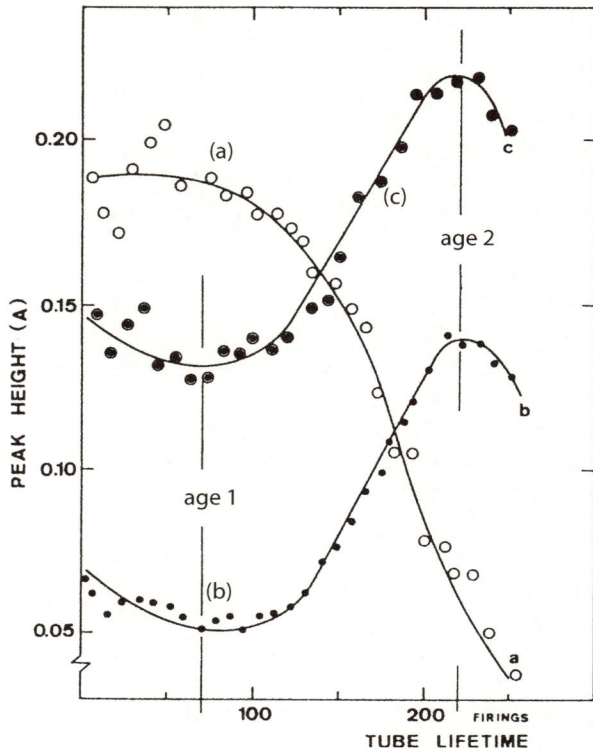

7.20
Nickel

In a standard official method [50], the plant material is prepared for analysis by either digestion with 60% *w/w* perchloric acid, 70% *w/w* nitric acid, 1:3 *m/v*, and digestion of the residue with 2 M hydrochloric acid, or by dry combustion at 500 °C followed by extraction of the residue with 6 M hydrochloric acid. The concentration of nickel in these extracts is determined by AAS at 232.0 nm employing either background correction, or by an AA spectrophotometric procedure involving formation of the nickel ammonium pyrrolidiniedithio-carbamate followed by chloroform extraction.

Differential pulse polarography [51] and adsorption voltammetry [52] have both been employed for the determination of nickel in plant tissues.

Uto and Sugawara [51] have described a procedure for the determination of nickel in the ng/ml range by differential pulse polarography. The method was used for the formation of water-soluble dithiocarbamate chelates, which were adsorbed onto the mercury electrode. An enhanced reduction current was obtained in an electrolysis solution composed of 0.1 M potassium chloride and 0.015 M ammonium oxalate/0.02 sodium hydroxide buffer in the presence of N-(dithiocarboxy)sarcosine. When evaluated using orchard leaf tissues, the method yielded recoveries of 1.36 nickel per gram, compared to certified values of 1.03 nickel per gram.

Braun and Metzger [52] showed that trace amounts of nickel obtained from natural environmental samples could be determined voltammetrically as nickel dimethyl glyoximate following adsorptive enrichment onto a rotating glassy carbon electrode, on which a thin mercury film has been deposited electrolytically. See Sect. 7.34.1.

7.21
Potassium

In a standard official method [53,54], potassium is determined in plant material by first digesting the sample with 60% perchloric acid: 70% m/m nitric acid 1:3 v/v followed by extraction of the residue with 2 M hydrochloric acid or by dry combustion at 500 °C followed by extraction with 6 M hydrochloric acid. Potassium in the extracts is determined flame-photometrically. There is no significant interference from other elements. See also Sects. 7.34.1 and 7.34.7.

7.22
Rubidium

See Sect. 7.34.7.

7.23
Ruthenium

Megarrity and Siebert [55] have described a method based on AAS for the determination of ruthenium in grass and animal faeces. In this method the sample was ashed at 350 °C with a mixture of potassium nitrate and potassium hydroxide, dissolved in dilute nitric acid and analysed via AA spectrophotometry using a carbon rod atomiser. The method is particularly free from interferences and is suitable for the determination of ruthenium at concentrations of between 5 and 50 µg per gram of dry matter. The atomic adsorption wavelength employed in this study was 349.7 nm.

A representative portion of the sample was dried at 80 °C and ground to pass a 1 mm sieve. A subsample was ground to fine powder using a Tema mill and 0.500 grams of this material weighed into a 25 ml glass beaker. Then, 0.06 grams of potassium nitrate and 0.02 grams of potassium hydroxide dissolved in 2 ml of water were added and the contents of the beaker dried at 100 °C prior to ashing at 350 °C for 16 hours.

The residue was suspended in 6 ml of 4 M nitric acid, heated on a water bath and diluted with water to 25 ml without filtration. A 5 µl aliquot was taken for analysis.

Uranium was the only interferent noted of those tested. The lowest level of ruthenium that could be detected was 5 $\mu g/g$ of dry matter. Between 98 and 100% of ruthenium at the 10–20 $\mu g/g$ level was recovered in this procedure.

7.24
Selenium

Selenium deficiency in soils occurs in some parts of the world [56] and it is standard practice in such areas to dose animals with selenium in order to correct this deficiency. Selenium-responsive diseases can appear when the level of selenium is lower than 0.03 $\mu g/g$ in the blood [57]. Levels of selenium in pasture plants associated with deficiency symptoms in animals are in the range 0.01–0.03 $\mu g/g$ [58].

The low levels of selenium involved are well beyond the analytical range of conventional flame AA spectroscopy, which has a detection limit of about 0.5 $\mu g/ml$ in solution when an air–hydrogen flame is used [59]. Although deficiency levels are slightly above the best reported concentration detection limits obtained with the carbon furnace atomisation technique [60], the matrix interference problems to be overcome appear formidable for this element [61,62].

Hydride generation techniques would satisfy these requirements and also give adequate sensitivity; this technique, as a separation process, apparently suffers no matrix interference problems during atomisation and does not necessarily require background correction for non-atomic absorption. Preliminary investigations into the generation of selenium hydride prior to atomisation led to a preference for the rapid method of generation [64] rather than the collection and storage method [63]. The use of sodium borohydride solution rather than zinc powder slurry was preferred, as the solution was easier to inject reliably and a preliminary reducing step was not required. The apparatus and procedure described by Duncan and Parker [64] was eventually chosen for this work, with some modifications.

A standard official method is based on fluorimetry [65–67]. In this method, the plant sample is digested with 60% m/m perchloric acid: 70% nitric acid, 3:1 v/v, and the residue is dissolved in 2 M hydrochloric acid. Any selenate present in this solution is converted to selenite by boiling. The concentration of selenate is then determined fluorimetrically as a dekalin extract of the complex formed with 2,3-diaminonaphthalene.

Olson [68] also studied this method and was able to determine down to 0.02 $\mu g/g$ selenium in plant materials.

Vijayakumar et al. [69] has described a method for the determination of trace quantities of selenium in plant tissues based on the interaction of selenium(IV)–iodine with an acid medium, leading to the liberation of iodine. This method was utilised for the indirect determination of selenium by AA spectrophotometry. The iodine is extracted into benzene and subsequently reductively stripped into an aqueous solution of ascorbic acid. After the extraction of the resulting iodine as tris(1,10-phenanthroline) cadmium(II) iodide into nitrobenzene, the cadmium content of the organic extract is determined by AAS. Beer's law is applicable up to 0.75 ppm of selenium. The few interferences are readily overcome. The chemical yield in the system is about 80% overall.

Bismuth, antimony, arsenic, arsenate, vanadate, thallium, lead, tin, dichromate, nitrate and iron all interfere in this procedure but the interferences by all of these except thallium, tin and dichromate can be overcome by suitable modifications to the method. Plant digests for analysis when prepared by nitric acid–perchloric acid digestion [70].

Cane sheath and neem leaf samples were used and the final acid digest was heated with concentrated hydrochloric acid to reduce any selenate to selenite. The solutions were finally made up to 100 ml and 5 and 10 ml aliquots were analysed. The samples did not contain any selenium, and recoveries were determined by the addition of 50 and 100 μg of selenium to the plant samples before decomposition. The results were all satisfactory. When these experiments were repeated with 1.0 and 2.5 μg of selenium as spikes, the recoveries were 80–120% at the 1 μg level and 96–120% at the 2.5 μg level (per five grams of sample).

Various workers have employed hydride generation AAS to determine selenium in acid digests of wheat [71], sugar beet [72], pasture [73] and plant material [74]. Thus, Clinton [73] described a method for the routine determination of selenium in plant samples at concentrations in the range 0.01–0.50 μg/g. Samples of mass one gram are digested with nitric and perchloric acid and then selenium hydride is generated from diluted digests by the controlled introduction of a solution of sodium borohydride. The selenium is subsequently atomised in a nitrogen–hydrogen-entrained air flame. Digests can be analysed at a rate of four per minute. For a pasture sample, which contained a 0.038 μg/g concentration of selenium, the relative standard deviation was 4.3%. The mean recovery of added selenium was 100.5%, with a relative standard deviation of 4.7%. The efficiency of hydride generation was 95%.

Plant samples were dried, ground and then redried for four hours at 90 °C before analysis. The dried sample (1 g) was placed in a borosilicate glass tube and digested with 1 ml of a mixed digestion acid (200 ml 72% m/v perchloric acid and 50 ml 68% m/v nitric acid) plus 5 ml redistilled 72% m/v nitric acid and two drops of kerosene to prevent frothing. The tubes were digested for three hours at 130 °C, and then 2 ml of redistilled 20% m/v hydrochloric acid is added when cold. After treatment with sodium borohydride, the solutions are evaluated at the selenium 196 nm resonance line.

The presence of up to 30 μg in the portion analysed does not interfere in this procedure, but 100 mg copper depresses the selenium signal by 9%. Therefore, one should be aware of copper interference in the case of plant material which is contaminated with copper spray residues.

Hydride generation inductively coupled plasma–mass spectrometry (ICP-MS) [75], gas chromatography–mass spectrometry (GC–MS) [76], gas chro-

matography inductively coupled plasma–mass spectrometry [76], isotope dilution mass spectrometry [77] and X-ray fluorescence spectrometry (XFS) [78]
have all been recently studied as methods for determining selenium species in
plant material.

The hydride generation ICP–mass spectrometric technique [75] had a sensitivity of 6.4 ng/g selenium in plant material and was applied to digests of
corn, kale and rice. In the isotope dilution mass spectrometric technique [77],
the samples were spiked with 76-selenium solution and digested on a heating
block at 150 °C with a mixture of nitric acid and hydrogen peroxide. Solid-
phase microextraction was used to extract selenium from plant material prior
to the gas chromatographic techniques [76]. See also Sects. 7.34.1 and 7.34.2.

7.25
Sodium

Sodium has been determined in plant material by a standard official method
[79]. In this method, the plant material is first digested with perchloric acid 60%
m/m: nitric acid 70% m/v (1:4 v/v), the residue is dissolved in hydrochloric acid,
and then it is analysed by AAS at the 589.0 nm emission line. See Sect. 7.34.1.

7.26
Strontium

See Sect. 7.34.7.

7.27
Tellurium

See Sect. 7.34.2.

7.28
Thallium

See Sect. 7.34.4.

7.29
Tin

Godar and Alexander [80] determined down to 0.5 mg/kg of tin in a 5 – 10 gram
sample of fruit and vegetables by a polarographic procedure. See Sect. 7.34.2.

7.30
Titanium

Abbasi [81] has described a spectrophotometric method employing *N-p*-methoxylphenyl-2-furohydroxamic acid as the chromogen for the determination of titanium in plant material. Traces of titanium are completely extracted as the above complex from acidic aqueous solutions obtained by ashing the plant material.

The golden yellow titanium complex is evaluated spectrophotometrically at 385 nm. Interference by iron, molybdenum, chromium, zirconium and tantalum is overcome by the addition of stannous chloride.

7.31
Uranium

Boomer and Powell [82] determined uranium in digests of plant material using ICP–MS. The lower limit of detection was 1 ng/ml for aqueous digests.

7.32
Vanadium

See Sect. 7.34.4.

7.33
Zinc

In an official method [83] for determining zinc in plant material, the sample is digested with perchloric acid 60%: nitric acid 70%, *m/v* 1:4, followed by 2 M hydrochloric acid. Alternatively, the plant material is dry ashed and the residue dissolved in 6 M hydrochloric acid. The extract is evaluated by AAS at the 213.9 nm emission line. See also Sect. 7.34.1, 7.34.4 and 7.34.7.

7.34
Multi-Cation Analysis

Most of the work published on the determination of cations in plant material is of course concerned not with the analysis of single elements present in the sample but instead a range of elements present in the sample.

Of these, the methods that predominate are those based on AAS and, to an increasing extent in recent years, coupled plasma-based methods and X-ray fluorescence spectroscopy. Each of these methodologies is reviewed below under separate headings.

7.34.1
Atomic Absorption Spectrometry (AAS)

The various methods located are reviewed in Table 7.4. It is seen that, in general, the detection limits achieved are better than 1 mg/kg of plant material; i.e., 1000 μg/kg or 1 μg/g.

Examples of various sample digestion procedures and methodologies are discussed in more detail below.

Lead and Cadmium by Nitric–Perchloric Digestion
and Ammonium Pyrrolidine Dithiocarbamate–Methyl Isobutyl Ketone Extraction

See [96]. In this method, 1–15 gram dry weight of samples were weighed into 250-ml Vycor beakers, covered with watchglasses and dried overnight at 120 °C. If necessary, three blanks were carried through the entire procedure. Concentrated nitric acid was added according to the formula:

$$\text{ml } HNO_3 = 10 + (5 \times \text{dry weight of sample in g})$$

The samples were then digested at room temperature for 2–12 hours. The solutions were boiled until the volume decreased to about one half of the original volume of nitric acid added, and concentrated perchloric acid was added according to the formula:

$$\text{ml } HClO_4 = 20 + (20 \times \text{weight of fat or oil present in g}) +$$
$$+ (7 \times \text{dry weight of sample in g})$$

Samples were then heated and maintained at boiling until only perchloric acid fumes remained (usually 0.5–1 hour). Solutions were boiled at a high temperature to a volume of 4–8 ml. The watchglasses and the sides of the beakers were washed with about 40 ml of water, and the solutions heated to about 90 °C and transferred to 125 ml separatory funnels.

The acidity of each digest was adjusted with concentrated ammonia solution in the presence of a drop of methyl violet (0.1% in water) until a blue-green hue just began to appear. The color of the indicator was stable only for several minutes and rapid adjustment was necessary.

After cooling, the solution in each separatory funnel was extracted and drained in turn as follows: APDC solution (5 ml 1%) was added with swirling (ten seconds). Immediately after, MIBK (70 ml) was added, the funnel was stoppered, and it was shaken manually for 60 seconds. The phases were separated (20 seconds) and the aqueous phase was drained, leaving about 1 ml. Stripping solution (nitric acid 3%–hydrogen peroxide 8%, 5 ml when Pb < 0.2 μg; 10 ml when 0.2 μg < Pb < 0.4 μg; and 20 ml when Pb > 0.4 μg) was pipetted into the funnel and the latter stoppered.

Table 7.4. Determination of metals in plant material and crops by atomic absorption spectrometry (from author's own files)

Elements	Sample	Sample digestion	Limit of detection	Comments	Reference
Copper, Manganese, Nickel	Plants	–	–	Flow injection on-line sorption preconcentration in a reactor, electrothermal AAS	[84]
Cobalt, Nickel, Copper	Plants	–	–	Ultrasonic slurry sampling, electrothermal AAS	[85]
Lead, Cadmium, Mercury, Selenium	Onions	–	–	Flame AAS	[86]
Calcium, Magnesium, Iron, Zinc, Manganese	Vegetables	–	–	Flame AAS	[87]
Copper, Manganese, Zinc	Wheat, Rice	–	–	Flame AAS	[88]
Copper, Manganese, Lead, Zinc	Vegetables	–	–	Flame AAS	[89]
Calcium, Magnesium, Sodium, Potassium, Iron, Copper, Zinc, Manganese	Cereals	–	–	Flame AAS	[90]
Aluminium, Chromium, Copper, Iron, Nickel, Lead	Vegetable oils	–	–	Flame AAS	[91]
Iron, Zinc, Calcium, Magnesium, Copper	Soy bean seed	–	–	Flame AAS	[92]
Sodium, Potassium	Wheat	–	–	Flame AAS	[93]
Sodium, Potassium Calcium, Magnesium, Manganese, Zinc	Plants	Acid digestion or dry ashing	–	Flame AAS	[94]

Table 7.4. (continued)

Elements	Sample	Sample digestion	Limit of detection	Comments	Reference
Arsenic, Aluminium, Iron, Zinc, Chromium, Copper	Plants	Digestion with 2 + 1 v/v concentrated nitric acid and sulfuric acid	As: 0.5 ng/g Al, Fe, Zn, Cr, Cu: 0.1 ng/g	Flame AAS	[95]
Lead, Cadmium	Spinach leaves	Concentrated nitric acid and perchloric acid	Pb: 20 ng Cd: 1 ng	Flame AAS	[96]
Copper, Manganese, Zinc	Plants, Barley	Dry ashing under oxygen with $KHSO_4$ and HNO_3	Cu: 0.25 mg/kg, Mn: 0.23 mg/kg, Fe: 1 mg/kg, Zn: 0.05 mg/kg	Flame AAS	[97]
Copper, Iron, Manganese, Zinc	Fruit, Kale, Potato, Spinach	Sulfuric acid	Cu: 0.1 mg/kg, Mn: 0.2 mg/kg, Fe: 1 mg/kg, Zn: 0.4 mg/kg	Flame AAS	[98]
Miscellaneous	Plants	–	Study of plant grinding method	Flame AAS	[99]

After shaking the funnels gently for five minutes, they were inverted on their stands to avoid seepage through the drainage spouts, left to stand overnight, shaken for another five minutes, and returned to an upright position. After an hour, the stopper of each was removed, the aqueous phase drained into a clean beaker, discarding the first 1–2 ml, and 2.00 ml were pipetted into a clean 25-ml pp bottle containing 6.00 ml of modification solution.

A dilution solution containing nitric acid (0.75%), ammonium dihydrogen phosphate (0.25%), ammonia (0.072%), hydrogen peroxide (2%) and MIBK (0.43%) was used for diluting samples and preparing standard lead and cadmium solutions. Standards were prepared fresh monthly.

Aliquots (20 µL) of samples and standards were pipetted into the graphite furnace, dried (125–150 °C), ashed (750 °C for 40 seconds), and atomised (2300 °C for 15 seconds). Absorbances of the samples were bracketed with those of standards. Blanks for both sets of solutions were measured and included in the final calculation. When the absorbances of the samples exceeded those of 60 ng/ml lead or 6 ng/ml cadmium standards, they were diluted to appropriate amounts with dilution solution.

Graphite tubes were initially conditioned by running through atomisation cycles for 15 aliquots of dilution solution, and were used for as many as 900 cycles. They were cleaned of carbon deposits every 100–200 cycles. Gradual erosion of the surface graphite during the lengthy use of the tubes caused 20–30% increases in actual temperatures for the same current settings, and adjustment of the latter was made every 300–500 firings.

Determination of "Acid-Insoluble" Lead

Samples were digested according to the digestion procedure to perchloric acid fumes, diluted to about 150 ml with water, and heated to near-boiling. Blanks containing lead standards equivalent to the "acid-soluble" lead content of the samples were simultaneously carried out through the entire procedure. The hot digests were filtered through membrane filters (cellulose acetate–nitrate, 0.3 µm porosity, 25 mm diameter), prewashed with nitric acid (5%). The precipitates were washed thoroughly with nitric acid (5%) and then water, and transferred with the filter into clean 50 ml Teflon–FEP beakers. Concentrated hydrofluoric acid (5 ml) and nitric acid (2 ml) were added and the solutions heated until the filters dissolved. Concentrated perchloric acid (7 ml) was added and the solutions were evaporated to perchloric acid fumes. If any precipitate was visible, hydrofluoric acid (7 ml) was again added and the solution evaporated as before. The solutions were diluted with water, transferred to a separatory funnel, and analysed according to the normal procedure.

Advantages of this method are:

1. Low detection limits, lead 20 ng, cadmium 1 ng;
2. Few interferences, and;
3. Good precision and accuracy.

Some results obtained on NBS standard reference plant materials are listed in Table 7.5.

In a further method, sample digests are prepared according to method 1(c) of the Analytical Methods Committee [100] using precautions described subsequently [101]. The resulting 100 ml of digest, which is in normally 5% v/v sulfuric acid, should not be colourless and should contain any suspended solids. At the same time, prepare two reagent blanks from the volume of acid used in sample oxidation.

Instrumentally each emission source is tuned to give maximum sensitivity-to-noise ratio according to the maker's instructions, at wavelengths of 327.4 nm for copper, 248.3 nm for iron, 279.5 nm for manganese and 213.9 nm for zinc. The instrumental conditions are adjusted to the fixed conditions described. Aspirate each solution in turn, taking a constant digital reading for a four-second integration time. The only significant interference in this method was the effect of calcium concentrations above 100 mg/l on the determination of manganese. Recoveries ranged from 92% zinc to 104% manganese. Some results obtained by this procedure are listed in Table 7.6.

Sulfuric Acid Digestion Procedure

Evans et al. [98] has described a method for the determination of total copper, iron, manganese and zinc in various plant materials. Organic matter is destroyed by wet oxidation and measurement is made directly upon the sulfuric acid digests by flame atomic absorption spectrophotometry. In the measurement, direct interferences from the inorganic species found in plant

Table 7.5. "Acid-soluble" lead and cadmium levels (μg/g) in NBS standard reference materials (from [96])

Material	Replicates	Sample size, g	Pb found $\pm$ s	Pb certified	Cd found $\pm$ s	Cd certified
Orchard	3	1	45.3 $\pm$ 0.7	45 $\pm$ 3	0.109 $\pm$ 0.002	0.11 $\pm$ 0.02
leaves	2	5	43.4 $\pm$ 1.0		0.104 $\pm$ 0.002	
Spinach	3	1	1.09 $\pm$ 0.06	1.2 $\pm$ 0.2		
leaves	1	9	1.18			
Tomato	3	1	5.95 $\pm$ 0.06	6.3 $\pm$ 0.3		
leaves	2	3	6.19 $\pm$ 0.18			
Bovine	3	2	0.33 $\pm$ 0.01	0.34 $\pm$ 0.08	0.26 $\pm$ 0.02	0.27 $\pm$ 0.04
liver			0.36			
	1	5				
	2	5			0.23 $\pm$ 0.01	
Pine	3	1	11.0 $\pm$ 0.6	10.8 $\pm$ 0.5		
needles	1	11	10.7 $\pm$			

Table 7.6. Replicate analysis for copper, iron, manganese and zinc on plant materials (from [100])

Element	Foodstuff materials*	Sample mass, g	Mean content, mg/kg	Range of content, mg/kg	Amount, μg	Repeatability, s_0 %	μg	Reproducibility, s %	μg	95% Confidence intervals (±), mg/kg†
Copper	Apple	10	0.29	0.23–0.33	2.89	9.3	0.27	10	0.30	0.069, 0.068
	Kale	1	4.99	4.40–5.59	4.99	7.4	0.37	8.0	0.40	0.095, 0.090
	Potato	2	2.70	2.49–3.09	5.41	5.1	0.28	6.6	0.36	0.071, 0.081
	Kale	2	5.29	4.70–6.01	10.6	5.1	0.54	8.6	0.91	0.14, 0.21
	Spinach	1	12.1	11.3–13.0	12.1	3.5	0.42	4.8	0.58	0.11, 0.13
	Flour	10	1.40	1.34–1.47	14.0	2.1	0.30	3.2	0.45	0.077, 0.10
Manganese	Apple	10	0.61	0.53–0.66	6.05	6.8	0.41	7.3	0.44	0.11, 0.10
	Potato	2	4.75	4.11–5.67	9.50	11	1.09	12	1.14	0.28, 0.26
	Kale	1	15.6	14.1–16.9	15.6	5.8	0.91	5.8	0.91	0.21, 0.21
	Kale	2	15.9	14.6–16.9	31.7	3.7	1.17	4.6	1.47	0.30, 0.33
	Flour§	10	5.40	5.19–5.82	54.0	1.9	1.00	3.6	1.9	0.26, 0.44
Iron	Potato	2	19.4	17.3–23.7	38.9	9.8	3.8	11	4.3	0.98, 0.97
	Apple	10	3.97	3.68–4.30	39.7	4.9	1.9	4.9	1.9	0.44, 0.44
	Kale	1	117	107–123	117	4.4	5.2	4.4	5.2	1.2, 1.2
	Flour	10	19.2	18.4–19.8	192	1.7	3.2	2.4	4.6	0.82, 1.04
	Kale	2	116	112–126	231	4.3	10	5.9	14	2.6, 3.2
	Spinach ‡ ≈	1	545	514–586	545	2.0	11	5.0	27	2.8, 6.1
Zinc	Apple§	10	0.58	0.42–0.71	5.8	7.4	0.43	16	0.93	0.11, 0.21
	Potato	2	9.46	8.72–10.9	18.9	6.0	1.14	7.8	1.48	0.29, 0.33
	Kale	1	33.1	30.9–35.1	33.1	4.5	1.5	5.0	1.6	0.39, 0.34
	Spinach	1	50.8	47.8–56.2	50.8	5.3	2.7	5.3	2.7	0.61, 0.61
	Flour‡	10	5.52	5.21–5.85	55.2	1.5	0.8	3.6	2.0	0.21, 0.45
	Kale	2	32.0	29.5–34.5	64.0	3.1	2.0	6.8	4.3	0.51, 0.97

* Each row represents 10 results obtained by 5 analysts, except for 1 gram of kale, for which 8 results were obtained by 4 analysts.

† These limits are for a single sample result, the first value derived from repeatability and the second derived from reproducibility. Concentrations are based on a 10-gram sample mass.

‡ 1% analytical significance.

§ 5% analytical significance.

≈ Measurement was made on a prepared solution diluted by a factor of 10 for copper and a factor of 5 for iron and manganese materials are shown to be a function of burner design and usage. Elimination of these interferences may be achieved for optimum conditions of use, thereby avoiding systematic bias. Isolation of the sources of variation during measurement reveals indirect interferences that enhance the standard deviation of measurement. The accuracy of the method when applied to foodstuffs is assessed for the levels of each element normally present. The standard deviations of the results are compared with those for measurement alone, and the further influence of indirect interference on the former is inferred.

Application of the total procedure to reference and retail foodstuffs indicates a satisfactory accuracy for the method, as indicated by agreement with certified or consensus values, with the exception of manganese, for which the means of the results are significantly higher. Comparison of the repeatability of the results with those from measuring standard solutions reveals that there is seldom statistical significance, except when there is an inhomogeneous elemental distribution in the samples examined. A similar comparison performed for the reproducibility shows that statistically significant differences may or may not occur for results from between-series measurements. The source of this contribution is believed to be the same as that giving the indirect interference for within-series measurement, i.e., repeatability, doubly reflected for between-series measurement. This is most marked for zinc, for which high absorbances are measured and which would be expected to be subject to greater effects from nebulisation variation than copper, iron or manganese. Because derived factors such as the 95% confidence intervals and the limit of detection should be based on the variation of results in survey analysis, these factors will be larger than might be anticipated. They will reflect, however, the effectiveness of the method for the four elements in routine practice.

Heanes [97] has described a method for determining copper, manganese and zinc in ashed plant extracts by flame AA spectrophotometry after cobalt and molybdenum have been assayed on separate aliquots of the same plant extracts by a spectrophotometric procedure [102]. (See Sect. 7.34.3). Ashing aids were necessary to maintain accuracy in the determinations. Concentrations of up to 3.5% m/m of silicon and calcium and 4% m/m of chlorine in the plants did not affect the determinations, but in some instances lower concentrations were determined in plant samples containing equal or higher levels of both added silica and calcium.

Potential interference was prevented during assays for copper by automatic background correction, for manganese and zinc by diluting the extract, and for all three analytes by selectively matching their chemical matrices with hydrochloric acid and potassium sulfate concentrations in respective working standards. Using these procedures, assays for the three elements were similar to determinations by flame AA spectrophotometry on samples digested by a standard wet digestion method. The ashing procedure is described below.

Sample ashed in the absence of the chemical ashing aids, potassium hydrogen sulfate and nitric acid, but in the presence of oxygen produced lower concentrations of copper, manganese and zinc than in comparable samples ashed with the chemical aids (Table 7.7). Varying the additions of potassium hydrogen sulfate from 0.5 to 1.0 gram and nitric acid from 1 to 3 ml had negligible effects on the determinations of the three elements in three samples that varied in elemental composition (Table 7.7).

Oxygen enrichment during ashing increased the concentrations of the elements determined, and slightly higher concentrations were sometimes assayed in samples ashed for 20 hours compared with 15 hours.

In general, these findings on the effects of ashing aids support those previously reported for determinations of cobalt and molybdenum in other aliquots of the same plant extract.

Thus, for all samples examined, the addition of the ashing aids improved the efficiency of ashing and the extraction of all five elements, i.e., copper,

Table 7.7. Effects of the ashing aids $KHSO_4$ and HNO_3 on the determination of Cu, Mn, and Zn in plant samples when oxygen is supplied (from [97])

Concentration in plant material*, mg/kg									
Ashing treatment		Barley glumes			Barley straw			Lucerne pasture I		
$KHSO_4$, g	HNO_3, g	Cu	Mn	Zn	Cu	Mn	Zn	Cu	Mn	Zn
0	0	2.3	46.6	8.4	4.6	35.9	20.9	7.6	30.7	39.3
0.5	0	2.9	47.8	9.0	5.9	37.8	22.2	8.8	33.8	42.0
0.5	1	3.4	51.5	9.7	6.8	39.5	27.0	9.3	34.5	42.9
0.5	2	3.3	51.3	9.9	6.8	39.6	27.2	9.4	34.5	43.1
0.5	3	3.3	51.3	9.8	6.8	39.5	27.2	9.4	34.6	42.9
0.75	0	3.2	48.6	9.2	6.5	39.8	27.0	9.3	34.4	42.4
0.75	1	3.4	51.5	9.8	6.8	39.7	27.2	9.3	34.4	42.9
0.75	2	3.3	51.4	10.0	6.6	39.8	27.1	9.4	34.3	43.1
0.75	3	3.4	51.6	10.0	6.8	39.7	27.2	9.4	34.6	43.1
1.0	0	3.3	49.1	9.5	6.6	39.5	27.0	9.3	34.4	42.9
1.0	1	3.3	51.3	9.7	6.7	39.7	27.3	9.3	34.5	42.8
1.0	2	3.4	51.6	10.0	6.8	39.8	27.0	9.4	34.3	43.2
1.0	3	3.4	51.4	9.8	6.8	39.6	27.2	9.3	34.5	42.9
LSD‡ ($P = 0.05$)		0.1	0.5	0.2	ns	0.5	0.3	0.2	0.3	ns

4 g of plant material were ashed in the presence of 100 ml/min of oxygen with $KHSO_4$ (0.75 g) and HNO_3 (1.5 ml) for 10 hours and H_2SO_4 and HNO_3 (0.75 ml each) for 10 hours
* Mean of three determinations
† Concentrations of the potentially interfering elements Si, Cl or Ca: barley glumes 4.6% m/m Si; barley straw 1.8% m/m Cl; lucerne pasture I 1.9% m/m Ca.
‡ LSD = Least significant difference ($P = 0.05$); the effects of the chemical treatments during ashing were not significant (ns); the nil nil $KHSO_4$–nilHNO_3 treatment for respective samples was excluded from the analyses of variance

manganese, zinc, cobalt and molybdenum. In this method, four-gram samples were ashed in the presence of 100 ml/min of oxygen with 0.75 g of potassium hydrogen sulfate and 1.5 ml of nitric acid for ten hours and 0.75 ml each of sulfuric and nitric acid for a further ten hours.

To three four-gram subsamples of nine different plant materials were added copper, manganese and zinc at concentrations of 10, 100 and 50 mg/kg, respectively. All samples were subsequently ashed and analysed for these three elements by the method described below. The mean recoveries varied from 96 to 100% for copper, 95 to 101% for manganese and 96 to 99% for zinc (Table 7.8).

Arsenic, Aluminium, Iron, Zinc, Chromium and Copper

Nitric Acid–Sulfuric Acid–Hydrogen Peroxide Digestion

Arafat and Glooschenko [95] have described a method for the simultaneous determination of these elements in plants which does not involve the use of perchloric acid.

In this method, a 0.5-gram sample of plant tissue was digested with a mixture of concentrated nitric and sulfuric acids (2 + 1 v/v) and hydrogen peroxide. Arsenic was determined by the hydride generation method. Aluminium, iron, zinc, chromium and copper were determined by direct flame AAS. The detection limits in dry plant material using 50 ml of aqueous solutions for analysis were 0.5 ng/g for arsenic and 0.1 μg/g for aluminium, iron, zinc, chromium and copper. The relevant standard deviations were 4, 6, 1, 11, 6 and 7%, respectively. All six metals were determined from the aliquot, with recoveries ranging from 93 to 118%. A study was made of the composition of the precipitate that settled out from the extracts. X-ray diffraction revealed the presence of α-aluminium oxide (corundum) and some quartz in the anti-bumping granules. α-Aluminium oxide was a source of contamination for the aluminium analysis.

The plant material was oven-dried for 24 h at 80 °C then pulverised in a Wiley cutting mill to 60 mesh (0.25 mm). An oven-dried milled sample (0.5 g) was weighed into a 125 ml Erlenmeyer flask and a few boiling chips, 10 ml of 16 M nitric acid and 5 M of 36 M sulfuric acid were added.

The mixture is digested for 15 minutes at 90 °C and then at 170 °C until the volume is reduced to 5 ml, avoiding dryness; 2 ml hydrogen peroxide (70% v/v) is then added and the solution heated until reaction ceases before it is made up to a standard volume. At the end of this procedure the acid content is about 10% v/v.

Aliquots of the digest are analysed for aluminium, chromium, iron copper and zinc by direct flame AAS. Together with the samples, blanks and standards covering the range from 0.1 to 30 μg/ml of each of the above metals in 10% sulfuric acid were also run.

Table 7.8. Recovery of Cu, Mn and Zn added to plant samples (from [97])

Sample	Concentration of Cu in plant material			Concentration of Mn in plant material			Concentration of Zn in plant material		
	Added sample concentration*, mg/kg	Mean analysis†, mg/kg	Mean recovery, %	Added sample concentration*, mg/kg	Mean analysis†, mg/kg	Mean recovery, %	Added sample concentration*, mg/kg	Mean analysis†, mg/kg	Mean recovery, %
Barley glumes	0	3.4		0	51.0		0	9.8	
	10	13.1	97	103	154.4	100	50	57.8	96
Barley straw	0	6.9		0	39.3		0	27.5	
	10	16.5	96	103	139.2	97	50	76.1	97
Wheaten hay	0	3.0		0	30.8		0	31.6	
	10	12.7	97	103	129.6	96	50	79.4	96
Lucerne pasture I	0	9.2		0	34.6		0	42.9	
	10	19.1	99	103	134.1	97	50	92.0	98
Medic pasture	0	7.3		0	17.1		0	26.4	
	10	17.2	99	103	117.5	97	50	74.2	96
Lupin straw	0	4.4		0	138		0	20.9	
	10	14.3	99	103	240	99	50	69.8	98
Subclover, ryegrass pasture I	0	7.1		0	167		0	25.8	
	10	17.0	99	103	271	101	50	75.4	99
Mixed legume and fog grass pasture	0	10.4		0	43.5		0	57.8	
	10	20.4	100	103	146.8	100	50	106.7	98
Kale	0	4.6		0	14.6		0	32.5	
	10	14.5	99	103	112.9	95	50	81.2	97

* Cu, Mn and Zn equivalent to 10 mg/kg Cu, 100 mg/kg Mn and 50 mg/kg Zn as $CuSO_4.5H_2O$, $MnSO_4.4H_2O$ and $ZnSO_4.7H_2O$, respectively, were added to 4-gram subsamples of plant material, which were then ashed by the recommended dry-ashing procedure.
† Mean of three determinations

Table 7.9. Results for determination of elements in *Sphagnum fuscum* (from [95])

Element	Lowest detectable level, μg/g	Adopted detectable level, μg/g	Results for ten replicate determinations	
			Mean μg/g	Relative standard deviation, %
As	0.00025	0.0005	0.55	4
Al*	0.05	0.1	26.1	6
Fe	0.05	0.1	26.0	1
Zn	0.05	0.1	41.1	11
Cr	0.05	0.1	0.1	6
Cu	0.05	0.1	27.2	7

*: Dinitrogen oxide flame

A variety of plant species were used to determine the applicability of this method, including mosses (*Sphagnum capillaceum, S. fallax*), lichen (*Caldonia* spp.) and higher plants (*Chamaedeephne calyculata*). The plant material dissolved completely in the acid and no difficulty was observed in the digestion. It was observed however, that some plant species dissolved more readily than others.

Table 7.9 presents the detection limits and statistical data for the determination of zinc, copper, iron, chromium, aluminium and arsenic in *Sphagnum fuscum,* based on ten replicate determinations. Table 7.9 also presents relative standard deviations and mean values, and recoveries ranged from 93% (copper) to 118% (arsenic).

7.34.2
Inductively Coupled Plasma–Atomic Emission Spectrometry (ICP–AES)

Schramel et al. [103] and Wolnik et al. [104] and Hahn et al. [105] have investigated the applicability of this technique to the determination of elements in plant materials.

Schramel [103] discusses the conditions for multi-element analysis of over 50 trace elements, giving detection limits. Wolnik [104] described a sample introduction system that extends the analytical capability of the inductively coupled argon plasma/polychromator to include the simultaneous determination of six elemental hydrides along with a variety of other elements in plant materials. Detection limits for arsenic, bismuth, selenium and tellurium range from 0.5 to 3 ng/ml and are better by at least an order of magnitude than those obtained with conventional pneumatic nebulisers, whereas detection limits for the other elements investigated remain the same. Results from the analysis of freeze-dried crop samples and NBS standard reference materials demonstrated the applicability of the technique. Results obtained by the analysis of a variety of plant materials are presented in Table 7.10.

Table 7.10. Spiked sample analysis (from [103])

Element	Spike amount, µg/g	NBS Rice flour Total amount, µg/g	% Recovery	NBS Wheat flour Total amount, µg/g	% Recovery	Corn Total amount, µg/g	% Recovery	Potatoes Total amount, µg/g	% Recovery	Soyabeans Total amount, µg/g	% Recovery
As	1.00	1.39	112	1.0	115	1.1	107	1.0	88.8 / 110 / 96.8	1.0	106
Bi	0.250	0.643	49.6	≤ 0.08	–	≤ 0.08	–	≤ 0.08	–	≤ 0.08	–
	1.00	1.0	99.4	1.0	115	1.0	119	1.0	95.4 / 96.2 / 116	1.0	96.0
Ge	0.250	0.25	96.5	≤ 0.03	–	≤ 0.03	–	≤ 0.03	–	≤ 0.03	–
	10.0	10	114	≤ 0.5	–	≤ 0.5	–	≤ 0.5	–	≤ 0.5	–
	1.25	≤ 0.5	–	≤ 0.5	–	≤ 0.5	–	1.25	208	≤ 0.5	–
Sb	2.50	2.50	51.7	–	–	–	–	–	–	–	–
	0.500	0.50	47.1	–	–	–	–	–	–	–	–
Se	1.00	1.31	96.2	1.87	76.8	1.18	98.0	1.27	118 / 101 / 109	3.36	96.0
Te	0.250	0.581	91.8	≤ 0.01	–	≤ 0.01	–	≤ 0.01	–	≤ 0.01	–
	1.00	1.0	104	1.0	103	1.0	103	1.0	98.2 / 102 / 109	1.09	107
	0.250	0.25	105	≤ 0.02	–	≤ 0.02	–	≤ 0.02	–	≤ 0.02	–

–: Not determined

Hahn et al. [105] used a hydride generation/condensation system with an ICP polychromator for the determination of arsenic, bismuth, germanium, antimony, selenium and tin in plant materials.

Detection limits range from 0.02 ng/ml for arsenic to 0.80 ng/ml for selenium, and precision values at 10 ng/ml are less than 6% relative standard deviation. Results of analyses of NBS standard reference materials (wheat flour, rice flour, spinach and orchard leaves) demonstrate the application of the method to the matrices. The layout of the apparatus is illustrated in Fig. 7.6.

In this method, portions (2 g) of the plant material were digested in 30 ml of digestion acid until fumes of nitric acid are produced then the solution is cooled, and diluted to 50 ml in 15% hydrochloric acid–10% sulfuric acid *v/v*. This solution was used for analysis. Recovery studies were made by spiking the digestion acid or sample just prior to the digestion procedure with varying amounts of standard solution. Quantification of the six elements was made from linear calibration curves verified by the method of standard additions.

Digestion recoveries obtained in this procedure ranged from 71% (50 ng/g selenium) to 110% (50% ng/g arsenic and 50 ng/g tin). Results obtained from various plant materials are listed in Table 7.11.

Pahlavanpour et al. [115] has described a method based on hydride generation and ICP–emission spectrometry for the determination of arsenic, antimony and bismuth in herbage.

This method is based on reduction of these elements to hydrides using sodium tetrahydroborate(III) and injection of the hydrides into an ICP for determination by atomic emission spectrometry.

In this method, the herbage is first dried at 50 °C for 48 hours and milled to pass though a 0.5 mm screen. Ground material (1 g) is combined with magnesium nitrate solution, which acts as an ashing acid, and the mixture is ignited at 200 °C for 30 minutes and then at 450 °C for 5 – 15 hours. The ignited residue is treated with potassium iodide and then dissolved in concentrated hydrochloric

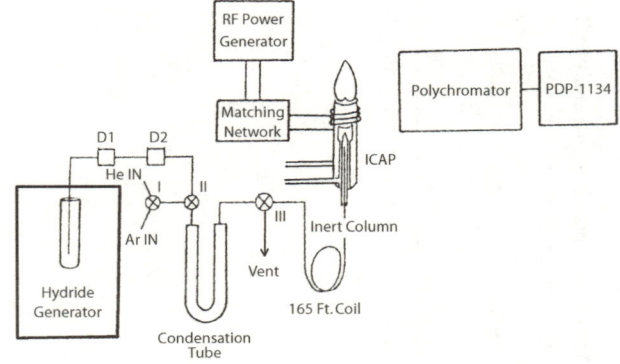

Figure 7.6. Hydride generation/condensation systems interfaced to an ICAP polychromator. D1 and D2: desiccant tubes; I, II and III: three-way valves. From [115]

Table 7.11. Sample results (from [105])

NBS Sample		As	Bi	Ge	Sb	Se	Sn
Rice flour	Certified, µg/g	0.41 ± 0.05				0.40 ± 0.10	
	Found[a], µg/g	0.44 ± 0.05	< 0.008	< 0.02	< 0.002	0.37 ± 0.06	< 0.02
	Added, µg/g	0.10	0.10	0.10	0.10	0.10	0.10
	% Recovery	99, 104	97, 97	108, 89	95, 80	89, 96	108, 71
	Added, µg/g	0.25	0.25	0.25	0.25	0.25	0.25
	% Recovery	103, 99	90, 87	91, 102	84, 92	97, 101	94, 114
Wheat flour	Certified, µg/g	(0.006)[b]				1.1 ± 0.2	
	Found[c], µg/g	0.006 ± 0.001	< 0.008	< 0.02	< 0.002	0.87 ± 0.06	< 0.02
Spinach	Certified, µg/g	0.15 ± 0.05			0.04 ± 0.01		
	Found[d], µg/g	0.17 ± 0.01	< 0.008	< 0.02	0.014 ± 0.003	< 0.003	< 0.02
Orchard leaves	Certified, µg/g	14 ± 2		0.15 ± 0	2.9 ± 0.1	0.08 ± 0.01	
	Found[e], µg/g	13 ± 1	< 0.008		2.8 ± 0.1	0.07 ± 0.01	0.18 ± 0.01

[a] Ten determinations, [b] Provisional value, [c] Three determinations, [d] Six determinations, [e] Three determinations

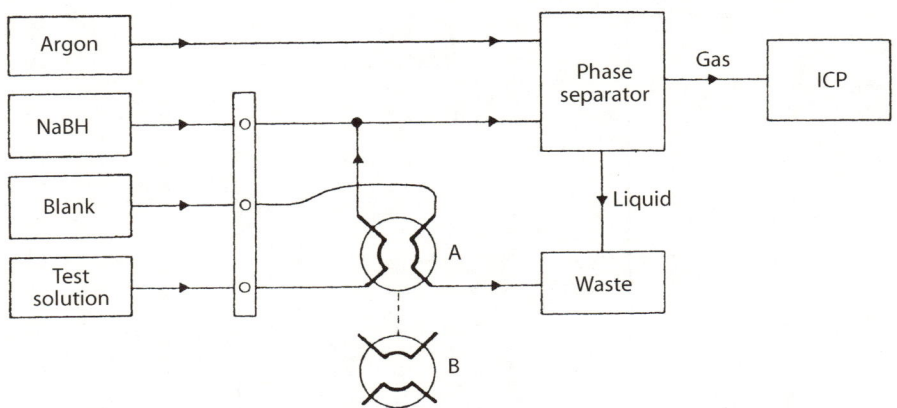

Figure 7.7. Schematic diagram of the continuous flow hydride generation system, showing the two positions of the four-port valve: (A) for analysis of the test solution and (B) for blank integrations used for changing the test solution. From [115]

acid. This solution is introduced into a hydride generator and then analysed in the system illustrated in Fig. 7.7.

Mean recoveries of these elements in herbage obtained by this procedure ranged from 98.7% for arsenic and 95.6% antimony to 95.6% for bismuth. The agreement with standard reference kale was acceptable.

7.34.3
Spectrophotometry

A classical example of the application of this technique is that of Heanes [102], who used it to determine cobalt and molybdenum in dry ashed plant material. The procedure uses a two-stage ashing process using nitric acid combined with either potassium hydrogen sulfate or sulfuric acid as ashing aids. Oxygen enrichment during ashing facilitated the complete oxidation of plant materials.

Modified spectrophotometric procedures are described for the quantitative determination of cobalt and molybdenum as the 2-nitrosonaphth-1-olate and toluene-3,4-dithiolate complexes in carbon tetrachloride. The extraction, chelation and phase separation steps permitted rapid sample handling, controlled interferences more effectively and provided accurate assays. The molar absorptivities for cobalt and molybdenum were 5.1×10^4 and 2.5×10^4 mol/l cm, respectively, and the detection limits for both elements were 4 ng/g.

In the digestion procedure, 6 ml of acidified 12.5% m/v potassium hydrogen sulfate is placed in a PTFE lined screw-cap bottle together with four grams of oven-dried plant material. The uncapped bottle is subjected to the following regime.

1. 120–175 °C with the doors ajar to allow for removal of water and nitric acid (usually one hour);
2. 175–350 °C until charring has moderated (usually one hour);
3. 350–550 °C for removal of sulfuric acid and organic by-products (usually 1.5 hours), and;
4. 550 °C with both doors closed for ten hours, in the presence of ducted oxygen at 100 ml/min.

The solution is then cooled to 150 °C and 3 ml of nitric acid 25% v/v:sulfuric acid 25% v/v solution is added. Subsequent reheating at 550 °C under oxygen produces a white ash, which is extracted with 10 ml 5.5 M hydrochloric acid. This suspension is diluted to 50 ml and the clear solution decanted from the silica precipitate prior to instrumental analysis.

Cobalt is determined in this extract by a spectrophotometric procedure involving the reaction of molybdenum with 2-nitrosonaphth-1-ol and evaluation of the carbon tetrachloride soluble complex formed at 307 nm.

Molybdenum is determined by a spectrophotometric procedure involving the reaction of molybdenum with dithiol and evaluation of the carbon tetrachloride complex at 680 nm.

There is distinct evidence that much-improved recoveries of cobalt and molybdenum from grains and grass can be obtained by using nitric acid in conjunction with potassium, hydrogen sulfate or sulfuric acid ashing acids and oxygen enrichment.

Negligible inference resulted from the presence of high levels of added silicon, calcium or chloride: all common constituents of plant materials.

Heanes [102] compared determinations of cobalt and molybdenum in various plant materials by wet and dry digestion procedures.

Close agreement ($P < 0.05$) between the results from the determinations of both elements obtained by both methods was observed for all materials, which varied appreciably in concentrations of cobalt, molybdenum and other elements (Table 7.12). Recoveries were generally high, ranging from 95 to 106% for cobalt and 93 to 99% for molybdenum.

7.34.4
Polarography

Cathode ray polarography has found limited application to trace elements in forage [106] and fruit and vegetables [107].

Nangniot [105] determined cobalt, cadmium, bismuth, copper, iron, manganese, molybdenum, nickel, lead, antimony, thallium, vanadium and zinc in forage. The forage sample is digested with nitric:perchloric acid 1:1 at 200–250 °C and then the solution diluted and filtered to remove silica. Following the addition of a supporting electrolyte, the solution is evaluated by square-wave cathode ray polarography.

Table 7.12. Comparison of Co and Mo concentrations in plant samples determined by the dry-ashing and wet digestion procedures (from [102])

Sample	Digestion procedure*	Concentration in plant material†	
		Co, ng/g	Mo, mg/g
Barley glumes	WD	21	0.36
	DA	30	0.38
Barley straw	WD	38	0.47
	DA	36	0.48
Wheaten hay	WD	65	0.22
	DA	55	0.22
Lucerne pasture 1	WD	245	0.46
	DA	235	0.46
Lucerne pasture 2	WD	188	0.98
	DA	179	0.98
Medic pasture	WD	50	0.23
	DA	50	0.22
Lupin straw	WD	131	0.61
	DA	142	0.59
Subterranean clover,	WD	285	0.51
ryegrass pasture 1	DA	274	0.54
Subterranean clover,	WD	201	0.30
ryegrass pasture 2	DA	189	0.30
Kale	WD	67	2.27
	DA	64	2.31

* Co and Mo analyses by wet-digestion procedure (WD), DA = dry-ashing.
† Mean of three determinations

7.34.5
Gas Chromatography

Bachmann [108] has shown that volatilisation can be used as a separation step to isolate metals from solid matrices. Thus lead can be isolated from plant materials in 100% yield when they are heated to 1000 °C. The lead is then reacted with hydrogen and volatile lead is swept into a gas chromatograph for quantitative evaluation [109]. Lead can be determined in ng to µg quantities in plant materials.

7.34.6
Neutron Activation Analysis

The only reference found to date for the application of this technique to plant materials is that of Moauro et al. [110], who applied it to the determination of 25 minor and trace elements.

7.34.7
X-Ray Fluorescence Spectroscopy (XFS)

Reuter [111] has discussed in detail a technique for the measurement of trace elements in plant materials by X-ray fluorescence spectrometry.

A numerical matrix correction technique is used to linearise fluorescent X-ray intensities from plant material in order to permit quantitation of the measurable trace elements. Percentage accuracies achieved on a standard sample were ±13% for sulfur and phosphorus and better than ±10% for heavier elements. The calculation employs all of the elemental X-ray intensities from the sample, relative X-ray production probabilities of the elements determined from thin film standards, elemental X-ray attenuation coefficients, and the areal density of the sample cm^2. The mathematical treatment accounts for the matrix absorption effects of pure cellulose and deviations in the matrix effect caused by the measured elements. Ten elements are typically calculated simultaneously: phosphorus, sulfur, chlorine, potassium, calcium, manganese, iron, copper, zinc and bromine. Detection limits obtained using a rhodium X-ray tube and an energy-dispersive X-ray fluorescence spectrometer are in the low ppm range for the elements manganese to strontium.

Measured and computed values of the matrix coefficient are shown in Table 7.13. The values agree within a few percent except for Fe and Mn in radishes, where the difference is 6%. A matrix correction factor of 2 means that the combined attenuation of the exciting and fluorescent X-rays is 50%. In radishes, about half of this figure is from the cellulose and the other half from the presence of 6% potassium.

In order to determine the effectiveness of the matrix correction procedure for elements lighter than manganese, a sample of known concentration was measured. The mean values and standard deviations of five separate measurements of NBS orchard leaves (SRM 1571) are shown in Table 7.14.

The precisions of the values for the orchard leaves improve with X-ray signal-to-noise ratio, which for this sample generally follows absolute concentration.

Table 7.13. Measured and calculated matrix correction coefficients (from [111])

Element	Cellulose Measured	Cellulose Calculated	Orchard leaves Measured	Orchard leaves Calculated	Radishes Measured	Radishes Calculated
Mn	1.46	1.43	1.87	1.78	2.18	2.05
Fe	1.35	1.33	1.65	1.60	1.92	1.81
Cu	1.16	1.17	1.32	1.31	1.45	1.41
Zn	1.14	1.14	1.25	1.25	1.36	1.34
As, $Pb_{L\alpha}$	1.08	1.08	1.16	1.15	1.22	1.20
Br	1.06	1.06	1.12	1.12	1.17	1.15
Rb			1.09	1.09		
Sr	1.05	1.05	1.08	1.08	1.12	1.11

Table 7.14. NBS Orchard leaves, SRM 1571 (from [111])

Element	XRF, ppm[a]	NBS, ppm[a]
P	2380 ± 180	2100 ± 100
S	2150 ± 380	2300[b]
K	15 700 ± 800	14 700 ± 300
Ca	19 800 ± 800	20 900 ± 300
Mn	91.1 ± 18.0	91 ± 4
Fe	314 ± 40	300 ± 20
Cu	10.4 ± 2.4	12 ± 1
Zn	25.6 ± 3.4	25 ± 3
As	12.0 ± 2.6	11 ± 2
Br	8.8 ± 1.6	10[b]
Rb	11.3 ± 5.2	12 ± 1
Sr	36.5 ± 4.0	37[b]
Pb	48.6 ± 3.8	45 ± 3

[a] Two standard deviations are reported.
[b] No standard deviation given

For elements heavier than calcium, the precision ranges from 5 to 10% of the amount present. For the elements present at higher concentrations – potassium and calcium – the precisions are 3 and 2%, respectively.

Hoffmann and Lieser [112] used XFS to determine a range of elements in leaves and grass and compared this technique with neutron activation analysis, AAS, ICP–atomic emission spectrometry, polarography and voltammetry.

Messerschmidt et al. [113] used total reflection XFS to determine arsenic and bismuth. These elements were first converted to hydrides which were recovered and analysed by the X-ray method.

Saleh [114] combined XFS and proton-induced X-ray emission (PIXE) to analyse flour for potassium, calcium, manganese, iron, copper, zinc, bromide, rubidium and strontium.

References

1. Maitani T, Xing DR, Ikeda C, Goda Y, Takeda M, Yoshihara K (1994) *Shokuhin Eiseigaku Zasshi* **36**:201.
2. Lisk DJ (1960) *J Agric Food Chem* **8**:121.
3. Thomas B, Roughan JA, Walters ED (1972) *J Agric Food Chem* **23**:1496.
4. Culver BD, Lech JF, Pradham NK (1975) *Food Technol Champaign* **29**:16.
5. Bibr B, Lerner J (1971) *J Agric Food Chem* **19**:1011.
6. Ministry of Agriculture, Fisheries and Food (1973) *The Analysis of Agricultural Materials – Cadmium in Plant Material, Method 9*, Technical Bulletin RB 427, HMSO, London, UK.

7. Ministry of Agriculture, Fisheries and Food (1973) *The Analysis of Agricultural Materials – Calcium in Plant Material, Method 12*, Technical Bulletin RB 427, HMSO, London, UK.

8. Ministry of Agriculture, Fisheries and Food (1973) *The Analysis of Agricultural Materials – Cobalt in Plant Material, Method 21*, Technical Bulletin RB 427, HMSO, London, UK.

9. AOAC (1975) *Official Methods of Analysis*, Association of Official Analytical Chemists, Washington, DC, USA, p. 37

10. Ministry of Agriculture, Fisheries and Food (1973) *The Analysis of Agricultural Materials – Copper in Plant Material, Method 25*, Technical Bulletin RB 427, HMSO, London, UK.

11. Andrus S (1955) *Analyst* **80**:514.

12. Simmons WJ, Loneragon JF (1975) *Anal Chem* **47**:566.

13. Simmons WJ (1978) *Anal Chem* **50**:870.

14. Schiller P, Cook GB, Beswick CK (1971) *Microchim Acta* **3**:420.

15. Ministry of Agriculture, Fisheries and Food (1973) *The Analysis of Agricultural Materials – Iron in Plant Material, Method 38*, Technical Bulletin RB 427, HMSO, London, UK.

16. Mortatti J, Krug FJ, Pessenda LCR, Zagatto EAG, Jorgensen SS (1982) *Analyst* **107**:659.

17. Jayman TCZ, Sivasubramanian S, Wijedasa MA (1975) *Analyst* **100**:716.

18. Sandell ER (1950) *Colorimetric Determination of Traces of Metals, Volume 3: Chemical Analysis*, Second Edition, Interscience, London, UK.

19. Cowling H, Benne EJ (1942) *J Assoc Off Agric Chem* **25**:555.

20. Jackson ML (1962) *Soil Chemical Analysis*, Constable, London, UK.

21. Ministry of Agriculture, Fisheries and Food (1973) *The Analysis of Agricultural Materials – Determination of Lead in Plant Material, Method 42*, Technical Bulletin RB 427, HMSO, London, UK.

22. Thomas B, Roughan JA, Walters ED (1972) *J Sci Food Agric* **23**:1493.

23. Marcus JR (1974) *J Assoc Off Agric Chem* **57**:970.

24. Stafilov T, Rizova V (1992) *Glas Hem Tehnol Maked* **11**:37.

25. Stephens SC, Littlejohn D, Ottaway JM (1985) *Analyst* **110**:1147.

26. Jahns G, Schunck W, Schwedt G (1983) *J Chromatogr* **259**:195.

27. Ministry of Agriculture, Fisheries and Food (1973) *The Analysis of Agricultural Materials – Magnesium in Plant Material, Method 45*, Technical Bulletin RB 427, HMSO, London, UK.

28. Ministry of Agriculture, Fisheries and Food (1973) *The Analysis of Agricultural Materials – Manganese in Plant Material, Method 47*, Technical Bulletin RB 427, HMSO, London, UK.

29. Gine MF, Zagatto EAG, Filho HB (1979) *Analyst* **104**:371.

30. Technicon Instruments Corp. (1977) *Digestion and Sample Preparation for the Analysis of Total Kjeldahl Nitrogen and/or Total Phosphorus in Food and Agricultural Products using the Technicon BD-20/40 Block Digestor*, Technicon Industrial Method No.369-75A, Technicon Instruments Corporation, New York, USA.

31. Marczenko Z (1964) *Anal Chim Acta* **31**:224.

32. PerkinElmer Corp. (1973) *Analytical Methods for Atomic Absorption Spectrophotometry*, PerkinElmer Corporation, Norwalk, CT, USA, p. 1–15.

33. Ministry of Agriculture, Fisheries and Food (1973) *The Analysis of Agricultural Materials – Total Mercury in Soil and Plant Material, Method 86*, Technical Bulletin RB 427, HMSO, London, UK.

34. Hon P-K, Lau O-W, Wong M-C (1983) *Analyst* **108**:64.

35. Van Loon JC (1980) *Analytical Absorption Spectroscopy: Selected Methods*, Academic, New York, NY, USA, p. 160.
36. Kimura Y, Miller VL (1962) *Anal Chim Acta* **27**:331.
37. Lugowska M, Rubel S (1987) *Chem Anal (Warsaw)* **32**:591.
38. Muscat VI, Vickers TJ Andren A (1972) *Anal Chem* **44**:218.
39. Tatton JO'G, Wagstaff PJ (1969) *J Chromatogr* **44**:284.
40. Bradfield EG, Stickland JF (1975) *Analyst* **100**:1.
41. Bradfield EG (1964) *J Sci Food Agric* **15**:469.
42. Kosta L, Byrne Ar (1969) *Talanta* **16**:297.
43. Wang S, Wai CM (1996) *Environ Sci Technol* **30**:3111.
44. Haller WA, Rancitelli LA, Cooper JA (1968) *J Agric Food Chem* **16**:1036.
45. Ministry of Agriculture, Fisheries and Food (1973) *The Analysis of Agricultural Materials – Molybdenum in Plant Material, Method 49*, Technical Bulletin RB 427, HMSO, London, UK.
46. Dick AT, Bingley JB (1947) *Aust J Exp Biol Med* **25**:193.
47. Hoenig M, Van Elsen YN, Van Canter R (1986) *Anal Chem* **58**:777.
48. Sturgeon RE, Chakrabarti CL (1977) *Anal Chem* **49**:90.
49. Sneddon J, Fuavao VA (1985) *Anal Chim Acta* **167**:317.
50. Ministry of Agriculture, Fisheries and Food (1973) *The Analysis of Agricultural Materials – Nickel in Plant Material, Method 52*, Technical Bulletin RB 427, HMSO, London, UK.
51. Uto M, Sugawara M (1985) *Fresen Z Anal Chem* **321**:68.
52. Braun H, Metzger M (1984) *Fresen Z Anal Chem* **318**:321.
53. Ministry of Agriculture, Fisheries and Food (1973) *The Analysis of Agricultural Materials – Potassium in Plant Material, Method 67*, Technical Bulletin RB 427, HMSO, London, UK.
54. Collins GC, Polkinhorne H (1952) *Analyst* **77**:430.
55. Megarrity RG, Siebert BD (1977) *Analyst* **102**:95.
56. Watkinson JH (1962) *Trans Jt Meet Comm IV V Int Soc Soil Sci*, p. 149.
57. Hartley WJ (1967) In: Muth OH (ed) *Symposium on Selenium in Biomedicine*, Avi, Westport, CT, USA.
58. Andrews ED, Hartley WJ, Grant AB (1968) *NZ Vet J* **16**:3.
59. Allan JE (1963) In: *4th Australian Spectroscopy Conference*, 20–23 Aug 1963, Canberra, Australia.
60. Baird RB, Gabrielian SM (1974) *Appl Spectrosc* **28**:273.
61. Ihnat M, Westerby RJ (1974) *Anal Lett* **7**:257.
62. Mesman BB, Thomas TC (1975) *Anal Lett* **8**:449.
63. Fernandez FJ (1973) *Atom Absorp Newslett* **12**:93.
64. Duncan L, Parker CR (1974) *Varian Techtron Technical Topic September 1974*, Varian Techtron Pty, Springvale, VC, Australia.
65. Ministry of Agriculture, Fisheries and Food (1973) *The Analysis of Agricultural Materials – Selenium in Plant Material, Method 70*, Technical Bulletin RB 427, HMSO, London, UK.
66. Levesave M, Vendette ED (1971) *Can J Soil Sci* **51**:85.
67. Hall RJ, Gupta PL (1969) *Analyst* **94**:292.
68. Olsen OE (1969) *J Assoc Off Anal Chem* **52**:627.
69. Vijayakumar M, Ramakrishna TV, Aravamudon G (1982) *Talanta* **29**:61.
70. Babbco AK, Pilpenko AT (1976) *Photometric Analysis: Methods for the Determination of Nonmetals*, Mir, Moscow, Russia, p. 236–299.
71. Rann CS, Hambly AN (1965) *Anal Chim Acta* **32**:346.

72. Molinari GP, Trevisan M, Nateli P, Del Re AAM (1987) *J Agric Food Chem* **35**:727.
73. Clinton OE (1977) *Analyst* **102**:187.
74. Norheim G, Haughan A (1986) *Acta Pharm Toxicol* **Suppl 59**:610.
75. Buckley WJ, Budac JJ, Godfrey DL (1992) *Anal Chem* **64**:724.
76. Meija J, Montes-Bayon M, Le Duc DL, Terry N, Caruso JA (2002) *Anal Chem* **74**:5837.
77. McLeod F, McGaw BA, Shard CA (1996) *Talanta* **43**:109.
78. Handley R (1960) *Anal Chem* **32**:1719.
79. Ministry of Agriculture, Fisheries and Food (1973) *The Analysis of Agricultural Materials – Sodium in Plant Material, Method 71*, Technical Bulletin RB 427, HMSO, London, UK.
80. Godar EM, Alexander OR (1946) *Ind Eng Chem Anal Edit* **18**:681.
81. Abbasi SA (1982) *Int J Environ Anal Chem* **11**:1.
82. Boomer DW, Powell MJ (1987) *Anal Chem* **59**:2810.
83. Ministry of Agriculture, Fisheries and Food (1973) *The Analysis of Agricultural Materials – Zinc in Plant Material, Method 81*, Technical Bulletin RB 427, HMSO, London, UK.
84. Ivanova E, Benkhedda K, Adams F (1998) *J Anal Atom Spectrom* **13**:527.
85. Takuwa DT, Sawula G, Wibetoe G, Lund W (1997) *J Anal Atom Spectrom* **12**:849.
86. Jirovetz L, Ecker G, Jacger W, Heiss T (1993) *Ernährung (Vienna)* **17**:265.
87. Vinas P, Campillo N, Lopez Garcia I, Hernandez Cordoba M (1993) *Anal Chim Acta* **283**:393.
88. Fu F, Zhang Z, Zheng J, Zhang M (1993) *Fenxi Huaxue* **21**:123.
89. De la Guardia M, Carbonelle V, Morales, Rubio A, Salvador A (1993) *Talanta* **40**:1609.
90. Rissova V, Stafilov T, Sivakov L (1993) *God Zv Zemjod Fak Univ Kiril Metodu – Skopje* **38**:359.
91. Martin-Polvillo M, Albi T, Guinda A (1994) *J Am Oil Chem Soc* **71**:347.
92. Wang Z, Wang Y, Song S, Wang F (1993) *Ji T Zhongguo Tiaoweipin* **5**:28.
93. Gelroth JA, Kadan RS (1987) *Cereal Foods World* **32**:443.
94. Allen SE, Parkinson JA (1969) *Spectrovision* **22**:2.
95. Arafat NM, Glooschenko WA (1981) *Analyst* **106**:1174.
96. Dabeka RW (1979) *Anal Chem* **51**:902.
97. Heanes DL (1981) *Analyst* **106**:182.
98. Evans WH, Dellar D, Lucas BE, Jackson FJ, Read JI (1980) *Analyst* **105**:529.
99. Stringari G, Pancheri I, Muller F, Faille O (1998) *Accredit Qual Assur* **3**:122.
100. Analytical Methods Committee (1960) *Analyst* **85**:643.
101. Evans WH, Read JI, Lucas BE (1978) *Analyst* **103**:580.
102. Heanes DL (1981) *Analyst* **106**:172.
103. Schramel P, Klase BJ, Hasse S (1982) *Fresen Z Anal Chem* **310**:209.
104. Wolnik KA, Fricke FL, Hahn MH, Caruso JA (1981) *Anal Chem* **53**:1030.
105. Hahn MH, Wolnik KA, Fricke FL, Caruso JA (1982) *Anal Chem* **54**:1048.
106. Nangniot P (1964) *J Electroanal Chem* **7**:50.
107. Wendt E (1937) *Lebensmittelindustr* **34**:101.
108. Bachmann K (1982) *Talanta* **29**:1.
109. Wahdat F, Shamsipoor (1977) *Fresen Z Anal Chem* **288**:191.
110. Moauro A, Triolo L, Avino P, Ferrandi L (1993) *Dev Plant Soil Anal* **33**:13.
111. Reuter III FW (1975) *Anal Chem* **47**:1763.
112. Hoffmann P, Lieser KH (1987) *Sci Total Environ* **64**:1.
113. Messerschmidt J, Von Bohlen A, Alt F, Klockenkämper R (1997) *J Anal Atom Spectrom* **12**:1251.
114. Saleh NS, Al-Saleh KA (1986) *Appl Phys Commun* **6**:195.
115. Pahlavanpour B, Thompson M, Thorne L (1981) *Analyst* **106**:467.

8 Determination of Organic Compounds in Plant Materials, Vegetables and Fruit

8.1
Carboxylic Acids and Ethers

Specific methods have been described for the determination in plants of pyruvic acid by gas chromatography (GC) [1] and phytic acid by spectrophotometric methods [2, 3].

Hueni and Uebersax [4] have used GC to determine low fatty acids in silage. In this method, fresh silage (100 grams) is mixed with 100–200 ml of water and allowed to stand for 30 minutes. The juice is expressed and 1.0 ml of hydrochloric acid is added per 10 ml of fluid. Insoluble material is removed by centrifugation and an aliquot of the supernatant liquid is injected directly into a gas chromatograph equipped with glass columns containing Porapak Q, the temperature programmed from 150 to 230 °C at 8 °C/min, and with a flame ionisation detector; nitrogen is used as carrier gas (30 ml/min).

A standard official method has been issued for the determination of C_1 to C_6 volatile carboxylic acids and lactic acid in silage juice [5, 6].

GC isotope ratio mass spectrometry [7] and GC using a caesium bromide thermionic detector [8] have been used to determine, respectively, carboxylic ethers in apples and tetraethyl pyrophosphate in chloroform–acetone extracts of crops in amounts down to 0.01 ppm.

8.2
2,4-Dichlorophenol

Supercritical fluid extraction with carbon dioxide has been used to determine 2,4-dichlorophenol in crops and straw [8].

8.3
N-Nitroso Compounds

Sen [9] has reviewed methods for the determination of nonvolatile N-nitroso compounds in crops, whilst Usero [10] used GC with a Hall detector to determine N-nitroso compounds.

8.4
Thiabendazole and S-Methylmethionine

Brandon [11] developed an enzyme immunoassay method using monoclonal antibodies to determine thiabendazole in apples and potatoes.

Kovatcheva [12] has described a method using an automated amino acid analyser for the determination of S-methylmethionine in cabbage, kohlrabi, celery and sweetcorn. The plant sample is first homogenised with 0.1 N hy-

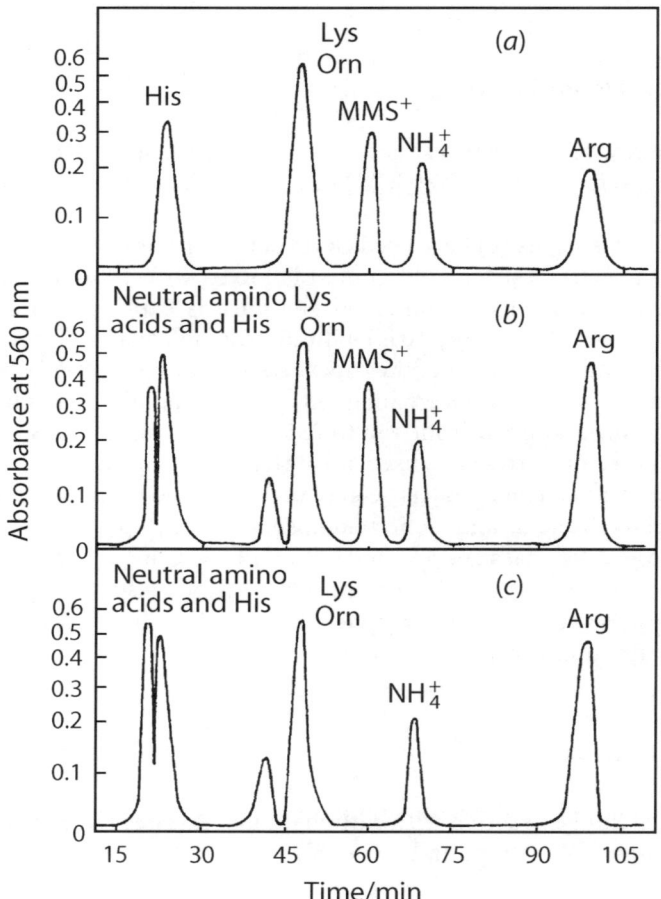

Figure 8.1. Chromatograms produced on a 100 mm column of an amino acid analyser with 0.3 N sodium citrate solution (pH 7.00) as eluting agent: (*a*) standard solution containing 1 μmol/ml of each amino acid; (*b*) 30 g of cabbage extract purified with Dowex 50-X8 in the ammonium form (0.97 mg of S-methylmethionine); and (*c*) the same sample as for (*b*) but after treatment at pH 10.0 for 30 minutes at 120 °C. *MMS* S-methylmethionine, *his* histidine, *lys* lysine, *orn* ornithine, *arg* arginine. From [12]

drochloric acid in 70% ethanol (3 + 1 m/v). This aqueous solution is passed down a column of Dowex 50-X8 cation exchange resin in the ammonium form and the amino acids eluted with 2 N ammonia solution and evaporated to dryness at 50 °C.

Following adjustment to pH 6.0, the solution is applied to a SP-Sephadex C-25 column in the sodium form. Amino acids are then eluted with 0.2 M citrate phosphate buffer, pH 8.0, and the effluent evaporated to dryness at 50 °C. The residue is dissolved in 0.1 N hydrochloric acid and applied to the amino acid analyser. Amino acids are separated by passing 0.2 M, pH 8 sodium citrate solution down the column. The S-methylmethionine content can then be obtained from the chromatogram, as illustrated in Fig. 8.1. The results obtained agree reasonably well with those obtained by thin-layer chromatography [13].

8.5
Miscellaneous Organic Compounds

Methods for the determination of these are reviewed in Table 8.1.

8.6
Mycotoxins

The trichothecene mycotoxins are a group of over 60 sesquiterpenoid compounds which are produced by a variety of imperfect fungi, including species of the genera *Cephalosporium, Fusarium, Myrothecium, Stachybotrys, Trichoderma* and *Trichothecium*. Structurally, trichothecenes are characterised by the 12,13-epoxytrichothec-9-ene ring system, with subgroup classes related to specific functionalities. As a group, the trichothecenes show a wide range of biological activity and have been involved in natural intoxication in humans and domestic animals following ingestion of mouldy grains. They are also toxic to plants and bacteria.

The naturally fluorescent mycotoxin *trans*-1-undecenyl (6-(10-hydroxy-6-oxo-*trans*-1-undecenyl) β-resorcylic acid μ-lactone) is a plant estrogenic mycotoxin which is suspected to cause infertility in dairy cattle and swine through the ingestion of mouldy feedstuffs. Other mycotoxins have been identified in cereals, grains, nuts and foods including aflatoxins, ochrotoxins, fusarium toxins, patulin, ergot alkaloids, atrinin, sterigmatocystin and penicillic acid.

Various techniques used for the determination of mycotoxins are reviewed below.

8.6.1
Liquid Chromatography (LC)

Dorner and Cole [27] determined aflatoxins in peanuts by LC with post-column iodination and modified micropump clean-up. β-Cyclodextrin enhances the

Table 8.1. Determination of miscellaneous organic compounds in plant material (from author's own files)

Organic compound	Sample type	Sample preparation	Analytical finish	Detection limit	Reference
Indole-2-ylacetic acid	Plants	Extraction conversion to silyl derivative	Capillary gas chromatography with NP detection [14,15], also HPLC [16]	–	[14–16]
Cyclopiazonic acid	Plants	–	Review of methods	125 ppb	[17]
Diosgenin; yamogenin	Fenugreek seeds	Hexane–ethyl acetate extraction	Infrared spectrometry	–	[18,19]
1-Naphthyl acetic acid, 2-(1-naphthyl acetamide)	Peaches	Acetone extraction, C_{18} clean-up	HPLC with UV detection	–	[20]
Geosmin (*trans*-1,10-dimethyl-*trans*-9-decalol)	Beans, beets	Freon 113 extraction	GC	–	[21]
Benomyl, carbendazim, ethoxyquin, thiabendazole	Apples, pears	–	HPTLC and HPLC	–	[22]
Ethoxyquin	Apples	Hexane extraction, 0:1 N hydrochloric acid extraction	GC with electron capture detection	0.2 ng (absolute)	[23]
Quinmerac acid	Cereals	–	Enzyme-linked immunoassay (ELISA)	–	[24]
Chlorocholine chloride	Grain	–	TLC	–	[25]
α-amylase	Barley, wheat	–	ELISA	–	[26]

fluorescence of aflatoxin B_1 and G_1 in aqueous systems, and this effect was used by Francis et al. [28] in developing a liquid chromatographic method for the determination of aflatoxins B_1, B_2, G_1 and G_2 in corn.

Rajakyla et al. [29] determined mycotoxins in grain by HPLC and thermo-spray LC–MS.

Shepherd [30] has compiled data on HPLC for the determination of the following mycotoxins in food: aflatoxins, ochrotoxins, fusarium toxins, patulin and engot alkaloids.

8.6.2
Thin-Layer Chromatography (TLC)

Rati et al. [31] used a TL chromatographic method to determine aflatoxin B_1 in corn and peanuts.

8.6.3
Supercritical Fluid Extraction (SCFE)–Direct Fluid Injection
Mass Spectrometry (DFI–MS)

Kalinoski et al. [32] has applied this method to the determination of tri-chothecene mycotoxins in wheat. The methods were based on chemical ionisa-tion MS and collision-induced dissociation tandem MS and enabled the rapid identification of ppm levels of several trichothecene mycotoxins. Supercritical carbon dioxide is shown to allow identification of mycotoxins with minimum sample handling in complex natural matrices such as wheat. Tandem MS tech-niques are employed for unambiguous identification of compounds of varying polarity, and "false positives" from isobaric compounds are avoided. Capillary column SCFC–MS of a SCF extract of the same sample was also performed, and detection limits in the ppb range appear feasible.

8.6.4
Differential Pulse Polarography (DPP)

A further method of determining the trichothecene mycotoxin deoxynivalenol in corn is based on DPP [33] and also on GC.

Palinisano et al. [33] found that the electrochemical activity of trichothecene toxins is related to the presence of an α,β-unsaturated keto group and conse-quently the polarographic method was selective for trichothecenes with keto groups. Differential pulse polarography gave a detection limit for a pure so-lution of deoxynivalenol of 0.029 μmol/l (8.6 ng/ml). This technique, coupled with an appropriate extraction procedure, was applied to the determination of deoxynivalenol in *Fusarium*-infected corn. A detection limit of about 50 ng/g (for 50 g of the original sample) was estimated.

The ground corn sample (50 g) was extracted with methanol–1% sodium chloride solution (55 + 45 v/v) (200 ml) and hexane (100 ml). The extract was further de-fatted with hexane (60 ml), re-extracted three times with chloroform (50 ml), and concentrated almost to dryness.

The residue was reconstituted with methanol–water (2 + 3) and purified by passage through a Sep-Pak C_{18} cartridge. Further purification of the extract was accomplished by preparative thin-layer chromatography carried out using chloroform–methanol–water (90 + 10 + 1) as the eluent. The deoxynivalenol area (checked by fluorescence quenching at the same R_F value of the standard) was scraped off and eluted with acetone. The extract was concentrated to dryness, reconstituted in 200 µl of methanol and divided into two equal fractions for gas chromatographic and polarographic analysis, respectively.

Differential Pulse Polarographic Analysis

A 3 ml aliquot of Britton-Robinson buffer solution–methanol (9 + 1) (supporting electrolyte) was transferred into the electrochemical cell and deoxygenated for 15 minutes with purified nitrogen, which was kept flowing over the sample solution during measurements. Polarograms were first run on each background solution and then on the sample solution by scanning the potential between –1.1 V and the potential of the electrochemical reduction of the supporting electrolyte. A scan rate of $1-5$ mV s^{-1}, a drop time of $1-2$ s and a modulation amplitude of $50-100$ mV were typically employed.

Gas Chromatography

A 100 µl aliquot of the extract that had been purified by thin-layer chromatography was evaporated, treated with Tri-Sil (50 µl) and allowed to stand for two hours to complete the reaction. The derivatised solution ($1-5$ µl) was analysed by GC under the following instrumental conditions: initial temperature, 150 °C (maintained constant for 0.5 minutes); final temperature, 280 °C (maintained constant for ten minutes); temperature programme, increasing at 10 °C per minute; injection-block temperature, 275 °C; flame ionisation detector temperature, 300 °C; and nitrogen flow-rate, 20 ml/min.

8.6.5
High-Performance Liquid Chromatography with Laser Fluorometric Detection

Disbold et al. [34] developed a laser fluorimetric method for the determination of zearalenone (6-(10-hydroxy-6-oxo-$trans$-1-undecenyl-β-resorcylic acid n-lactone))-infected corn. By combining laser fluorimetry with high-pressure liquid chromatography, these workers were able to detect and quantitate the naturally fluorescent mycotoxin zearalenone in contaminated corn samples.

Experiments with zearalenone standards show that the linear fluorimeter response covers four orders of magnitude with a detection limit of 300 pg zearalenone injected onto a C_{18} reverse-phase column. The corn samples are first purified using a small silica gel column. The recovery from this step is 86% over the range from 5 ppb to 2.5 ppm. Based on the magnitude of the zearalenone signals compared to the flatness of the baseline for zearalenone-free corn samples, a limit of 5 ppb is placed on the detection of zearalenone by this procedure.

Analysis is performed on an aqueous methanol (25:75 v/v) extract of the ground corn sample. After the addition of ammonium sulfate, the extract is partitioned with hexane, then partitioned with methylene dichloride. This extract is evaporated to dryness and the residue taken up in benzene–acetonitrile (98:2 v/v). Final analysis is carried out by HPLC with a laser fluorometric detector.

Various other workers have reviewed the sample preparation and preservation [35] and analytical determination of [36–38] mycotoxins in cereals and Inhat [39] and Beaver [41] have reviewed gas chromatographic methods for the determination of mycotoxins.

8.7
Volatile Organic Compounds

8.7.1
Volatiles in Plant Materials

Volatile organic compounds (VOCs) are an important group of chemicals that permeate our environment. The concentrations of VOCs in vegetation are one factor that must be considered in an assessment of the environment's exposure to these chemicals. The major route for plant uptake of volatile hydrophobic compounds is through sorption of the compounds directly from air [42]. This uptake by vegetation is suggested to be species-dependent [43]. Several models are used to predict bioconcentration factors based on the partitioning of organic compounds between air and an organic phase (as octanol) and between air and an aqueous phase [40, 44–46]. Documented determinations resulting in bioconcentration factors (BCFs) are limited to tetrachloroethene in pine needles [48] and 1,2,4-trichlorobenzene in soybeans [47].

The uptake of tetrachloroethene in pine needles is reported to be more complex than the published model [45]. In that study, pine needles in a chamber were exposed to elevated levels of tetrachloroethene, and its concentration in the needles were predicted via its K_{oa} and lipid content. When the needles had been exposed to the much lower environmental levels, the concentration of tetrachlorethene was much greater than predicted. It was suggested that an additional compartment in the needles bioconcentrated the tetrachlorethene in excess of theory but had a limited capacity to absorb tetrachlorethene.

Hiatt [48] has investigated the behaviour of volatile organic compounds and tests the use of lipid content and K_{oa} to predict BCF in an uncontrolled environmental setting. The urban environmental variations encountered in these experiments introduce errors when estimating BCFs, but highlight a major factor affecting the equilibria of VOCs between leaves and air.

The environmental site investigated for this study was at the US Environmental Protection Agency's National Exposure Research Laboratory on the University of Nevada Las Vegas (UNLV) campus. The VOCs monitored in this study were those that are persistent in the environment, that tended to bioconcentrate in an organic phase (hydrophobic), and that could easily be detected in air and leaves. Nine different plant species found on the UNLV campus were analysed by using vacuum distillation coupled with GC–MS to determine concentrations of VOCs. Grab air samples were taken before and immediately after collection of leaves to determine the concentration of VOCs in air.

The bioconcentration of volatiles in the leaves of some species can be predicted using the partition coefficients between air and octanol (K_{oa}) and by only considering VOC absorption in the lipid fractions of the leaves. For these leaves, the bioconcentration factors agreed with existing models. Leaves of some species displayed a bioconcentration of volatiles that greatly exceeded theory. These hyper-bioconcentration leaves also contain appreciable concentrations of monoterpenes, suggesting that a terpenoid compartment should be considered for the bioconcentration of organic compounds in leaves. Adding an additional terpenoid compartment should improve the characterisation of volatile organic compounds in the environment. The uptake of VOCs from air by leaves is rapid, and the equilibration rates are seen to be quicker for compounds that have higher vapour pressures. The release of VOCs from the leaves of plants is slower for hyper-bioconcentration leaves.

Keymeulen et al. [49] determined various volatile chlorinated hydrocarbons in plant leaves by GC–MS. The method consists of solvent extraction with pentane followed by GC–MS.

Kaupp and Sklarz [50] reported a clean-up method for the determination of polyaromatic hydrocarbons in plant samples including maize leaves. The two-step clean-up consisted of gel permeation chromatography on a porous styrene–divinylbenzene copolymer, followed by further clean-up on silica gel.

Thermal desorption techniques have been used to determine plant volatiles [51].

8.7.2
Volatiles in Fruit and Vegetables

Scudamore [52] determined 2-aminobutane in potatoes by HPLC. This substance is used to control certain potato tuber diseases. The amine was distilled from potatoes, dansylated and determined using reverse-phase HPLC with fluorescence detection. Recovery of 2-aminobutane by distillation was about 95% from standard solutions and 92% from treated potatoes. The lower limit of detection was below 0.2 µg/kg.

Supercritical fluid chromatography has been used to characterise aroma compounds in strawberries [53] and HPLC has been used to determine furaneol and related compounds in strawberries [54].

Steele et al. [55] used a dynamic heated headspace purge and trap extraction technique with selected ion monitoring capillary GC–MS to measure styrene monomer levels in amounts down to 2 ng/g in tomatoes and milk.

Cairns et al. [56] employed GC–MS techniques using chemical ionisation and electron impact to confirm the presence of ethylene dibromide in fruits and grains determined by electron capture detection gas chromatography. Interferences from both the solvent and coextractables have been minimised to permit a determination using only two ions detected in the correct experimental intensity ratio belonging to the monobromoethylene carbonium fragment ion. A complementary measurement made under negative ion methane chemical ionisation determined that bromide was present.

Ohta and Osajima [57] recovered the volatile compounds in onions using a cold trap apparatus.

8.7.3
Volatiles in Grains

The UK Panel on Fumigant Residues in Grain [58] have developed an electron capture gas chromatographic method for the determination of volatile fumigants such as carbon tetrachloride, chloroform and other chloroaliphatic compounds down to 6 mg/kg of carbon tetrachloride in fumigated wheat, and maize was determined by this method.

8.8
Insecticides and Pesticides

8.8.1
Plant Materials

Some examples of the types of procedures used to determine insecticides and pesticides in plant materials are given below.

Ueji [59] determined carbaryl and propoxur (o-isopropylphenyl N-methylcarbamate) in crops in amounts down to 0.0005 ppm. The carbamates were reacted with trifluoroacetic anhydride solution in ethyl acetate by heating in the dark at 50 °C. This reaction was quantitative and reproducible and the stability of the N-trifluoroacetyl derivatives was high. The derivatives of these insecticides were subjected to GC with electron capture detection on five-foot columns packed with Chromosorb W coated with 5% OV-17 or OV-25, 2% poly(ethanediol adipate). The most efficient stationary phases were OV-17 and OV-25, 5% OV-17 being particularly good. To determine m-tolyl methylcarbamate residues in unpolished rice grain or rice straw, a sample of powdered rice

(10 g) or chopped straw (5 g) was extracted with dichloromethane in a Soxhlet apparatus, the extract was partitioned between acetonitrile and hexane, and the acetonitrile phase was diluted with 4% sodium chloride solution and extracted with dichloromethane. This extract was passed through a column of Florisil (5 g), water-saturated dichloromethane being used as eluent, and the eluate was cleaned up on a column of activated alumina (10 g), with acetone–hexane (1:9) as eluent. The residue from the evaporation of the eluate was dissolved in ethyl acetate, and, after the trifluoroacetylation reaction, the solution was analysed by gas chromatography. Recoveries ranged from 91.2 to 98.8% for samples fortified with m-tolyl methylcarbamate at the 0.1 – 0.4 ppm level.

Westlake et al. [60] determined m-S-butylphenyl methyl-(phenylthio)car-bamate (RE 11775) in vegetation by a chromatographic procedure. The sample is extracted with dichloromethane, chloroform or acetonitrile, followed by clean-up (if necessary) on a column of Florisil, silica gel or alumina. The purified residue is submitted to GC on either a stainless steel column (3 ft × 0.25 ″) packed with 5% OV-225 on Gas-Chrom Q (60 – 80 mesh) and operated at 242 °C, with nitrogen as carrier and a flame photometric detector operated in the S mode, or on a glass column (3 ft × 6 mm od) with identical packing and operated at 195 °C, with hydrogen as carrier gas (100 ml/min) and an electrolytic conductivity nitrogen detector. Recoveries of added RE 11775 from water, soil and mud samples were about 100% and from grass and Lucerne about 80%. Down to 0.01 and 0.1 ppm could be determined in grass and in lucerne, respectively.

Numerous other methods have been described and are previewed in Table 8.2.

8.8.2
Fruit and Vegetables

HPLC features highly in methods for the determination of insecticides in fruit and vegetables. Thus, Clark et al. [79] developed a method for the determination of ethyl and methyl parathion residues on vegetable material using reverse-phase HPLC with series UV–electrochemical detection. Sample preparation techniques involving acetonitrile extraction of the vegetable were developed which avoided the usual preliminary column fractionations and which allowed the parathions to be recovered with an average of 95% recovery at concentrations of less than 50 ng/g for the plant material. Relative standard deviations of about 5% were obtained using five different plant samples. The selectivity of electrochemical detection meant that it was not necessary to chromatographically resolve the plant components from the pesticides which were electrochemically active and it allowed rapid analysis. Series detection proved useful in distinguishing various components in the samples from pesticides, in distinguishing various pesticides, and in comparing the operating characteristics of the two detectors.

Table 8.2. Determination of insecticides and pesticides in plant materials (from author's own files)

Insecticide type	Sample type	Sample preparation	Analytical finish	Detection limit	Reference
Carbamate type					
Oxamyl (methyl-N,N'-dimethyl-N-[(methyl-carbamoyl)oxy]-1-thio oxamidate)	Plants including: leaves, potato, tomato and wheat	Ethyl acetate extraction of sample	Addition of excess copper sulfate, back-titration with standard EDTA to 1-(2-pyridylazo)2-naphthol	–	[61]
Mesurol (N-methyl type carbamate)	Plants, blueberries	Pervaporation of trifluoroacetyl and pentafluorobenzyl derivatives	GC, GC-MS	–	[62,63]
Carbamoyl oximes, carbamothioc acids, dithiocarbamates and phenyl urea types	Vegetation	–	HPLC with post-column photolysis for the generation of fluorophores	–	[64]
Carbamate type	Crops	C$_8$ HPLC with post-column basic hydrolysis and electrochemical detection	–		[65]
Carbaryl, propoxur (isopropyl phenyl N-methyl carbamate)	Crops	Reaction with trifluoroacetic anhydride	GC with ECD	0.005 ppm	[66]
M-S-butylphenyl methyl (phenylthio)carbamate	Vegetation, grass	Dichloromethane or chloroform extraction	GC with FPD	0.1 ppm	[67]
Methomyl [S-methyl-N-(methylcabamoyl)oxy-thioacetimidate]	Crops	Dichloromethane extraction	GC	0.5 ppm	[67]

Table 8.2. (continued)

Insecticide type	Sample type	Sample preparation	Analytical finish	Detection limit	Reference
Permethrin type					
Permethrin, cypermethrin, deltamethrin, fenvalerate	Crops	GC with ECD		–	[68]
Cypermethrin and its degradation products	Leaves	–	GC with ECD and GC–MS, and achiral and chiral HPLC	–	[69]
Organophosphorus types					
Organophosphorus types	Plants	Solvent extraction from freeze-dried sample	GC with ECD	–	[70]
Organophosphorus types	Crops	–	GC	–	[71]
Organophosphorus types	Plants	–	Laser MS	–	[171]
Miscellaneous types					
Diuron	Crops	Methanol extraction, methylene dichloride partitioning	Silica gel HPLC with photoconductivity detection	–	[72]
Dichlorbenil	Crops	Steam distillation	HPLC with pre-column preconcentration	–	[73]
Azinphos-methyl and its metabolites	Leaves	–	GC with flame ionisation and dual-flame detection	–	[74]
Pyridate and its metabolite	Plants	Solvent extraction	HPLC and MS–MS	–	[75, 76]
Miscellaneous	Plants	Soxhlet extraction	Study of extraction conditions	–	[77]
20 miscellaneous pesticides	Leaves	–	Static SIMS, polar pesticides most easily detected	–	[78]

Figure 8.2 shows a representative chromatogram of a turnip green sample which was obtained by using series ultraviolet and amperometric detection.

Baker and Bottomly [80] developed a multi-residue method for the determination of synthetic pyrethroids in fruit and vegetables. After extraction with hexane–acetone, the pyrethroids are separated from coextractives by a partition process and chromatography on a silica gel column and quantitatively determined by electron-capture gas–liquid chromatography and/or HPLC using an ultraviolet spectrophotometric detector.

This separation is illustrated in Fig. 8.3.

The HPLC system takes only about 30 minutes to screen for all of the pyrethroids, but suffers from the disadvantage that complete resolution of all

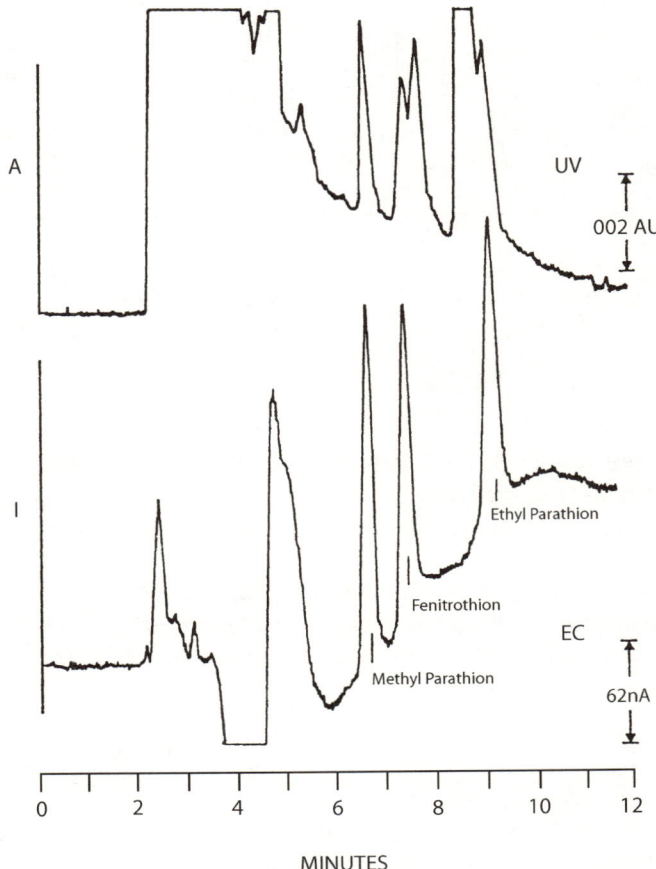

MINUTES

Figure 8.2. Chromatograms of a turnip green sample using series ultraviolet and electrochemical detection: 40.04 ng of methyl parathion and 44.52 ng of ethyl parathion injected; mobile phase: 64% acetonitrile, 36% 0.05 M ammonium acetate; pH 5.0, flow rate, 1 ml/min, UV detection at 270 nm; EC detector at –0.97 V *vs*. Ag/AgCl. From [79]

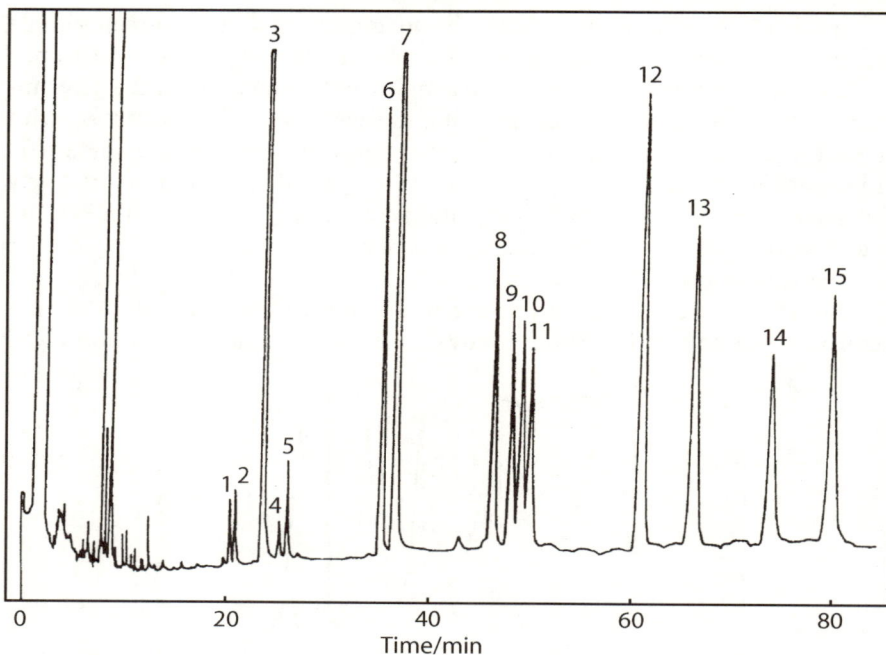

Figure 8.3. Separation of synthetic pyrethroids (0.8 ng of each) on a 25 m OV-101 WCOT capillary column. Peaks: *1* = *cis*-methrin; *2* = bioresmethrin; *1 + 2* = resmethrin; *4* and *5* = phenothrin; *6* = *cis*-permethrin; *7* = *trans*-permethrin; *8, 9, 10* and *11* = cypermethrin; *12* and *13* = fenvalerate; and *14* and *15* = deltamethrin. From [80]

the compounds is not achieved. Deltamethrin, fenvalerate, bioresmethrin and resmethrin are not resolved completely. However, this technique can be used for initial screening, with gas–liquid chromatography being used for confirmation of identity.

Gas chromatography is another favoured method, and has been used for example to determine organophosphorus pesticides in fruit and vegetables and also in many other methods of pesticide analysis. See Table 8.3.

8.8.3
Grains and Cereals

Again, HPLC and GC seem to be the methods of choice, particularly in more recently published methods when they are interfaced with a mass spectrometric detector.

Thus Bottomly and Baker [133] have described a multi-residue method for the determination of organophosphorus and synthetic pyrethoid pesticides and carbaryl in grain. After extraction with acetone–methanol, the pesticides

Table 8.3. Determination of insecticides and pesticides in fruit and vegetables (from author's own files)

Insecticide type	Sample type	Sample preparation	Analytical finish	Detection limit	References
Carbamate type					
Thiocarbamate type	Apples	Solid-phase extraction	HPLC, UV detection and GC flame detection	–	[81]
Thiocarbamate type	Apples	Solid-phase extraction	HPLC, UV detection and GC flame detection	–	[82]
Methomyl, methomyl oxime carbamate types	Fruit	Solvent extraction	HPLC	–	[83]
30 N-methyl carbamates	Vegetables	–	Review of HPLC methods	–	[84]
Seven carbamate types and their metabolites	Fruit and vegetables	Extraction then gel permeation chromatography	HPLC with pre-column hydrolysis and reaction with o-phthaldehyde to form fluorophores	–	[85]
15 Carbamates and their metabolites	Fruit and vegetables	Ethyl acetate extraction–clean-up on C₈ Sep-Pak cartridge	Derivatisation with 1-fluoro-2,4-dinitrobenzene–GC with ECD		[86]
Carbamate type and its metabolites	Fruit and vegetables	Solvent extraction, clean-up on amino-bonded silica	HPLC with post-column hydrolysis and addition of o-phthaldehyde to form fluorophores		[87]
21 Carbamate types and their metabolites	Fruit and vegetables	–	On-line HPLC		[88]
Organochlorine types					
Organochlorine pesticides and metabolites	Fruit and vegetables	Steam distillation	Study of extraction conditions		[89]
Organochlorine and organophosphorus type	Fruit and vegetables	–	MS		[90]

Table 8.3. (continued)

Insecticide type	Sample type	Sample preparation	Analytical finish	Detection limit	References
Organophosphorus type					
Fonophos(o-ethyl-9-phenylethyl phospho-nodithioate)	Vegetables	–	MS	–	[91]
Organophosphorus type	Apples	Chlorofuran extraction, Na$_2$SO$_4$–Florisil–Celite clean-up	GC with ECD detection Online size exclusion chromatography–silica gel. HPLC–GC with flame photo-metric detection	ng/g [93]	[92]
Organophosphorus type	Vegetables	SCFE with methanol-modified carbon dioxide	–	–	[94]
Organophosphorus type	Fruit and vegetables	Microextraction with water:ethyl acetate	Study of extraction conditions	–	[95]
Organophosphorus type	Carrots	Packed C$_{18}$ column SFC with carbon dioxide	GC with ECD		[96]
Organophosphorus type	Fruit and vegetables	Collaborative study on extraction studies	–		[97]
Organophosphorus type	Vegetables	–	Review of methods		[98]
Organophosphorus type	Vegetables	–	TLC		[98]
Dichorvos, dimethoate, parathion, malathion	Fruit and vegetables (carrots, beans, tomatoes, peas)	Extraction with toluene:acetone (1:1)	GC with alkali flame ionisation and photometric detection		[99–102]

Table 8.3. (continued)

Insecticide type	Sample type	Sample preparation	Analytical finish	Detection limit	References
Malathion, dimethoate, omethoate, chlorfenvinphos	Vegetables (lettuce, tomatoes, cabbage)	Dichloromethane extraction	Reductive amperometric detection. GC with alkali flame ionisation and photometric detection	< 0.1 mg/kg	[103]
Dimethoate, omethoate	Peaches, apples	Solvent extraction, solvent partitioning clean-up	GC with FID	–	[104]
Pyrethroid type					
Pyrethroids, organo-phosphorus types	Vegetables	Ethanol extraction, partitioning with toluene, Florisil clean-up	GC	–	[105]
Natural pyrethrins, synthetic pyrethroids	Fruit and vegetables	Acetone or acetonitrile extraction, hexane partitioning, Florisil clean-up	Capillary GC with ECD	–	[106]
Pyrethroids	Fruit and vegetables, grain	Acetone:petroleum ether extraction, Florisil on Florisil/charcoal clean-up	Packed column GC with ECD detection	–	[107]
Pyrethroids	Fruit and vegetables	Methanol extraction, toluene partitioning, Florisil/charcoal clean-up	Capillary column GC with ECD detection	–	[108]
Miscellaneous pesticides/insecticides					
Linuron,3,4-dichloroaniline	Potatoes	–	C_{18} HPLC with amperometric detection	–	[109]
Abametin	Fruit and vegetables	Acetonitrile extraction, hexane partitioning aminopropyl solid-phase extraction clean-up	Formation of fluorescent derivative and HPLC	–	[110]
Propachlor	Maize, potatoes	Water extraction, steam distillation, acetic acid extraction	GC with FID	–	[111]

Table 8.3. (continued)

Insecticide type	Sample type	Sample preparation	Analytical finish	Detection limit	References
o-Phenylphenol, imazalil, thiabenda-zole	Citrus fruits	Acetone extraction	Silica gel HPLC, scanning at 300 nm	–	[112]
Paraoxon	Vegetables	–	HPLC with chemiluminescence-based flow sensor	–	[113]
Profenofos	Tomatoes	–	GC with CD and GC–MS	–	[114]
Paraquat, diquat	Potatoes	–	High-performance capillary-zone electrophoresis with UV detection	–	[115]
Trifluralin	Carrots	–	GC with ECD	–	[116]
Fluazifop-butyl and breakdown products	Potatoes	–	GC	–	[117]
Linuron, trifluralin	Carrots	Hexane–ethyl acetate extraction, Florisil clean-up	GC with ECD, also HPLC with UV detection	–	[118]
Buprofezin	Vegetables	Ethyl acetate extraction, gel permeation chromatography clean-up	Capillary GC with SIMS detection	–	[119]
Buprofezin	Vegetables	Ethyl acetate extraction	GC with NP detection	–	[120]
Vamidothion and its oxidation metabolites	Apples	Acetone extraction, transfer to methylene chloride	Capillary GC–MS with secondary ion monitoring	–	[121]
Dichlorprop	Apples	–	Polarisation fluorescence immunoassay	–	[122]
Azodrin ([3-(dimethoxy-phosphinyl) N-methyl-cis-crotonamide])	Strawberries	Solvent extraction	GC–MS of trifluoroacetyl derivatives	–	[123]

Table 8.3. (continued)

Insecticide type	Sample type	Sample preparation	Analytical finish	Detection limit	References
2-Chloroethyl phosphamic acid	Pineapple, apples, cherries, onions	Solvent extraction and methylation	GC, with potassium thermionic detection	–	[124]
Miscellaneous pesticides/insecticides	Fruit and vegetables	Gel permeation chromatography clean-up	GC-MS with ion trap detection	–	[125]
Miscellaneous pesticides/insecticides	Fruit and vegetables	–	Spectrometric square wave polarography	–	[126]
Miscellaneous pesticides/insecticides	Vegetables	Investigation automated gel permeation chromatographic clean-up	–	–	[127]
Miscellaneous pesticides/insecticides	Vegetables	Ethyl acetate extraction, C_{18} Sep-Pak clean-up, solvent partitioning	Derivatisation with 1-fluoro 2,4 dinitrobenzene, GC with ECD	–	[128]
Miscellaneous pesticides/insecticides	Fruit and vegetables	Extraction gel permeation chromatography, Nuchar-Celite clean-up	HPLC with post-column hydrolysis and addition of o-phthaldehyde to produce fluorophores	–	[129]
Miscellaneous pesticides/insecticides	Spinach	Collaborative testing of methodology	–	–	[130]
Miscellaneous pesticides/insecticides	Apples	–	HPLC	–	[131]
Miscellaneous pesticides/insecticides	Fruit and vegetables	Reviews of methodology	GC	0.001 – 0.05 ppb	[132]

GC: gas chromatography, ECD: electron capture detection, SCFE: supercritical fluid extraction are separated from coextractives by a partition process with dichloromethane and chromatography on an acidic aluminium oxide column. Quantitative determination is made by packed column gas–liquid chromatography using an electron capture detector for the organochlorine pesticides and a flame photometric detector for the organophosphorus pesticides. HPLC using an ultraviolet spectrophotometric detector is used for the determination of synthetic pyrethoids and carbaryl. Capillary column gas–liquid chromatography is used to confirm the identities of suspected residues of organochlorine and synthetic pyrethoid pesticides.

Recoveries of organochlorine, organophosphorus and synthetic pyrethoids determined in barley and wheat were respectively 81 – 128% when present at the 0.04 – 0.05 mg/kg level, 70 – 129% when present at the 0.1 – 10 mg/kg level and 63 – 146% when present at the 1 μ/kg level.

Various committees set up by the UK Society for Analytical Chemistry have carried out very detailed studies on the application of GC with flame photometric detection to methanol extracts of grain for the determination of malathion and dichlorvos [134] and organophosphorus in pesticides [135].

Other methods for the determination of insecticides in grain are reviewed in Table 8.4.

8.9
Herbicides in Plant Materials, Vegetables and Grain

Methods for the determination of herbicides usually involve solvent extraction of the sample with methanol, acetone or an acetonitrile followed by LC or GC.

Other techniques include TLC, SCFC and enzyme immunoassay (Table 8.5).

8.10
Fungicides in Vegetables, Fruit and Grain

Most of published methods for the determination of fungicides in vegetables, fruit and cereals are based on GC (Table 8.6).

Thus, Caverley and Unwin [158] have described a rapid and sensitive technique for the determination of residues of the fungicides, furalaxyl and metalaxyl in plants. Plants are macerated with acetone and after filtration and dilution with water, partitioned with chloroform. The extracts are subjected to GC with a nitrogen-specific detector after removal of chloroform and dissolution in acetone. Recoveries are generally better than 80%, with detection limits of 0.1 mg/kg for lettuce.

Furoxyl is [methyl-N-(2,6-dimethylphenyl)-N-(2-furoyl)alaninate] and melaloxyl is [methyl-N-(2,6-dimethylphenyl)-N-(2-methoxy-diacetyl)alaninate].

Farrow et al. [159] determined the systemic fungicide carboxin (2,3-dihydro-6-methyl-5-phenylcarbamoyl-1,4-oxathiin), which is used in barley and wheat seed treatment of cereals.

Table 8.4. Determination of insecticides and pesticides in grain and cereals (from author's own files)

Insecticide type	Sample type	Sample preparation	Analytical finish	Detection limit	Reference
Carbamate	Grain, fruit and vegetables	Solvent extraction clean-up on amino-bonded silica	HPLC–post-column reaction with o-phthaldehyde	–	[88]
Chlorophenoxy carboxylic acid type	Cereal, berries	–	Pesticide methylated, then GC–MS	–	[136]
Organonitrogen type	Grain	Florisil sorbent trap	GC	–	[137]
Deltamethrin	Wheat	–	TLC, sulfuric acid–potassium permanganate spray detection	–	[138]
Biosmethrin	Grain	–	Enzyme immunoassay	–	[139]
Pyrethroid type	Cereals	Nonpolar solvent extraction–solid-phase extraction	Differential pulse polarography of 3-methoxybenz-aldehyde produced from pesticides by alkaline hydrolysis	–	[140]
Chlorpyrifos-methyl, methoprene	Grain	–	Enzyme immunoassay	–	[141]
Bromoxynil, ioxynil	Cereals	Solvent extraction clean-up by solvent partitioning	Perfluorination–GC with ion trap MS detection	–	[142]
Agrochemicals	–	–	Flow-injection thermospray system	–	[143]
Triazine, urea, N-phenyl carbamate, thiocarbamate, substituted aniline and uracil types	Corn, potatoes, carrots	–	GC–MS	0.1 ppm	[144,145]

Table 8.5. Determination of herbicides in plant materials, grain and vegetables (from author's own files)

Herbicide type	Sample type	Sample preparation	Analytical finish	Detection limit	Reference
Phenylurea type	Crops	Methanol extraction clean-up on solid-phase extraction cartridges	HPLC with detection at 242 nm	–	[146]
Substituted urea and carbamate type	Plants	Acetone extraction	TLC and GC	0.001–0.5 ppm	[147]
Substituted phenylurea types	Grain	Methanol extraction	LC with UV detection	–	[151]
Sulfonylurea type	Plants	SCFE	SFC	–	[152]
Atrazine	Corn	–	Enzyme immunoassay	–	[153]
Atrazine, propazine, simazine	Root crops	50% aqueous methanol extraction, water:chloroform partitioning	GC with N-specific detection	0.02 ppm	[154]
Imazethepyr	Plants	–	Fibre-optic immunosensor	–	[155]
Metobromuron	Plant leaves	–	Column chromatography on alumina and spectrophotometry of azo derivatives	–	[156]
Nitrofen (2,4-dichlorophenyl-4′-nitrophenyl ether)	Vegetables	Acetonitrile extraction then petroleum ether extraction	GC with ECD	10 ng/g	[157]

Table 8.6. Determination of fungicides in fruit and vegetables and grain (from author's own files)

Fungicide type	Sample type	Sample preparation	Analytical finish	Detection limit	Reference
Vegetables and Fruit					
Fungicides, anti-sprouting agent	Potatoes	–	GC and HPLC	–	[161]
Fungicides	Vegetables	Acetone extraction, adsorption on diatomaceous earth cartridges	–	–	[162]
Fluzazole and its phenyl metabolite	Cereals, fruit	(1) Ethyl acetate extraction clean-up on silica or Florisil, or (2) gel permeation chromatography	GC with NP detection or capillary GC-MS	1–2 ppm	[163]
Dinocap	Fruit (apples, pears, grapes)	Acetone extraction, C_{18} clean-up and silica gel solid phase columns	HPLC with UV detection	0.1 ppm	[164]
Vinclozolin, captan	Fruit and vegetables	On-line microextraction clean-up on silica gel deactivated with 10% water	GC with ECD	–	[165] [166]
Chlorothanonil	Tomatoes, cucumber	Solvent extraction	GC with ECD and MS detection	–	
Carbendazim	Crops, grain	Methanol extraction partitioning into methylene chloride solid-phase silica gel clean-up	HPLC on amino-bonded column with fluorescence detection at 285 nm (315 nm)	–	[167]
Chloraniformethan (N-[2,2,2-trichloro-1-(3,4-dichloro-anilo)ethyl] formamide]	Grain, cucumber	Soxhlet extraction with methanol	TLC then GC with ECD	1 ng (absolute)	[168]
Binapacryl [2-(1-methyl-n-propyl)-4,6-dinitrophenyl-2-methyl crotonate]	Fruits (apple, cherry, pear, peach, plum)	Sample plus celite, dimethyl formamide and hexane: diethyl ether (4:1 v/v), homogenised and solvent phase filtered off	GC with tritium foil ECD	200 pg (absolute)	[169]
Dithiocarbamate type, zineb, maneb, mancozeb, thiram	Fruit and vegetables	Sample treated with tin chloride–hydrochloric acid in a sealed vessel. Headspace injected into a GC	Headspace GC, flame photometric detection	0.2 mg/kg	[170]

Table 8.7. Determination of plant growth regulators and retardants (from author's own files)

Type	Sample type	Sample preparation	Analytical finish	Detection limit	Reference
Dalapon (2,2-dichloropropionic acid), regulator	Plants	Ethyl ether extraction	GC with ECD	< 10 ppm	[168]
Triazole type, pyrimidine type (retardants)	Plants	Ethylene dichloride extraction	Silica gel chromatography with N-thermionic detection	–	[169]
Abscisic acid, indole-3-yl acetic acid (regulator)	Plants	–	HPLC	–	[170]
Triazole type, pyrimidine type (retardant)	Plants	Methylene dichloride partitioning	LC on silica and C_{18} columns using NP thermionic detection	–	[171]

The carboxin is extracted from the sample with acetone in a Soxhlet extraction apparatus and, after concentration of the extract, is determined via gas–liquid chromatography using a nitrogen-selective detector. The presence of carboxin is confirmed by the use of a sulfur flame photometric detector. Recoveries ranged from 73 to 80% (barley) and 73 to 78% (wheat).

Initial attempts by Baker et al. [160] to determine the nonsystemic triphenyl tin-based fungicide fentin by HPLC were unsuccessful due to a lack of sensitivity and interference by coextractives. They therefore decided to investigate the applicability of spectrofluorimetry to this determination. Different extraction procedures are described for vegetables and cocoa products. For potatoes, for example, the grated potato is dried with anhydrous sodium sulfate and then Soxhlet-extracted with dichloromethane.

Spectrofluorimetry of the hydroxyl flavone complex of fentin enabled down to 0.5 ng/kg of fungicide in potato, celery and sugar beet to be determined.

8.11
Growth Regulators

A limited amount of work has been carried out on the determination of plant growth regulators and retardants (Table 8.7).

References

1. McHan F, Horvat RJ (1987) *J Agric Food Chem* **35**:241.
2. Oberleas D, Harland BF (1986) *Phytic Acid Symposium Applied Phytic Acid* 77 1986.
3. Winter M, Brondl W, Herrmen K (1987) *Z Lebensm Unters Forsch* **184**:11.
4. Hueni K, Uebersax P (1973) *Landw Forsch* **26**:125.
5. Ministry of Agriculture, Fisheries and Food (1973) *The Analysis of Agricultural Materials – Volatile Fatty Acids (C_1–C_{16}) and Lactic Acid in Silage Juices, Method 85*, Technical Bulletin RB 427, HMSO, London, UK.
6. Jones DW, Kay JJ (1976) *J Agric Food Chem* **27**:1005.
7. Karl V, Dietrich A, Mosandi A (1994) *Phytochem Anal* **5**:32.
8. Thomson CA, Chesney DJ (1992) *Anal Chem* **64**:848.
9. Sen N, Kubecki S (1987) *J Food Add Contam* **4**:357.
10. Usero JL, Angels d'Arco M, Izquierdo C (1987) *Casado Tec Lab* **11**:98.
11. Brandon DL, Binder RG, Wilson RE, Montague WC (1993) *J Agric Food Chem* **41**:996.
12. Kovatcheva EG (1979) *Analyst* **104**:79.
13. Kovatcheva EG, Popova YS (1977) *Nahrung* **21**:465.
14. Hunter WJ (1986) *J Chromatogr* **362**:430.
15. Kling GJ, Perkins LM, Capillo PE, Eisenberg BA (1987) *J Chromatogr* **407**:377.
16. Rivera VG, Morgan PW, Stipanovic RD (1986) *J Chromatogr* **358**:243.
17. Lansden JA (1938) *J Assoc Off Anal Chem* **69**:964.
18. Jeffries TM, Hardman R (1976) *Analyst* **101**:122.
19. Jeffries TM, Hardman R (1972) *Planta Med* **22**:78.
20. Schmidt ER, Kynclova E, Soevegjarto F (1992) *Ernaehrung (Vienna)* **16**:274.
21. Acree TE, Lee CY, Butts RM, Barnard J (1976) *J Agric Food Chem* **24**:430.

22. Corti P, Dreassi E, Politi N, Aprea C (1992) *Food Addit Contam* **9**:243.
23. Winell B (1976) *Analyst* **101**:883.
24. Baumann RA, Hart de Klejin VM (1993) *Meded Fac Handbouwkd Toegepaste Riol Wet (Univ Gent)* **58**:173.
25. Breuggemann J, Ocber HD (1986) *Chem Mikrobiol Technol Lebensm* **10**:113.
26. Masoic P, Zawistowski J, Zawistowski U, Howes N (1993) *J Cereal Sci* **17**:115.
27. Dorner JW, Cole RJ (1988) *J Assoc Off Anal Chem* **71**:43.
28. Francis OJ, Wore GM, Larman AS (1987) *J Assoc Off Anal Chem* **70**:842.
29. Rajakyla E, Laasasenaho K, Sakkers PJD (1987) *J Chromatogr* **384**:391.
30. Shepherd MJ (1986) In: Cole RJ (ed) *Modern Methods in the Analysis and Structural Elucidation of Mycotoxins*, Academic, Orlando, FL, USA, p. 293.
31. Rati ER, Prema V, Shantha T (1987) *J Food Sci Technol* **24**:90.
32. Kalinoski HT, Udseth HR, Wright BW, Smith RD (1986) *Anal Chem* **58**:2421.
33. Palmisano F, Visconti A, Bottalico A, Lerario P, Zambonin PG (1981) *Analyst* **106**:992.
34. Disbold GJ, Karny N, Zare RN (1979) *Anal Chem* **51**:67.
35. Dickens JW, Whitaker TB (1986) *Mod Methods Anal Struct Elucidat Mycotoxins* 29.
36. Francis OJ, Ware JM, Carman AS (1988) *J Assoc Off Anal Chem* **71**:41.
37. Gorst Delman CP (1986) In: Cole RJ (ed) *Modern Methods in the Analysis and Structural Elucidation of Mycotoxins*, Academic, Orlando, FL, USA, p. 95.
38. Shotwell OL (1986) In: Cole RJ (ed) *Modern Methods in the Analysis and Structural Elucidation of Mycotoxins*, Academic, Orlando, FL, USA, p. 51.
39. Inhat M (1974) *J Assoc Off Anal Chem* **57**:368.
40. Neshelm S, Trucksees MW (1986) In: Cole RJ (ed) *Modern Methods in the Analysis and Structural Elucidation of Mycotoxins*, Academic, Orlando, FL, USA, p. 239.
41. Beaver RW (1986) In: Cole RJ (ed) *Modern Methods in the Analysis and Structural Elucidation of Mycotoxins*, Academic, Orlando, FL, USA, p. 205.
42. Travis CC, Hattemer-Frey HA (1988) *Chemosphere* **17**:277.
43. Buckley EH (1982) *Science* **216**:520.
44. Paterson S, Mackay D, Bacci E, Calamari D (1991) *Environ Sci Technol* **25**:866.
45. Frank H, Frank W (1989) *Environ Sci Technol* **23**:365.
46. Paterson S, Mackay D, McFarlane C (1994) *Environ Sci Technol* **28**:2259.
47. Paterson S, Mackay D, Gladman A (1991) *Chemosphere* **23**:539.
48. Hiatt MH (1998) *Anal Chem* **70**:851.
49. Keymeulen R, Voutetaki A, Von Langenhove H (1995) *J Chromatogr* 699:223.
50. Kaupp H, Sklarz M (1996) *Chemosphere* **32**:849.
51. Esteban JL, Martinez-Castro I, Sanz J (1993) *J Chromatogr* **657**:155.
52. Scudamore KA (1980) *Analyst* **105**:1171.
53. Polesello S, Lovati F, Rizzolo A, Rovide C (1993) *J High Res Chromatogr* **16**:555.
54. Sanz C, Perez AG, Richardson DG (1994) *Food Sci* **59**:139.
55. Steele DH, Thornburg MJ, Stanley JS, Miller RR, Brooke R, Cushman JR, Cruzan G (1994) *J Agric Food Chem* **42**:1661.
56. Cairns T, Siegmund EG, Doose GM, Hundley HK, Barry T, Petzinger G (1984) *Anal Chem* **56**:2138.
57. Ohta H, Osajima Y (1992) *J Fac Agric Kyushi Univ* **37**:125.
58. Panel on Fumigant Residues in Grain (1984) *Analyst* **99**:570.
59. Ueji M, Kanazawa J (1973) *Jpn Analyst* **22**:16.
60. Westlake WE, Monika I, Gunther FA (1972) *Bull Environ Contam Toxicol* **8**:109.
61. Singhal JP, Khan S, Bansal OP (1978) *Analyst* **103**:872.
62. Greenhalgh R, Marshall D, King RR (1976) *J Agric Food Chem* **24**:266.
63. Coburn JA, Ripley BD, Chen ASY (1976) *J Assoc Off Anal Chem* **59**:188.

64. Miles CJ, Moye HA (1987) *Chromatographia* **24**:628.
65. Krause RT (1988) *Chromatographia* **442**:333.
66. Ueji M, Kanazawa J (1973) *Jpn Analyst* **22**:16.
67. Reeves RG, Woodham DW (1974) *J Agric Food Chem* **22**:76.
68. Sukul P (1994) *Toxicol Environ Chem* **44**:217.
69. Clan JJ (1992) *Int J Environ Anal Chem* **49**:189.
70. McEachern PR, Foster GD (1993) *J Chromatogr* **632**:119.
71. Sasaki K, Suzuki T, Saito Y (1987) *J Assoc Off Anal Chem* **70**:460.
72. Zahnow EW (1987) *J Agric Food Chem* **35**:403.
73. Schmidt M, Hamman R, Kettrup A (1988) *Int J Environ Anal Chem* **33**:1.
74. Allmaier G, Goergl A, Schmid ER, Wagner K (1986) *J High Res Chromatogr* **9**:762.
75. Alawi MA (1986) *Chromaographia* **22**:40.
76. Jaklin J, Krenmayr P, Varmuza K, Heegeman W, Landvoight W (1988) *Fresen Z Anal Chem* **330**:704.
77. Mliellet A (1986) *Ann Faisil Expert Chim Toxicol* **72**:245.
78. Ingram JC, Groenewold GS, Appelhans AD, Delmore JE, Olsen JE, Miller DL (1997) *Environ Sci Technol* **31**:402.
79. Clark GJ, Goodin RR, Smiley JW (1985) *Anal Chem* **57**:2223.
80. Baker PG, Bottomley P (1982) *Analyst* **107**:206.
81. Howard AL, Brave C, Taylor LJ (1993) *J Chromatogr Sci* **31**:323.
82. Howard AL, Thomas CLR, Taylor LT (1994) *Anal Chem* **66**:1432.
83. Alawi MA (1981) *Fresen Z Anal Chem* **309**:8.
84. McGarvey BD (1993) *J Chromatogr* **642**:89.
85. Chaput D (1988) *J Assoc Off Anal Chem* **71**:542.
86. Brauckholt S, Their HP (1987) *Z Lebensm Unters Forsch* **184**:91.
87. De Kok A, Hiemstra M, Vreeker CP (1987) *Chromatographia* **24**:469.
88. De Kok A, Hiemstra M (1992) *J Assoc Off Anal Chem* **75**:1063.
89. Parrendo M, Larson B (1993) *J Trace Microprobe Te* **11**:133.
90. Pulpiw HM (1993) *J Assoc Off Anal Chem* **76**:1369.
91. Stan HJ, Abraham B, Behla L, Kellert M, Mitterapiingsbl GDCL (1976) *Deutsch Chem Fachgr Lebensm Gerrichte Chem* **30**:146.
92. Neicheva A, Kovacheva E, Marudov G (1987) *J Chromatogr* **437**:249.
93. De Paoli M, Barbina MT, Mondini R, Pezzoni A, Valentina A, Grob K (1992) *J Chromatogr* **626**:145.
94. Wuchner K, Ghijsen RT, Brinkman UAT, Grob R, Mattieu J (1993) *Analyst* **118**:11.
95. Steinwandter H (1992) *Fresen J Anal Chem* **343**:887.
96. Yarita T, Nomura A, Horimoto Y, Yamada J (1994) *Microchem J* **49**:145.
97. Steinwandter H (1994) *Fresen J Anal Chem* **348**:688.
98. Kumar R, Sharma CB (1987) *J Liq Chromatogr* **10**:3637.
99. Society of Analytical Chemistry (1997) *Analyst* **102**:858.
100. Abbot DC, Crisp S, Tarrant KR, Tatton JO'G (1970) *Pest Sci* **1**:10.
101. Sissons DJ, Telling GM (1970) *J Chromatogr* **47**:328.
102. Sissons DJ, Telling GM (1970) *J Chromatogr* **48**:468.
103. Smart NA, Hill ARC, Roughan PA (1978) *Analyst* **103**:770.
104. Ferreira JR, Falcao MM, Tainha A (1987) *J Agric Food Chem* **35**:506.
105. Wan HB, Wong MK, Lim PY, Mok CY (1994) *J Chromatogr* **662**:147.
106. Nakamura Y, Tonogai Y, Tsumura Y, Ito Y (1993) *J Assoc Off Anal Chem* **76**:1348.
107. Pang G-F, Fan C-L, Chao Y-Z, Zhao T-S (1994) *J Assoc Off Anal Chem* **77**:738.
108. Pang G-F, Fan C-L, Chao Y-Z, Zhang T-S (1994) *J Chromatogr* **667**:348.
109. Maruyaina M (1992) *Fresen J Anal Chem* **343**:890.

110. Chamkasem N, Papathakis ML, Lee S (1993) *J Assoc Off Anal Chem* **76**:691.
111. Wenzel KD, Mohnke M, Grahl R (1985) *Fresen Z Anal Chem* **322**:423.
112. Dellacassa E, Martinez R, Moyna P (1993) *J Planar Chromatogr* **6**:326.
113. Roda A, Rauch P, Ferri P, Girotti S, Ghini S, Carrea G, Bovara R (1994) *Anal Chim Acta* **294**:35.
114. Ismail SMM, Ali HM, Habiba RA (1993) *J Agric Food Chem* **41**:610.
115. Wigfield YY, McCormack KA, Grant R (1993) *J Agric Food Chem* **41**:2315.
116. Mortimer RD, Black DB, Dawson BA (1994) *J Agric Food Chem* **42**:1713.
117. Clegg BS (1987) *J Agric Food Chem* **35**:269.
118. D'Amato A, Semeraro I, Bicchi C (1993) *J Assoc Off Anal Chem Int* **76**:657.
119. Valverde-Garcia A, Fernandez-Alba AR, Herrara JC, Rodlan E (1994) *J Assoc Off Anal Chem Int* **77**:1041.
120. Valverde-Garcia A, Gonzales-Pradas E, Aguilera del Real A (1993) *J Agric Food Chem* **41**:2319.
121. Tsumura Y, Matsuki H, Tunogai Y, Nakamura Y, Kato S, Ito Y (1993) *J Food Protect* **56**:437.
122. Garcia Sanchez F, Navas A, Alonso F, Lovillo J (1993) *J Agric Food Chem* **41**:2215.
123. Lawrence JF, McLeod HA (1976) *J Assoc Off Anal Chem* **59**:637.
124. Bache CA (1970) *J Assoc Off Anal Chem* **730**:53.
125. Cairns T, Luke MA, Chiu KS, Navarro D, Sigmund G (1993) *Rapid Commun Mass Spec* **7**:1070.
126. Wandte E (1987) *Lebensmittelindustrie* **34**:161.
127. Gretch FM, Rosen JD (1987) *J Assoc Off Anal Chem* **70**:109.
128. Braukoff S, Their HP (1987) *Z Lebensm Unters Forsch* **184**:91.
129. Chaputi D (1988) *J Assoc Off Anal Chem* **71**:542.
130. Thier HP, Speccht W, Gilsbach W (1993) *Cereal Foods World* **38**:62.
131. Dreyfuss MF, Lotfi H, Marquet P, Debord J, Daquet JL, Lachatre G (1994) *Analusis* **22**:273.
132. Cochrane WP (1979) *J Chromatogr Sci* **17**:124.
133. Bottomley P, Baker DG (1984) *Analyst* **109**:85.
134. Society of Analytical Chemistry (1973) *Analyst* **98**:19.
135. Society of Analytical Chemistry (1980) *Analyst* **105**:515.
136. Meemkan HA, Rudolph P, Fuerst P (1987) *Deut Lebensm Rundchem* **83**:239.
137. King JW, Hopper ML, Luchtefeld RG, Tyalor SL, Orton WL (1993) *J Assoc Off Anal Chem* **76**:857.
138. Appaiah KM, Sreenivasa MA, Nagarajai KV (1993) *Indian Food Packer* **47**:61.
139. Hill AS, McAdam DP, Edward SL, Skerritt JH (1993) *J Agric Food Chem* **41**:2011.
140. Corbini G, Biondi C, Proietti D, Draessi E, Corti P (1993) *Analyst* **118**:183.
141. Edward SL, Hill AS, Ashworth P, Matt J, Skerritt JH (1993) *Cereal Chem* **70**:748.
142. Sanchez-Brunarte C, Garcia-Valcarcel AI, Tadeo JL (1994) *Chromatographia* **38**:624.
143. Hayward MJ, Snodgran TJ, Thomas ML (1993) *Rapid Commun Mass Spec* **7**:85.
144. Lawrence JF (1976) *J Agric Food Chem* **24**:1236.
145. Prestel D, Weisgerber I, Klein W, Forte K (1976) *Chemosphere* **5**:137.
146. Lagana A, Marino A, Fago G, Prado-Martinz B (1994) *Analusis* **22**:63.
147. Cohen IC, Wheals BB (1969) *J Chromatogr* **43**:233.
148. Farrington DS, Hopkins RG, Ruzicka JHA (1977) *Analyst* **102**:377.
149. McNally MEP, Wheeler JR (1988) *J Chromatogr* **435**:63.
150. Wigfield YY, Grant R (1993) *Bull Environ Contam Toxicol* **51**:171.
151. Lawrence JF (1974) *J Agric Food Chem* **22**:137.
152. Anis NA, Eldefrawi ME, Wong RB (1993) *J Agric Food Chem* **41**:843.

153. Corbaz R, Artho D, Ceschini M, Hausermann M, Plantefere JC (1969) *Bectr Tabakforsch* **5**:80.
154. Kvalvag J (1974) *Analyst* **99**:666.
155. Caverly DJ, Unwin J (1981) *Analyst* **106**:389.
156. Farrow PG, Hoodless RA, Hopkinson A (1975) *Analyst* **105**:249.
157. Baker PG, Farrington DS, Hoodless RA (1980) *Analyst* **105**:282.
158. Martindale RW (1988) *Analyst* **113**:1229.
159. Di Muccio A, Dommarco R, Attard Barbini D, Santilo A, Girolimetti S, Ausili A, Ventriglia M, Generali T, Vergori L (1993) *J Chromatogr* **643**:363.
160. Guinivan RA, Gagnon MR (1994) *J Assoc Off Anal Chem Int* **77**:728.
161. Schenke FJ, Hennessy MK (1993) *J Liq Chromatogr* **16**:755.
162. Steinwandter H (1994) *Fresen J Anal Chem* **348**:692.
163. Valverde-Garcia A, Gonzalez-Pradas E, Aguilera-Del Real A, Urena-Amate MD, Camacho-Ferre F (1993) *Anal Chim Acta* **276**:15.
164. Regis-Rolle SD, Bauville GM (1993) *Pestic Sci* **37**:273.
165. Hoodless RA, Sargent M (1976) *Analyst* **101**:161.
166. Baker PG, Hoodless RA (1973) *Analyst* **98**:172.
167. Society of Analytical Chemistry (1981) *Analyst* **106**:782.
168. Getzendaner ME (1969) *J Assoc Off Anal Chem* **52**:824.
169. Reed AN (1988) *J Chromatogr* **438**:393.
170. Guinn G, Brummett DL, Beier RC (1986) *Plant Physiol* **81**:997.
171. Morrelli JJ, Hercules DM (1986) *Anal Chem* **58**:1294.

9 Determination of Organometallic Compounds in Plants and Crops

Mercury- and tin-containing fungicides have been used to control diseases in fruit, vegetables and cereals.

9.1
Organomercury Compounds

Mercurial fungicides have been used to control scab in apples. Gutenmann and Lisk [1] have described a modified Schöniger combustion of dried apple tissue, which replaces wet ashing prior to the determination of mercury. Loss of mercury by volatilisation is eliminated in the closed combustion flask. Apple tissue is dried on cellophane overnight under vacuum, then burned in an oxygen-filled flask with a balloon attached for pressure control. Mercury is determined spectrophotometrically after extraction of the absorbed solution with dithizone. About 12 samples can be burned and analysed in one day. Recovery of mercury from apples in the 0.3 to 0.6 ppm range averaged 83.6%. Up to 0.18 ppm of mercury was found in apples treated with mercurial fungicides for scab control.

Liang et al. [2] eliminated matrix interference in the determination of methyl mercury in biological materials, by using a solvent extraction technique involving no critical cleaning steps. Recoveries were close to 100% with a relative standard deviation of less than 5%.

9.2
Organotin Compounds

Nangniot and Martens [3] determined triphenyltin acetate fungicide in vegetable matter by a method based on the hanging drop mercury electrode.

Gauer et al. [4] described a gas chromatographic method for the determination of the residues of tricyclohexylhydroxystannane and its dicyclohexyl metabolite on strawberries, apples and grapes that have been treated with Pictran miticide. Crop samples were treated with aqueous hydrobromic acid to form bromo derivatives of the organotin compounds, and these derivatives were extracted into benzene. When the residue levels were less than 1 μg/l, the derivative solution was cleaned up on a column of silica gel. The derivatives were determined by gas chromatography at 200 °C on a column packed with 2% of OV-225 on Chromosorb GAW–DMCS or at 100 °C on a column packed with 0.5% of OV-225 on glass beads with helium as a carrier gas. Background interference was minimised by use of a halide-sensitive Coulson detector. Recovery of 1 mg/l of added tricyclohexylhydroxystannane was 80–95%, and at 0.1 mg/l it was 78 to 89%. Conditions are also described for the gas chromatographic determination of cyclohexylstannane acid, another possible degradation product of Pictran.

Triphenyltin has been determined in potatoes in amounts down to 0.1 ng/l by extraction with ethanolic potassium hydroxide, derivatisation with sodium tetraethyl boron and solid-phase microextraction gas chromatography IC–PMS [5].

Newsome [6] determined methyl mercury compounds in wheat flour and ground oats by extraction with benzene–formic acid followed by purification and gas–liquid chromatography. Interfering substances were removed from the extracts by column chromatography on silicic acid and partitioning with cysteine acetate solution. The method is sensitive in the 0.01–0.9 ppm range with a recovery of generally better than 95%.

References

1. Gutenmann WH, Lisk DJ (1960) *J Agric Food Chem* **8**:306.
2. Liang L, Horvat M, Cernichiari E, Gelein B, Balogh S (1996) *Talanta* **43**:1883.
3. Nangniot P, Martens PH (1961) *Anal Chim Acta* **24**:276.
4. Gauer WD, Seiber JN, Crosby DC (1974) *J Agric Food Chem* **22**:252.
5. Vercauteren J, De Meester A, De Smaele T, Vanhaecke F, Moens L, Dams R, Sandra P (2000) *J Anal Atom Spectrom* **15**:651.
6. Newsome WH (1971) *J Agric Food Chem* **19**:567.

10 Determination of Anions in Plants and Crops

10.1
Borate

Plants

Ogner [1] has described an automated analyser method for the determination of boron-containing anions in plants. This is based on the formation of a fluorescent complex between these anions and carminic acid at pH 7. The plant tissues are ashed at 550 °C and the residue dissolved in 0.5 N hydrochloric acid prior to adjustment to pH 6–7 with sodium carbonate solution. The solution is excited at 470 nm and fluorescence intensities measured at 585 nm. Interferences by the reaction of some cations with carminic acid are overcome by passing the solution through an ion exchange column to exchange the cations for sodium ions. Analytical recoveries of boron anions were in the range 98–104%. The detection limit of the method was 5 µg/l boron.

Plant Extracts

Lopez Garcia et al. [2] have described a rapid and sensitive spectrophotometric method for the determination of boron complex anions in plant extracts and waters which is based on the formation of a blue complex at pH 1–2 between the anionic complex of boric acid with 2,6-dihydroxybenzoic acid and crystal violet. The colour is stabilised with polyvinyl alcohol. At 600 nm the calibration graph is linear in the range 0.3–4.5 µg boron per 25 ml of final solution, with a relative standard deviation of ±2.6% for µg/l of boron. In this procedure to determine borate in plant tissues, the dried tissue is treated with calcium hydroxide, then ashed at 400 °C. The ash is digested with 1 N sulfuric acid and heated to 80 °C, neutralized with cadmium hydroxide and then treated with acidic 2,6-dihydroxybenzoic acid and crystal violet, and the colour evaluated spectrophotometrically at 600 nm. Most of the ions present in natural waters or plant extracts do not interfere in the determination of boron complex anions by this procedure. Recoveries of boron from water samples and plant extracts were in the range of 97–102%.

10.2
Bromate

Vegetables

Osborne [3] has described a method for the determination of bromate in flour using aqueous extraction and photometric flow injection analysis.

10.3
Bromide

Vegetables

Roughan et al. [4] have described a gas chromatographic method for the determination of bromide/total bromine as methyl bromide fumigant in lettuce, cucumber and tomatoes.

In this method, the bromide ion is converted to 2-bromoethanol by reaction with ethylene oxide in acetonitrile diisopropyl ether under acidic conditions. The 2-bromoethanol thus produced is determined by gas chromatography using an electron capture detector.

Mass spectrometry was used on a non-routine basis to identify the chromatogram peaks. The mean recovery for dried vegetable substrates is 97% for a wide range of bromide levels, equivalent to approximately 20 – 1000 mg/kg on a fresh mass basis. The method can be used to determine bromide down to 0.1 mg/kg of substrate fresh mass.

Beernaert and Vandezande [5] developed a gas chromatographic method for the determination of inorganic bromide in vegetables.

Grain

A method has been described [6] for the determination of bromide in grain. In this method, bromide ion is allowed to react selectively with ethylene oxide to form 2-bromoethanol, which is separated and determined by gas chromatography using an electron capture detector. In calibration tests of this method carried out over seven laboratories on standard grains and maize containing 50 mg/kg bromide, interlaboratory standard deviations of between ±6.4 and ±6.2 mg/kg were obtained. Mean rates of recovery were in the range of 92 – 109%.

Shiga et al. [14] have described a method for determining total inorganic bromide in crops by HP–LC.

10.4
Chloride

Plant Extracts

Davey and Bembrick [7] have described a method for the determination of chloride in water extracts of plants based on measurement of the EMF developed between two silver–silver chloride electrodes in a cell with a liquid junction and suitable electrolyte.

See also Sect. 10.11.

10.5
Cyanide

Plant Material

Cyanide in trace amounts is found in a large number of plants, mainly in the form of substituted glycosides. Affected plants include grasses, fruit kernels, pulses, linseed and cassava.

Harris et al. [8] has described methods for the determination of cyanide in these materials based on either spectrophotometry using *p*-phenylene diamine pyridine or gas chromatographically following conversion of cyanide to cyanogen bromide. Cyanide is extracted from the sample by digestion with phosphoric acid. Recoveries were in the range 96–99% (spectrophotometric method) and 90–96% (gas chromatographic method).

10.6
Fluoride

Vickery and Vickery [9] have investigated the interference by aluminium and iron in the ion-selective electrode method for the determination of fluoride in plant extracts. They demonstrated that plant ashes may contain sufficient of these two elements to seriously interfere in the determination of fluoride when using the fluoride-selective electrode. In the presence of these metals, the known additions method gives erroneous results, as did that involving the attempted formation of complexes with ethylene diamine tetraacetic acid (disodium salt) or 1,2-cyclohexylenedinitrilotetraacetic acid.

Good recoveries of fluoride ion were obtained in the presence of aluminium, iron, magnesium or silicate using sodium citrate as the complexing agent. Greater than 90% recovery of fluoride was obtained in the analysis of ashes of commercial tea high in aluminium (2000 mg/kg) and iron (2800 mg/kg).

Villa [10] determined fluoride in vegetative matter using an ion-selective electrode. Fluoride was extracted from dried vegetation by stirring with 0.1 N perchloric acid for 20 minutes at 20 °C. The fluoride content of the extract was determined at pH 1 using the method of standard additions, thus eliminating the need to decomplex fluoride prior to analysis. The method was applicable over the range 4–2000 mg/kg fluoride in vegetative matter such as grass, apples, pine needles, alfalfa and sorghum.

10.7
Nitrate and Nitrite

Plant Materials

Hemmi et al. [11] has described a differential pulse polarographic procedure for the determination of nitrate in environmental samples such as silage, grass, plants, snow and water. This method utilizes the catalytic reaction between nitrate and uranyl ion in the presence of potassium sulfate. The differential pulse polarographic peak is proportional to the nitrate ion concentration from 1 to 50 µmol/l. The detection limit for nitrate in water is 8×10^{-7} mol/l. Using this procedure, between 1 and 70 mg/g nitrate was found in vegetation samples.

Hepher and Alexander [12] developed a miniature distillation unit for the determination of nitrate in plant material.

Bradfield and Cooke [13] give details of a procedure for the determination of chloride (as well as nitrate phosphate and sulfate) in aqueous extracts of plant materials by an ion chromatographic technique with indirect ultraviolet detection. Recoveries ranged from 84 to 108%. This technique is discussed in further detail in Sect. 10.11.

Garcia-Gutierrez [15] has described an azo-coupling spectrophotometric method for the determination of nitrite and nitrate in vegetables.

Nitrite is determined spectrophotometrically at 550 nm after treatment with sulfanilic acid and N-1-naphthylethylenediamine to form the azo dye. In another portion of the sample, nitrate is reduced to nitrite by passing the solution buffered to pH 9.6 through a cadmium reductor, and total nitrite is determined as above.

Vegetables were boiled with water and calcium carbonate, treated with freshly precipitated aluminium hydroxide, and filtered prior to analysis as above.

Tanaka et al. [16] have described a spectrophotometric method for the determination of nitrate in vegetable products. This procedure is based on the quantitative reaction of nitrate and 2-*sec*-butylphenol in sulfuric acid (5 + 7), and the subsequent extraction and measurement of the yellow complex formed in alkaline medium. The column reaction is sensitive and stable and absorbances measured at 418 nm obey Beer's law for concentrations of nitrate-nitrogen between 0.13 and 2.5 µg/ml. In this procedure, the vegetable matter is digested at 80 °C with a sodium hydroxide silver sulfate solution, concentrated sulfuric acid and 2-*sec*-butylphenol are added, and after 15 minutes of standing time the nitrated phenol is extracted with toluene. Finally, the toluene layer is back-extracted with aqueous sodium hydroxide and evaluated spectrophotometrically at 418 nm. The standard deviation of the whole procedure was 1.4%, and analytical recoveries ranged between 91 and 98%.

Schuster and Lee [1] compared various methods for the determination of nitrate and nitrite in vegetables and grains.

10.8
Phosphate

See Sect. 10.11.

10.9
Sulfate

Plants

Krug et al. [1] used flow injection turbidimetry to determine sulfate in natural waters and plant digests. They described an improved flow injection system with alternative streams of reagents. Samples were injected into an inert carrier comprising 0.3% EDTA disodium salt and 0.2 mol/l sodium hydroxide. The inert carrier is mixed with 5% barium chloride containing 0.05% polyvinyl alcohol to form a barium sulfate suspension. The range of the method can be extended to low concentrations by continuously adding sulfate to the sample carrier stream. System performance is improved by automatic pumping of the reagent stream and an alkaline EDTA solution at high flow rate. All operations were controlled by an electronically operated proportional injector commutator. No baseline drift was observed even after analysis of 3000 samples. The method is capable of analysing 120 samples per hour with a relative standard deviation of less than 1% for sulfate concentrations in the ranges 1–30 mg/l or 5–200 mg/l (plant digests). Analytical recovery was 97–102%. Plant samples were digested with a nitric acid–perchloric acid mixture without further treatment [2].

Grain

Basargin et al. [20] has described a spectrophotometric procedure for the determination of sulfate in grain. This method is based on the formation of a coloured complex with an absorption maximum at 640 nm between sulfate and 3,6-bis-(4-nitro-2-sulfphophenylazo)chromotropic acid. Down to 2 mg/l of sulfate in grain can be determined by this procedure with a relative error of ±1.3%. Borate, chloride, nitrate, perchlorate, arsenate and chromate do not interfere.

10.10
Vanadate

Ibbasi [2] determined metavanadate in plant extracts by a method based on the formulation of a violet colour with vanadium (V) on the addition of a chloroform solution of N-(p-N-N-dimethylanilo-3-methoxy-2-naphtho)hydroxamic acid to the acidified (4–6 mol/l) plant extract. This solution was evaluated spectrophotometrically at 570 nm. The detection limit was 0.05 μg vanadium at a dilution ratio of 1×10^7. Very few interferences occur in the procedure.

10.11
Multi-Anion Analysis

Bradfield and Cooke [13] described an ion chromatographic method using a UV detector for the determination of chloride, nitrate, sulfate and phosphate in water extracts of plants and soils. Plant materials are heated for 30 minutes at 70 °C with water to extract anions. Soils are leached with water

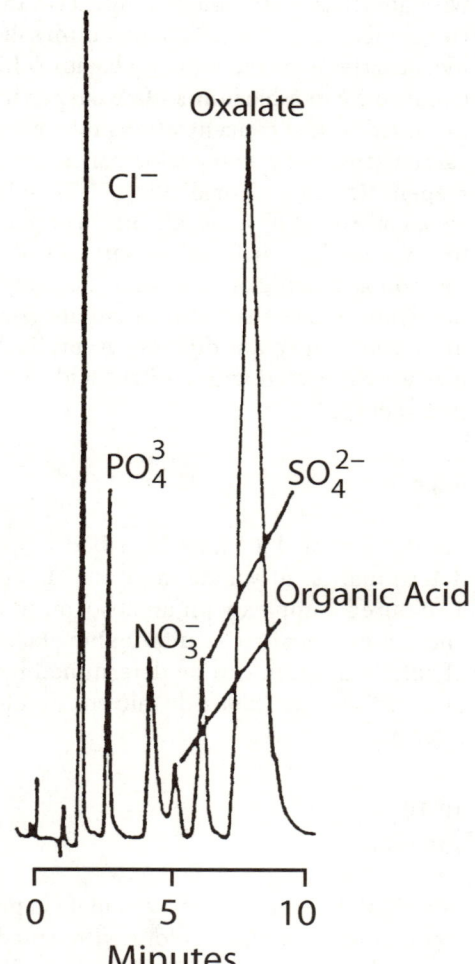

Figure 10.1. Ion chromatograph of an aqueous extract of fresh spinach. From [13]

and Dowex 50-X4 resin added to the aqueous extract, which is then passed through a Sep-Pak C_{18} cartridge, and the eluate is then passed through the ion chromatographic column. The best separation of these ions was obtained using a 5×10^{-4} mol/l potassium hydrogen phthalate solution in 2% methanol at ph 4.9. A reverse-phase system was employed. Retention times were 5.5, 7.9, 12.6 and 18 minutes for chloride, nitrate, phosphate and sulfate respectively. Recoveries ranged from 84 to 108%, with a mean of 97%.

Figure 10.1 shows an ion chromatograph of an aqueous extract of spinach.

References

1. Ogner G (1980) *Analyst* **105**:916.
2. Lopez-Garcia I, Cordoba MN, Sanchez-Pedreno C (1985) *Analyst* **110**:1259.
3. Osborne BG (1987) *Analyst* **112**:137.
4. Roughan JA, Roughan PA, Wilkins JPG (1983) *Analyst* **180**:742.
5. Beernaert H, Vandezande A (1986) *Med Fac Landbouww Univ Gent* **51**:191.
6. Panel 2 of the Committee for Analytical Methods for Residues of Pesticides and Veterinary Products in Foodstuffs of the Ministry of Agriculture, Fisheries and Food (1976) *Analyst* **101**:386.
7. Davey BG, Bembrick MA (1969) *Proc Soil Sci Soc Am* **33**:385.
8. Harris JR, Herson GHJ, Hardy MJ, Curtis DJ (1980) *Analyst* **105**:974.
9. Vickery B, Vickery MI (1970) *Analyst* **101**:445.
10. Villa AE (1976) *Analyst* **104**:545.
11. Hemmi H, Hasebe K, Ohzeki K, Kawbara T (1998) P319, Hakodate Technical College, Tokura-cho 226, Hakodate 042, Japan.
12. Hepher MJ, Alexander RH (1987) *Lab Practice* **36**:76.
13. Bradfield EG, Cooke DT (1985) *Analyst* **110**:1409.
14. Shiga N, Shimamura Y, Matano O, Goto S (1986) *Nippon Noyaku Gakk* **11**:585.
15. Garcia-Gutierrez G (1973) *Infeion Quim Analet Pura Apl Ind* **27**:171.
16. Tanaka A, Nose N, Iwasaki H (1982) *Analyst* **107**:190.
17. Schuster BE, Lee K (1987) *J Food Sci* **52**:1632.
18. Krug FJ, Zagatto EAG, Reis BF, Bahia OF, Jacintha O, Jorgensen SS (1983) *Anal Chim Acta* **145**:179.
19. Krug FJ, Filho HB, Zagatto EAG, Jorgensen SS (1977) *Analyst* **102**:503.
20. Basagin NN, Men'shikova NL, Belova VS, Myassishcheva LG (1968) *Z Anal Khim* **23**:732.
21. Abbhasi SA (1981) *Int J Environ Stud* **18**:51.

11 Contaminant Contents of Soil and Crops

Information on the levels of various metals present in farmland is given in the first column of Table 11.1. Depending on the history of the land, various studies have shown that the toxic metal levels can vary over a wide range; for example lead concentrations of between 3 and 710 mg/kg have been found against a maximum acceptable level for farmland of 100 mg/kg. While some of this metal is naturally occurring – in other words it may have a geological origin – the remainder represents manmade pollution such as airborne pollutants and the use of agrochemicals or sewage sludge as a fertiliser.

A proportion of the metals in the soil enter grass and crops grown on the soil and this represents a potential hazard to farm animals and man, who eats the animals and also crops grown on the land.

It is to be expected that a relationship exists between the metal content of soils and the metal content of the crop. Table 11.2 gives data on the maximum metal contents observed in soils (taken from Table 11.1) and the maximum metal contents determined for various crops, including corn, wheat and rice flours, apples, potatoes, broccoli and kale. A plot of maximum metal contents (mg/kg) in soil and crops respectively, shows the relationship between these parameters (Fig. 11.1). Metal contents in crops in the range 0.01 – 1000 mg/kg increase with increasing soil metal content in the range 1 – 100 000 mg/kg.

This relationship is, understandably, not as precise as would be expected due to the many variables involved. It does, however, provide benchmark data.

The relationship is:

$$\log C_s = \log C_c + 2.1 \pm 0.8$$

in other words

$$\log \frac{C_c}{C_s} = -2.1 \pm 0.8$$

where C_s is the concentration (in mg/kg) of the metal in the soil and C_c is the concentration (in mg/kg) of the metal in crops.

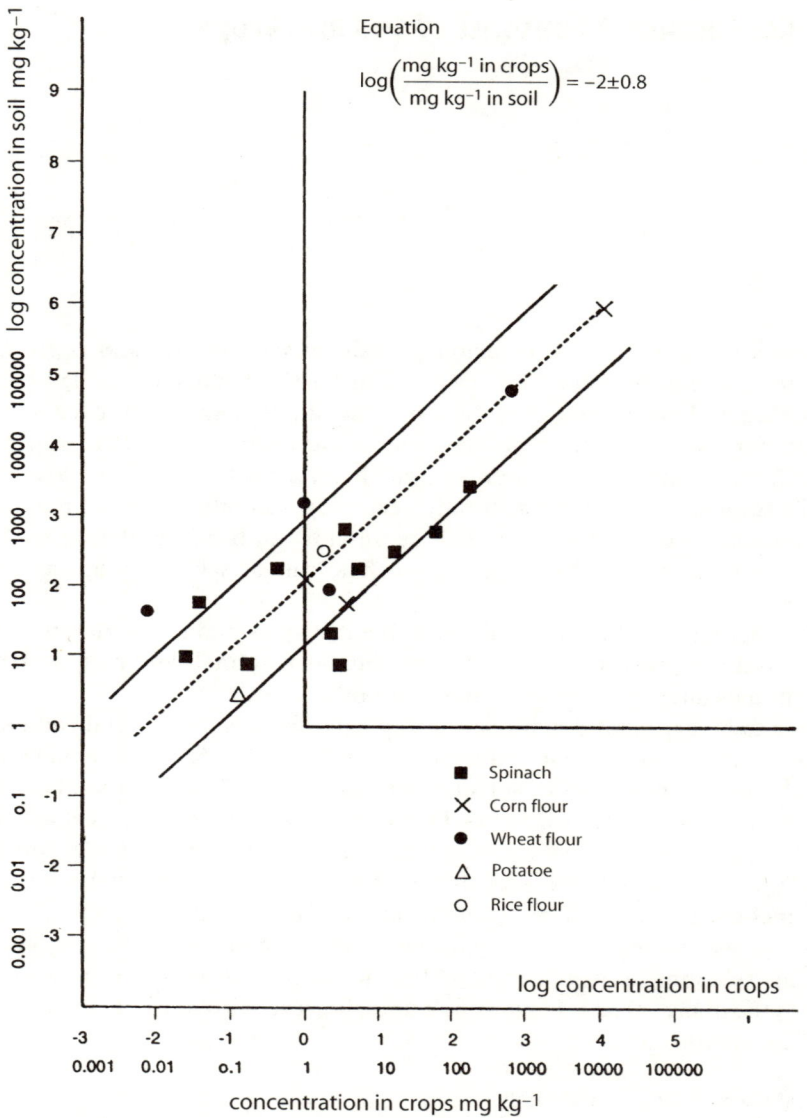

Figure 11.1. Relationship between metal content of soil and metal content of crops grown in soil (from author's own files)

This is in fair agreement with the ratio proposed by O'Connor [1], admittedly for organic compounds, of:

$$\frac{C_c}{C_s} = < 0.01 \quad \text{i.e.} \quad \log \frac{C_c}{C_s} = -2$$

Table 11.1. Relationship between maximum metal contents of soils and crops (from author's own files)

	Soils		Cornflour		Wheatflour		Rice flour		Apples		Potatoes		Spinach		Kale		Overall–crops	
	mg/kg	log mg/kg	mg/kg	log mg/kg	mg/kg	log mg/kg	mg/kg	log mg/kg	mg/kg	log mg/kg	mg/kg	log mg/kg	mg/kg	log mg/kg	mg/kg	log mg/kg	mg/kg	log
Zn	600	2.78	37	1.57	10.6	1.02	5.2 / 19.4	0.72 / 1.29	0.71	-0.15	10.9	1.04	56.2	1.75	35	1.54	56.2	1.75
Cu	240	2.38	7.5	0.87	2.0	0.30	-	-	0.3	0.52	3.1	0.49	13.9	1.14	6	0.78	13.9	1.14
Ni	74	1.87	3.6	0.56	0.06	-1.22	-	-	-	-	-	-	7.5	0.87	1.1	0.04	7.5	0.87
Co	21	1.32	-	-	0.013	-1.89	-	-	0.007	-2.15	0.06	-1.22	1.67	0.22	0.06	-1.22	1.67	0.22
Cd	5.1	0.71	0.10	-1.00	0.04	-1.40	-	-	-	-	-	-	2.5	0.40	1.1	0.04	2.50	0.40
Pb	710	2.85	0.61	-0.22	0.016	-0.80	-	-	0.10	-1.00	1.20	0.08	3.2	0.50	2.9	0.46	3.2	0.50
Cr	171	2.23	2.5	0.40	0.04	-1.4	20	1.30	0.66	-0.18	0.06	-1.22	3.86	0.59	0.35	-0.46	3.86	0.59
Mn	2,750	3.44	-	-	8.5	0.93	9	0.95	43	0.63	5.7	0.76	173	2.24	16.9	1.23	173	2.24
Fe	77,500	4.89	-	-	20	1.30	-	-	-	-	24	1.38	586	2.77	126	2.10	586	2.77
Sb	66	1.82	-	-	<0.002	<-0.27	-	-	-	-	-	-	0.04	-1.40	-	-	0.04	-1.40
Bi	40	1.60	-	-	1.0	0.00	-	-	-	-	-	-	<0.008	<-2.10	-	-	<0.008	<-2.10
Hg	3.3	0.52	-	-	-	-	-	-	-	-	0.15	-1.82	-	-	-	-	0.15	-0.82
Vn	200	2.30	-	-	-	-	-	-	-	-	-	-	-	-	0.4	-0.40	0.40	-0.40
Se	111	2.04	0.53	-0.28	1.9	0.28	0.30 / 1.87	-0.52 / 0.27	0.04	-1.40	-	-	0.78	-0.11	0.14	-0.85	1.9	0.28
Ag	5.0	0.70	-	-	0.02	-1.7	-	-	0.004	-2.40	0.03	-2.52	0.15	-0.82	0.03	-2.52	0.15	-0.82
Sn	7.1	0.85	-	-	<0.02	<-1.7	-	-	-	-	-	-	<0.02	<-1.70	-	-	<0.02	-1.70
As	1,375	3.14	-	-	1.0	0.00	0.45	-0.35	-	-	-	-	0.17	-0.77	-	-	1.00	0.00
Mo	434	2.64	-	-	-	-	1.6	0.20	-	-	-	-	-	-	-	-	1.6	0.20

Table 11.2. Metal pollutant levels in soil (from author's own files)

	Found in soil, mg/kg	Acceptable level, mg/kg
Heavy metals[a]		
Zinc	11 – 600	300
Copper	1 – 240	100
Nickel	3 – 74	100
Cobalt	0.2 – 21	50
Cadmium	0.04 – 5.1	5
Lead	3 – 710	100
Chromium	9.2 – 171	100
Total heavy metals	28 – 1820	755
Major constituents: Heavy metals		
Manganese[a]	188 – 2750	400
Iron[b]	400 – 77 500	–
Others		
Arsenic	6 – 1375	
Titanium	0.5 – 8600	
Aluminium	9990 – 85 000	
Barium	500 – 700	
Minor constituents		
Antimony	0.6 – 66	
Bismuth	0.1 – 4.0	
Mercury	0.03 – 3.3	
Vanadium	4 – 200	
Beryllium	2.7	
Selenium	0.01 – 111	
Silver	0.05 – 0.5	
Uranium	1.2 – 4.0	
Tin	1.5 – 7.1	
Molybdenum	0.2 – 437	
Tungsten	1.2 – 18	
Hafnium	0.6 – 4.1	
Lanthanum	15 – 47	

[a] Heavy metals for which acceptable levels exist
[b] Heavy metals for which acceptable levels do not exist

Thus, if a soil contained 1000 mg/kg of a metal then the concentration would be 8 mg/kg, with a range of 1.3 to 50.2 mg/kg in crops; corresponding median values of the levels of metal in soil were 10 000 mg/kg and 100 000 mg/kg, respectively, at the 80 and 800 mg/kg levels. These levels would definitely be of environmental concern (see the maximum levels of manganese (2750 mg/kg), iron (77 500 mg/kg), arsenic (1375 mg/kg), titanium (8600 mg/kg) and aluminium (85 000 mg/kg) found in crops, as reported in Table 11.1.

Thompson and Thoresby [2] have carried out a comparison of soils and plants grown in these soils. They obtained arsenic levels of 103 to 467 µg/g soil and 1.3 to 6.1 µg/g in the corresponding plants. This corresponds to log concentration in crops/log concentration of arsenic in soil values of between −1.85 and −1.9, which is in good agreement with the value of −2.1 ± 0.8 quoted previously.

References

1. O'Connor GA, Chaney RL, Ryan JA (1991) *Rev Environ Contam Toxicol* **121**:129.
2. Thompson AJ, Thoresby PA (1977) *Analyst* **102**:9.

Subject Index